Assessing Ecological Risks of Biotechnology

BIOTECHNOLOGY

BIOTECHNOLOGY SERIES

1. R. Saliwanchik	*Legal Protection for Microbiological and Genetic Engineering Inventions*
2. L. Vining (editor)	*Biochemistry and Genetic Regulation of Commercially Important Antibiotics*
3. K. Herrmann and R. Somerville (editors)	*Amino Acids: Biosynthesis and Genetic Regulation*
4. D. Wise (editor)	*Organic Chemicals from Biomass*
5. A. Laskin (editor)	*Enzymes and Immobilized Cells in Biotechnology*
6. A. Demain and N. Solomon (editors)	*Biology of Industrial Microorganisms*
7. Z. Vaněk and Z. Hošťálek (editors)	*Overproduction of Microbial Metabolites: Strain Improvement and Process Control Strategies*
8. W. Reznikoff and L. Gold (editors)	*Maximizing Gene Expression*
9. W. Thilly (editor)	*Mammalian Cell Technology*
10. R. Rodriguez and D. Denhardt (editors)	*Vectors: A Survey of Molecular Cloning Vectors and Their Uses*
11. S.-D. Kung and C. Arntzen (editors)	*Plant Biotechnology*
12. D. Wise (editor)	*Applied Biosensors*
13. P. Barr, A. Brake, and P. Valenzuela (editors)	*Yeast Genetic Engineering*

Assessing Ecological Risks of Biotechnology

Edited by

Lev R. Ginzburg
State University of New York
Stony Brook, New York

Butterworth–Heinemann
Boston London Singapore Sydney Toronto Wellington

Recognizing the importance of preserving what has been written, it is the policy of Butterworth–Heinemann to have the books it publishes printed on acid-free paper, and we exert our best efforts to that end.

Editorial and production supervision by Science Tech Publishers, Madison, WI 53705.

Library of Congress Cataloging-in-Publication Data
Assessing ecological risks of biotechnology/edited by Lev R. Ginzburg.
p. cm.—(Biotechnology ; 15)
Includes bibliographical references.
ISBN 0-409-90199-7
1. Microbial biotechnology—Safety measures. 2. Microbial biotechnology—Environmental aspects. 3. Genetic engineering—Safety measures.
4. Ecology. I. Ginzburg, Lev R. II. Series.
TP248.27.M53A87 1991
660′.62′0289—dc20 90-1719
CIP

British Library Cataloguing in Publication Data
Assessing ecological risks of biotechnology.
1. Biotechnology. Ecological aspects
I. Ginzburg, Lev R. II. Series
660.6

ISBN 0-409-90199-7

Butterworth–Heinemann
80 Montvale Avenue
Stoneham, MA 02180

10 9 8 7 6 5 4 3 2 1

Printed in the United States of America

CONTRIBUTORS

H. Reşit Akçakaya
Applied Biomathematics
Setauket, New York

Duane F. Berry
Crop and Soil Environmental Sciences
Virginia Polytechnic Institute and State University
Blacksburg, Virginia

Richard Condit
Department of Biology
Princeton University
Princeton, New Jersey

Maryln K. Cordle
Office of Agricultural Biotechnology
Office of the Assistant Secretary for Science and Education
USDA
Washington, D.C.

Monica A. Devanas
Department of Biological Sciences
Rutgers University
New Brunswick, New Jersey

Daniel E. Dykhuizen
Department of Ecology and Evolution
State University of New York at Stony Brook
Stony Brook, New York

L.E. Ehler
Department of Entomology
University of California
Davis, California

Lev R. Ginzburg
Department of Ecology and Evolution
State University of New York at Stony Brook
Stony Brook, New York

Robert A. Goldstein
Electric Power Research Institute
Palo Alto, California

Charles Hagedorn
Crop and Soil Environmental Sciences
Virginia Polytechnic Institute and State University
Blacksburg, Virginia

Susan S. Hirano
Agricultural Research Service, USDA
Department of Plant Pathology
University of Wisconsin
Madison, Wisconsin

Rogier A.H.G. Holla
Department of Law
European University Institute
Florence Ferrovia, Italy

Conrad A. Istock
Department of Ecology and Evolutionary Biology
University of Arizona
Tucson, Arizona

Peter Kareiva
Department of Zoology
University of Washington
Seattle, Washington

Junhyong Kim
Department of Ecology and Evolution
State University of New York at Stony Brook
Stony Brook, New York

Richard E. Lenski
Department of Ecology and Evolutionary Biology
University of California
Irvine, California

Morris Levin
Maryland Biotechnology Institute
University of Maryland
College Park, Maryland

Robin Manasse
Department of Zoology
University of Washington
Seattle, Washington

James A. Moore
Agricultural Engineering Department
Oregon State University
Corvallis, Oregon

John H. Payne
Biotechnology, Biologics, and Environmental Protection
Animal and Plant Health Inspection Service, USDA
Hyattsville, Maryland

Marvin Rogul
Maryland Biotechnology Institute
University of Maryland
College Park, Maryland

Mark Sagoff
Institute for Philosophy and Public Policy
University of Maryland
College Park, Maryland

Daniel Simberloff
Department of Biological Science
Florida State University
Tallahassee, Florida

Guenther Stotzky
Department of Biology
New York University
New York, New York

Anthony D. Thrall
Electric Power Research Institute
Palo Alto, California

Christen D. Upper
Agricultural Research Service, USDA
Department of Plant Pathology
University of Wisconsin
Madison, Wisconsin

Alvin L. Young
Office of Agricultural Biotechnology
Office of the Assistant Secretary for Science and Education
USDA
Washington, D.C.

Lawrence R. Zeph
U.S. Environmental Protection Agency
Office of Toxic Substances
Washington, D.C.

CONTENTS

PREFACE

This volume presents a comprehensive analysis of ecological risk assessment for biotechnology as viewed predominantly by scientists doing research in this area, but also by regulators, philosophers, and research managers. The emphasis is on the ecological risks associated with the release of genetically engineered organisms into the environment. The 17 chapters of the book are organized into four parts.

Part I discusses the ecological experience gained from previous biological introductions. In the last few decades, humans have introduced many nonnative species into a number of different ecosystems. Some of these introductions were planned, as in the case of organisms released for biological control of pests, and some were accidental. Chapters 1 and 2 summarize the history of invasions resulting from such introductions, describe the properties that seem to characterize invading species, give an account of ecological impacts of invasions by nonnative species, and attempt to synthesize the experience gained from these introductions for predicting the outcome of a different type of introduction–releasing genetically engineered organisms.

Most applications of biotechnology involve microorganisms. Even in cases where the engineered organism is, for example, a plant, microorganisms are often used as vectors for manipulating the genetic material being engineered. Therefore, the ecology and the genetics of microorganisms are of great importance in assessing the risks of biotechnology. The second part of this book explores these characteristics of microbial communities. Particular emphasis is given to the transport of microorganisms, since one of the major ecological concerns about biotechnology is the danger of the spread of genetically engineered organisms to ecosystems other than the one to which they are released. Similarly, the transfer of genes to other organisms is an ecological concern. The last chapter of Part II deals with the stability of genes and the exchange of genetic material between different species.

Risk assessment involves predicting certain outcomes and estimating the probability associated with each of these outcomes, as well as assessing their consequences. Such quantitative estimates can only be made with mathematical models. The third part of this book reviews mathematical models that can be used for ecological risk assessment at four different levels. At lower levels, the effect of genetic engineering on fitness of individuals is explored. Chapters 10 and 11 utilize models at genetic and population levels to predict the effect of introducing genetically engineered organisms into a biological system.

Part IV concerns the regulation of biotechnology, current research trends, and social values. Chapters 12 through 15 present an overview of the regulatory processes employed by governmental agencies in the United States and in Europe and give the point of view of regulators. Chapter 16 is written by research managers, i.e., people managing the research efforts funded by industrial organizations. It discusses projects currently underway and gives an indication of what kinds of research will emerge in the next few years. Finally, the book ends with a philosopher's point of view about what should be regulated and why.

The aim of this book is to give readers an understanding of the complexity of biotechnology risk assessment. The multitude of scientific issues addressed in the first 11 chapters of the book, as well as the immediate difficulties that regulators are faced with, show that the field is still in an early stage of development. However, this also means that there are many opportunities for scientists to make a contribution to this growing field. I hope this book will be useful in pointing out these areas that need immediate attention.

Lev R. Ginzburg

Assessing Ecological Risks of Biotechnology

PART I

Experience with Introduced Organisms

CHAPTER
1

Keystone Species and Community Effects of Biological Introductions

Daniel Simberloff

Whether an introduced species will survive and how it will affect its new community appear at first to be idiosyncratic and unpredictable matters. This is perhaps best seen by comparing the trajectories of several apparently similar species introduced into the same region, or of native species that seem to be relatively innocuous components of their community and related introduced species that have appeared to wreak havoc.

The house sparrow (*Passer domesticus*) and the tree sparrow (*P. montanus*) were both introduced into the United States in the 19th century (Long 1981); in 1853, the first successful introduction of the house sparrow occurred, after two similar propagules had failed. Aided by secondary introductions, the species rapidly spread and ultimately occupied the entire area of the contiguous United States and adjacent Canada, and Mexico. It is now one of our most common birds. The tree sparrow was first released in 1870 in St. Louis and has generally remained restricted to the vicinity of that city; it probably numbers about 25,000 individuals. Although the precise community roles of these two birds are not known, the house sparrow certainly causes agricultural damage and has displaced several native insectivores, while there are no reports that the tree sparrow affects the resident community. There is no obvious explanation for this striking difference. Both species are often associated with human habitats and use similar foods. More remarkably, in some other parts of the world where the tree sparrow has been introduced, it has spread widely and become very numerous, even in areas where the house sparrow was also introduced (such as southeastern Australia). And in some areas where both are native, such as southern Asia, the tree sparrow occurs in larger numbers.

Another American example concerns the introduction to the southern part of the United States of the South American fire ant, *Solenopsis invicta*, and its native congener, *S. geminata*, (Simberloff 1985). The full impact of the introduced species on the community has yet to be determined, but it is known as a major predator of other insects (Porter et al. 1988) and consumes much native vegetation. The sheer number and size of its colonies suggest it must play a major community role. The native species, which is in the same subgenus, seems to be a relatively minor component of its communities, and there is no clear reason for this difference. Finally, four mongoose species (*Herpestes*) have been introduced in various parts of the world, and three of them have had little obvious effect, while the fourth (*H. auropunctatus*) has been a scourge in all four areas of introduction (Ebenhard 1988). Simberloff (1985), Ehrlich (1986 and 1989), and Ashton and Mitchell (1989) give other examples of related groups of species with dramatically different impacts.

1.1 QUANTITATIVE ESTIMATES

Most introduced species probably do not survive, though there is no way to estimate the exact fraction because it is very unlikely that one would witness the arrival and disappearance of a few individuals. Nevertheless, the sheer number of sightings of transient birds that do not subsequently establish populations suggests that the number of propagules of less obvious but equally dispersive taxa (e.g., insects and plants) must be staggering, and only a small number of these are known to produce ongoing populations. For example, North America, with approximately 100,000 native insect and mite species, currently has about 1,700 known introduced species, a figure unlikely to rise above 3,000 with more exhaustive study (Knutson et al. 1986; Simberloff 1986). It is hard to estimate how many species actually had propagules reach North America, but there were probably thousands. In Britain, of *known* introductions, 240 of 409 insect species failed, as did 166 of 488 vascular plant species (Williamson and Brown 1986). I cannot prove this, but I believe that propagules of several times as many species must have immigrated, and subsequent failure to record them means they failed. In the Hawaiian islands, 679 animal species (almost all insects) were purposely introduced between 1890 and 1985 for biological control purposes. Of these, 436 failed to establish populations (Funasaki et al. 1988).

Biological control introduction attempts are especially likely to be recorded, so there is a higher likelihood that we would know about these attempts than for other introductions. Also, the selection process (consideration of habitat, host organisms, and propagule size) that leads to a biological control introduction almost certainly ensures that a smaller fraction of such introductions fail than of unplanned introductions. In sum, data are scarce, but strongly suggest most introductions fail.

If information is insufficient to document the fate of many introductions, there are even fewer adequate data to demonstrate effects of surviving introductions on the resident community (Simberloff 1981). Many observations are suggestive, as introductions often constitute natural experiments. But the sort of controlled experimental evidence that would be most convincing is usually lacking. For example, the North American mink (*Mustela vison*), first imported to fur farms in Britain in 1929 (Lever 1977), began to spread rapidly in the mid-1950s, at approximately the same time that otters were precipitously declining (Chanin and Jeffries 1978). This synchronicity led many people (e.g., Lever 1977) to suggest a causal relation between the two events, a view many otter hunters held very strongly (Chanin and Jeffries 1978). Similarly, in Sweden the mink was introduced as farm escapees in the late 1920s and began to increase rapidly at approximately the same time that the otter declined there, again leading to suggestions that the introduction of one caused the decline of the other (Erlinge 1972). However, in both countries, careful examination of hunting records shows that the otter decline slightly preceded the rapid increase in mink numbers. The mink may nevertheless affect the otter—for example, it seems to have more catholic food habits, permitting it to colonize areas that are not prime otter habitat and perhaps to restrict otters to optimal habitat. However, overall, the otter seems to affect the mink much more than vice-versa. Pollution, particularly of organochlorine pesticides such as dieldrin, seems to be the main cause of the otter decline (Erlinge 1972; Chanin and Jeffries 1978).

Technically, perhaps every surviving introduced species can be said to affect the community in some way, if only by changing species composition. Perhaps this latter effect would be the minimum possible change. Game species are often introduced to fill "empty niches," but Ebenhard (1988) finds only one example where the *total* game supply was increased—the moose (*Alces alces*) in Newfoundland uses coniferous forests without apparent effects on other game. However, even in this example the presence of a new species is probably at the expense of shifts in abundance elsewhere in the system. Though niches are often said to be "vacant" (Lawton 1984; Price 1984), it is notoriously difficult to establish that they are (Herbold and Moyle 1986), and such observations probably reflect differences in community organization or in defining niches rather than truly unused resources. Ultimately, resources are metabolized in some way, if only by bacteria, so it is difficult to imagine that an introduced species would not affect energy flow and nutrient cycling as well as composition. Of course, the total amount of energy or nutrient flowing through a particular ecosystem component need not be changed by the invader, but the topology of the cycle or food web surely is.

In any event, an invader that fills an "empty" niche can still have an enormous impact. In 1788, the first English settlers in Australia brought 7 cattle, 7 horses, and 44 sheep. By 1974, cattle alone numbered 30 million. The cattle produce at least 300 million cowpats each day, which would destroy six million acres of pasture annually if they were not removed (Waterhouse 1974). Native kangaroos and other marsupial herbivores produce much smaller, drier pellets, which are used by some 250 native dung beetles (scarabeids), but these beetles cannot use the introduced droppings fully. The unprocessed cowpats have various ecological effects, such as fostering growth of plant species not favored by cattle, providing a breeding habitat for various insects, and simply drying and covering the ground. One type of African dung beetle was unintentionally introduced before 1900, but in 1967 four other types were released. At least three of these quickly became established over wide areas and rapidly removed vast numbers of cowpats. The beetles also increased soil fertility and permeability by working the dung into the soil. There can be no doubt that these species altered many ecosystem properties. This saga is not yet complete, however, because today the cane toad, *Bufo marinus*, introduced in 1935 to control root beetles in sugarcane, is eating the introduced dung beetles, as well as reducing populations of many native species (Floyd and Easteal 1986). To reduce the cane toad numbers, Waterhouse (1974) suggested releasing giant African dung beetles of the genus *Heliocopris* that, when swallowed whole, would bore their way through the toad's body wall, but this has not yet been done.

An introduced species might affect one particular resident species, for example by predation, herbivory, competition, or vectoring a disease. Such direct effects have been the predominant focus of introduction studies. The effect may subsequently reverberate through the community, as changes in populations of the affected species in turn affect other species. For example, the American chestnut (*Castanea dentata*), a major element in many eastern deciduous forests in North America, has been almost eliminated from its original range by chestnut blight, a fungal disease (*Cryphonectria parasitica*) probably introduced from Asia around 1900 (McCormick and Platt 1980). In less than fifty years the pathogen spread over 91 million ha of hardwood forests and virtually eliminated the tree (von Broembsen 1989). Other trees, especially oaks and hickories, have replaced chestnuts in these forests (McCormick and Platt 1980; Krebs 1985). Vitousek (1986) suggests there has been little long-term effect on ecosystem structure, though the cited reference (McCormick and Platt 1980) deals almost exclusively with tree species composition. At least 60 moth species feed on North American chestnuts; seven are apparently host-specific to *C. dentata* (Opler 1978). Some of these may now be extinct. Predators and especially parasites on these moths are not very well known, but it is quite conceivable that several of these can ultimately be affected. Quimby (1982) suggests that the oak wilt disease *Ceratocystis fagacearum* increased on native oaks in the northeastern part of the United States because the susceptible red oak, *Quercus rubra*, increased in the absence of chestnut. So the full impact of chestnut's disappearance may be varied and recondite.

The chestnut example shows that the initial impact on one species produces ripple effects throughout an entire ecosystem and may well affect communities in more ways than just species composition alone. Such indirect effects have not been well studied. Most literature on introductions has focused on direct effects on particular species, with at most cursory consideration of the propagation of indirect effects. This situation mirrors one in ecology in general, in which direct pairwise effects such as predation, competition, and parasitism have historically received the most attention. This situation is now being redressed with a burst of interest in indirect effects (e.g., Kerfoot and Sih 1987), and one hopes that the ecology of introductions will participate in this development.

In 1981 I sought evidence in the literature that introduced species had directly caused competitive extinctions as envisioned in the equilibrium theory of island biogeography (MacArthur and Wilson 1967) and found rather few such scenarios. I examined ten review papers summarizing at least 854 introductions and pointed out that very few studies came close to satisfying stringent criteria for demonstrating an effect analogous to those enunciated by Kitching and Ebling (1967) for proving exclusion by predation. Most literature descriptions of effects of introductions are quite impressionistic. I found just three competitive extinctions that could unequivocally be assigned to these introductions, and only 71 extinctions in total (mostly caused by predation or habitat change). More surprisingly, for 678 of these 854 introductions (79.4%) there seemed no clear demonstration of any effect on the community or ecosystem.

Herbold and Moyle (1986) differed with this conclusion on the grounds that there was insufficient study in virtually all of these cases to conclude there were no effects, and that in some sense the assumption of "no effect" in this matter is less conservative than the assumption of an effect, in the absence of adequate data. They are undoubtedly correct on these points. I would also add that there are so many possible effects on the resident community that even careful studies probably do not rule out all of them. All that can be said is that the sorts of data normally presented in reviews of introductions simply do not allow an estimate of the frequency of any kind of effect, and that the effects are probably only noted when they are quite dramatic. Further, as Herbold and Moyle observe, an introduction may have an effect well after it occurs, yet the later the effect, the less likely one is to note it in a cursory examination or to link it to the introduction event.

Two other attempts to review the introduction literature are vulnerable to the same criticisms, but are perhaps useful as rough indicators of the frequencies of major and relatively immediate effects of introductions. Dritschilo et al. (submitted for publication) studied the literature on "ecological damage" by 913 vertebrates, invertebrates, and plants successfully introduced into California. Only 23 (2.5%) were listed as causing "major ecological damage," while 788 (86.3%) were characterized as causing "no ecological damage." Ebenhard (1988) studied literature records on 1,559 successful introductions of 330 species of mammals and birds. In 20% of the mammal introductions a plant or habitat effect through herbivory was recorded, while another 17% caused ecological effects by predation. For birds, the analogous figures are 0.4% and 1.4%, respectively. Competition is notoriously difficult to see (Simberloff 1981; Herbold and Moyle 1986; Ebenhard 1988); Ebenhard finds competition between introduced and native species in 2.9% of the mammal introductions and 3.1% of the avian introductions. At least some effect was suggested in 40% of the mammal introductions and 5% of the bird introductions. Ebenhard notes that more thorough study would increase these figures, but that researchers are less likely to publish on introductions without notable consequences.

Again I would emphasize that most reports are inadequate to assess subtle effects, especially delayed ones. And assignment of even major effects in such "natural experiments" (Diamond 1986) can easily be in error. The historical reconstruction that gives good post-facto rationalizations of how an introduced species affected a particular ecosystem is far from trivial. An example is the well-known *Opuntia-Cactoblastis* story (Krebs 1985). The prickly pear, *Opuntia*, from the southern United States was planted as a hedge plant in eastern Australia after introduction in 1839. By 1880 it was a recognized pest and by 1925 occupied almost 250,000 km^2. About half of this area was impenetrably dense growth 1–2 m high. In 1925, eggs of *Cactoblastis cactorum*, a moth from northern Argentina, were introduced. The caterpillars damage the cactus by burrowing inside the pads and by introducing bacterial and fungal infections. The moth quickly destroyed vast expanses of *Opuntia*, which now exists in much lower densities, and cactus populations in some pasturelands have even become extinct (Myers et al. 1981). Because the moth eggs are clumped, even in woodlands many plants are at least temporarily free of the moth (Monro 1967; Myers et al. 1981; Osmond and Monro 1981), and some entire marginal populations of the cactus may not contain the moth (Taylor 1990). It is possible that a clever researcher who did not know the history would be able to figure out that the moth controls the plant and is responsible for its absence from 250,000 km^2, but this deduction would not be trivial and would surely take detailed study and experiment.

The literature surveys, then, may be summarized as follows: At least some ecological effect, if only a direct effect on individual species in the target community, seems likely in at least 20% of all surviving introductions. The true figure may well be much higher but is obscured by inadequate study. Indirect effects are likely, but are far less frequently noted in surveys.

1.2 DETERMINANTS OF MAJOR ECOLOGICAL EFFECTS

There seems to be no single, sure-fire criterion for predicting a major ecological effect by an introduced species. This lacuna probably reflects a more general one—the organization of communities and ecosystems is itself quite poorly understood (cf. Paine 1988; Peters 1988), and few "rules" or "laws" have been found to characterize them, short of the laws of thermodynamics. Communities seem so idiosyncratic that, even though we can learn through experiment and exhaustive observation how particular ones operate, such knowledge does not tell us all we would want to know about other, unstudied communities, even though they may superficially appear similar. We usually cannot predict with great precision the effect of a particular extinction or other perturbation, so it is not surprising that we cannot do better at predicting the effect of an introduced species. In fact, we might expect the effects of an introduced species to be more difficult to predict than many other kinds of perturbations because the perturbing element is alive and dynamic on a number of time scales. The population size and structure of an introduced species can vary over the short term, so that any model of its short-term effects would have to entail a population dynamics model. And the population can evolve by natural selection, drift, or hybridization (Ebenhard 1988) so that key properties change. Several papers have focused on the amount and causes of observed evolution (or lack thereof) in introduced species (e.g., Johnston and Selander 1964 and 1971; Selander and Johnston 1967; Pietsch 1970; Pankakoski and Nurmi 1986), but there seems to be no study of how such evolution will modify ecological effects, except for the literature of evolution of benignity of disease (e.g., Fenner and Ratcliffe 1965; Allison 1982; Levin et al. 1982; Ewald 1983).

However, the absence of a single, readily predicted "key" to the ecological effects of introductions does not mean that no prediction is possible. The concept of "keystone" species provides guidance in predicting community and ecosystem effects of introduced species, though the predictions will be qualitative and general, like the concept itself. One may at least be able to pinpoint those introductions particularly prone to have major impacts. Paine (1969) suggested that certain predator species, though they may be relatively unimportant as energy transformers and need not be very numerous, can greatly modify the species composition, physical structure, and integrity of entire complex systems. He termed these keystone species. His two examples both concerned predatory marine invertebrates—the starfish, *Pisaster ochraceus*, in the Pacific rocky intertidal zone and tritons (*Charonia* spp.) on the Great Barrier Reef. The starfish, preferring to eat mussels, prevents these from dominating space and eliminating other species (Paine 1966). The tritons might, by favoring as prey the crown-of-thorns starfish (*Acanthaster planci*), prevent this species from destroying large patches of living coral reef (though other hypotheses than destruction of the tritons are possible for crown-of-thorns, population explosions [Ford 1988]). In either case the key is that the keystone species prevents a particular prey species from otherwise exerting a dominating influence. The concept of keystone species has been generalized—for example, Gilbert (1980) defined "keystone mutualists" as organisms, typically plants, that provide critical support to a large number of animal species that are, in turn, critical to the existence of other species. The tree *Caesuria corymbosa* fills this role by maintaining several frugivores that are themselves critical to seed dispersal of a number of other plants (Howe and Westley 1988).

Introduced species have played so many different keystone roles that it is necessary to classify them in order to get an overview. I divide them arbitrarily into three categories. First, some species themselves constitute a new structural habitat with diverse microhabitats for many other species. Second are species that create but do not comprise a new structural habitat. Finally, several introduced species have affected many other species (e.g., by direct predation), but have not greatly affected the structural habitat, at least initially. The lines separating these categories are fuzzy; for example, many introduced tree and shrub species constitute habitats for other species, but they also modify the soil habitat and affect other plants through shading, allelopathy, and other means. An introduced predator can also ultimately modify the plant community, and thus the structural habitat, by removing pollinators or seed-dispersers.

1.3 SPECIES THAT CONSTITUTE NEW HABITATS

Along rivers in the arid southwestern United States, salt-cedar (*Tamarix* spp.) and Russian olive (*Eleagnus angustifolia*) trees have established forests where none existed previously (Knopf and Olson 1984; Vitousek 1986) with dramatic and far-reaching effects. Salt-cedars were introduced in the 19th century. Once they are established, their deep roots allow them to maintain themselves in situations where most plants would not survive. For example, the Glen Canyon Dam regulated flooding on the Colorado River and allowed salt-cedar, among very few species, to become established on shores that had previously been swept clear of plants. Now there are numerous small forests of substantial *Tamarix* trees probably providing habitat for animals. *Tamarix* transpiration causes massive water loss from desert water bodies (Vitousek 1986). For example, in the 1930s to 1940s, *Tamarix* invaded Eagle Borax Spring in Death Valley, and by the 1960s its water use had drained the surface of a large marsh. Subsequent removal of the *Tamarix* led to reappearance of surface water and recovery of the original biota. Russian olive, introduced as an ornamental before 1900 and escaped from cultivation by the 1920s, tends to be upstream from salt-cedar, and many birds typical of native riparian vegetation are absent from Russian olive, which has formed monocultures and replaced the original species in some areas (Knopf and Olson 1984). Russian olive provides no habitat for entire guilds, such as cavity-nesters. This species has displaced others in entire marshlands in South Dakota; much of the eastern part of this state may be a Russian-olive forest within 50 years (Knopf 1986; Olson and Knopf 1986).

Although mangroves cover intertidal soft substrate in tropical sheltered bays and estuaries in most of the world, similar Hawaiian habitats were occupied by *Hibiscus tiliaceus* until 1902, when the American Sugar Company planted red mangrove (*Rhizophora mangle*) seedlings from Florida in Molokai (Walsh 1967). A second introduction in 1922 brought other mangroves from the Philippines to the archipelago. Now, by natural dispersal and perhaps deliberate plantings, mangroves (especially *Rhizophora*) have spread to many parts of the islands, completely replacing *Hibiscus* and forming forests up to 20 m tall in some areas. The arboreal arthropod fauna seems mostly composed of cosmopolitan "tramp" species, though there is substantial folivory (Simberloff, unpublished observations). Use of these habitats by vertebrates has not, to my knowledge, been studied, while research on the aquatic web housed by the roots in Hawaii is preliminary (Walsh 1967). Since healthy mangrove swamps drop almost 4.5 tons of leaves annually per acre, and the roots form critical habitat for fishes and shrimp (Carey 1982) and accumulate sediment (Holdridge 1940), the consequences of this introduction to energy flow, nutrient cycling, and succession must be staggering, if largely unstudied.

1.4 SPECIES THAT MODIFY EXISTING HABITAT

European marram grass (*Ammophila arenaria*) was introduced to California in 1896 to stabilize sand dunes (Slobodchikoff and Doyen 1977). It has a number of morphological adaptations that trap sand and retard drift. However, as dunes stabilize, marram grass itself declines and species characteristic of stabilized sand invade and replace it. Native dune plants of California are also adapted to loose sand, and tend to be replaced by marram grass (Barbour et al. 1976), which alters the light environment (Mooney et al. 1986) and dune topography and orientation (Barbour and Johnson 1988) and severely depresses dune arthropod populations and species richness (Slobodchikoff and Doyen 1977).

In 1884 water-hyacinth (*Eichhornia crassipes*) was brought from South America to New Orleans to decorate tables at the Cotton Exposition. A Floridian attendee brought some back to his pond near Palatka, FL, from which they escaped into the St. Johns River. By 1893, navigation problems were severe enough that removal programs were instituted in Florida (annual costs reached $3,000,000 by 1984). By the mid-1970s, water-hyacinth covered 90,000 acres of Florida lakes and streams with a dense mat, shading and killing the native aquatic plants and greatly reducing fish, turtle, alligator, and waterfowl populations (Ehrenfeld 1970; Schardt 1985). Rafts of water-hyacinth uprooted native vegetation and contributed to filling in Florida waters at a rate of over four tons of sediment per acre annually (Schardt 1985).

Through modification of soil characteristics, other plants also modify habitat by their presence and facilitate colonization by other exotic plants at the expense of native species. The subsequent invaders can themselves have major impacts. In the early 1900s, the lower North Spit of Humboldt Bay in California was covered chiefly by sand, with a few small plants hugging the ground (Miller 1988). In 1908 the operator of a fog-signal station brought seeds of the yellow bush lupine (*Lupinus arboreus*) from San Francisco to prevent drifting sand, an operation repeated nearby in 1917 by the U.S. Army Corps of Engineers. The plant spread at the rate of about five acres per year through 1984 and is partly responsible for the displacement of 65% of the original unforested dune vegetation at this site. A likely scenario for this effect (A. Pickart, personal communication) is that the lupine, which hosts nodulating bacteria, survives in the nutrient-poor dune mat and modifies the environment, perhaps by moisture retention and shade, to make it favorable for the establishment of other introduced species. These, in turn, increase nitrogen and organic matter in the soil, which ultimately leads to the degradation of the original community.

Vitousek (1986) presents a similar picture of invasion by the Atlantic shrub *Myrica fava* of young volcanic regions on the island of Hawaii. These sites are nitrogen poor and have no native symbiotic nitrogen fixers. *Myrica* forms near-monocultures and fixes nitrogen, which in turn can facilitate the introduction of further exotics. A possible result is the acceleration of succession and alteration of energy flow and nutrient cycling.

In California the introduced African ice plant, *Mesembryanthemum crystallinum*, has had equally devastating effects on the native vegetation, also by changing the soil (Vivrette and Muller 1977; Macdonald et al. 1989). This species is an annual and accumulates salt throughout its life; then, when it dies, the fog and rain leach the salt into the soil, where it suppresses growth and germination of native species. It forms a thick carpet that shades out other vegetation, and when disturbances open small areas, these are invaded by introduced weeds rather than native plants.

Introduced plants can also modify an entire ecosystem by altering fire regimes. *Melaleuca quinquenervia* was introduced from Australia to south Florida around 1900 as a potential timber tree. It can survive on almost any soil in south Florida, but achieves its main impact by displacing less fire-resistant species in areas susceptible to periodic burning (Ewel 1986). It has numerous morphological and chemical adaptations to fire, and these have helped it to replace cypress in many areas (Myers 1984). Several introduced plants act as "fire enhancers" in Hawaii (Vitousek 1986). Among these are two species of *Andropogon* that formed dense ground cover in some areas after feral goats were removed. The fires burned 100 times as much area when the plant thrived after goat removal, and the *Andropogon* species were more able than the native species to reestablish themselves after fire. These observations suggest that the goat was a keystone herbivore before its removal, while the *Andropogon* acted as keystone species later.

In the 18th and 19th century, much of the northeastern North American coast consisted of mud flats and salt marshes, rather than the current rocky beach. This change is due to population growth of the exotic European periwinkle snail, *Littorina littorea* (Bertness 1984; Dean 1988). Introduced (probably for food) in Nova Scotia around 1840, it has slowly worked its way south. The periwinkle eats algae on rocks and also rhizomes of marsh grasses. When it is experimentally excluded, algae and then mud quickly cover the rocks, after which grasses invade the mud. In addition to changing the entire physical structure of the intertidal, the periwinkle has many direct effects on other species. For example, almost all hermit crabs in parts of New England now occupy periwinkle shells, allowing one to speculate that the crabs are much more common because of the introduction (Dean 1988). The periwinkle displaces and depresses the population of its native congener, *L. saxatilis* (Yamada and Mansour 1987). It prevents *Fucus* germlings and barnacle cyprids from establishing (Lubchenco and Menge 1978; Lubchenco 1983; Petraitis 1983). And the periwinkle has competitively excluded the native mud snail, *Ilyanassa obsoleta*, from many habitats, such as salt marshes and eel grass beds, where it was formerly abundant (Brenchley and Carlton 1983). Oddly, *I. obsoleta*, itself introduced on the Pacific coast of North America, has competitively displaced a native mud snail, *Cerithidea californica*, there (Race 1982).

Introduced feral pigs (*Sus scrofa*) have modified entire communities and ecosystems. In 1920, approximately 100 individuals brought to North Carolina for hunting escaped from an enclosure and by the 1940s had invaded the Great Smoky Mountains National Park. They "root" primarily in high-elevation deciduous forests in the summer, when they greatly reduce understory cover and number of species as well as modify species composition (Bratton 1975). By selectively feeding on plant species with starchy bulbs, tubers, and rhizomes, they have locally extinguished such species (Ebenhard 1988). Further, they have greatly modified soil characteristics by thinning the forest litter, mixing organic and mineral layers, and creating bare ground, changes that in turn increased concentrations of nitrogen and potassium in soil solution and accelerated leaching of many minerals from the soil and litter (Singer et al. 1984). Where pigs are common, rooting has nearly eliminated two litter species, the southern red-backed vole (*Clethrionomys gapperi*) and the northern short-tailed shrew (*Blarina brevicauda*) (Singer et al. 1984). In Hawaii feral pigs cause soil changes similar to those in the Great Smokies (Vitousek 1986). In addition, they aid in the dispersal of exotic plants (Loope and Scowcroft 1985) and selectively eat particular native species (Stone 1985). Their rooting and defecation favor exotic invertebrates (Stone 1985).

In 1929, coypus or nutrias (*Myocastor coypus*) were brought to Britain to be used in fur farms; some had escaped by 1932. Gradually spreading, they greatly modified aquatic habitats by feeding on marsh and water plants and boring into banks (Lever 1977; Usher 1986; Gosling 1989; Macdonald et al. 1989). They eat young reed and marsh shoots, cut up rhizomes, and often trample marsh vegetation. They drove particularly favored food plants to local extinction and destroyed vast areas of reed-beds, threatening the habitat of many marsh bird species. A series of elimination campaigns begun in 1960 gradually restricted the range of these animals and appear to have finally eradicated them completely (Gosling 1989; Usher 1989).

1.5 KEYSTONE SPECIES THAT DO NOT INITIALLY CHANGE HABITATS

The fynbos shrublands of the southwestern cape in South Africa comprise many endemic plants in the Proteaceae, of which over 170 species have their seeds dispersed by native ants (Bond and Slingsby 1984). The Argentine ant, *Iridomyrmex humilis*, has recently invaded this community, and has a history elsewhere of replacing other ants (e.g., Crowell 1968). The Argentine ant is slow to find seeds of one fynbos species, does not move them as far as native ants do, and does not store them in underground nests as the natives do. The result is much lower germination rates, and the prospect for the entire community is slow decline and even extinction of many species by gradual loss of seed reserves. Thus the Argentine ant acts as a keystone species through its interference with dispersal mechanisms. In Hawaii the same introduced ant severely depresses native ground-dwelling arthropods in high-elevation shrublands, including pollinators of endemic shrubs and herbs that are obligate outcrossers (Medeiros et al. 1986).

The South American ant, *Wasmannia auropunctata*, introduced to Santa Cruz in the Galapagos Islands in the early twentieth century, has spread to five islands (Lubin 1984). It depressess most of the other ant species, including the few endemic ones; only species living deep in the soil persist in areas where it is in high density. In addition it has greatly lowered populations of a scorpion and two spiders, and reduced the numbers of insect species and individuals caught in sticky traps. Finally, *Wasmannia* tends coccids for their "honeydew," and these species have increased densities.

In recent decades, particularly in the 1970s, the forest bird species of Guam have declined precipitously and contracted their ranges. Several species are probably extinct now, and the remainder are so rare that Ralph and Sakai (1979) describe the island as "the most massive avian desert we have ever seen . . . ." Consequences to other species, such as plants with fruits that might have been eaten by birds or insects that might have been prey to birds, must be enormous, though these propagated effects are unstudied. The large Australian brown tree snake, *Boiga irregularis*, was accidentally introduced to Guam, probably in military cargo in the late 1940s or early

1950s (Savidge 1987). Only one other snake is found on the island, a rarely observed, small, blind, vermiform species. The tree snake's density and activity can be accurately estimated because it climbs on electrical transmission lines and short-circuits them (Fritts et al. 1987). Savidge (1984, 1985, and 1987) found that birds disappeared from various parts of Guam exactly when *Boiga* populations increased there, and that local mammals also serving as prey for the snakes declined precipitously at the same time. Although her arguments are cogent, they have not convinced all parties, as a different mechanism cannot be ruled out in this uncontrolled natural experiment. Jenkins (1983), Diamond (1984), and Diamond and Case (1986) suggest that use of organochlorine pesticides by the U.S. military during and after World War II and later by farmers is the main cause of the decline. The timing of pesticide use is suggestive, although Grue (1985) shows that the single report of residues in tissue and guano of a Guam bird species gave concentrations far lower than those associated elsewhere with mortality or reproductive failure. If the snake *is* the cause of bird decline, it certainly must be tallied as a keystone species.

The Nile perch (*Lates niloticus*) was deliberately introduced into Lake Victoria in the late 1950s to improve the economics of the fishing industry. Its record in this regard is mixed, but it has had a catastrophic effect on the native haplochromine fishes (Hughes 1986; Payne 1987). The more-than-200 small endemic species had evolved in the absence of such a large, voracious predator and had evolved traits, like mouth-brooding and swim-bladders that cannot be rapidly adjusted, that made them easy targets for the perch. Many populations have already declined greatly, particularly in areas away from sheltering rocks. Whereas originally the native fishes comprised most of the perch diet, they are now so rare that the perch feed primarily on their own young and a small prawn.

1.6 SYNERGISMS AND INVASIONS BY ENTIRE COMMUNITIES

The key to the effects of many of the most devastating biological introductions is not the activity of a single species, but rather those of an entire complex of species whose coevolutionary adaptations facilitate their survival and growth. Crosby (1986) details how whole Eurasian communities have successively invaded and transformed the biota of the temperate areas all over the world, with plants, pathogens, and animals interacting synergistically to produce an invincible juggernaut. The crucial feature, according to Mack (1989), is that temperate grasslands outside Eurasia lacked large, hoofed grazers and were dominated by caespitose grasses intolerant of grazing or trampling. These are quickly eliminated when disturbance increases, as by the introduction of cattle or sheep. On the other hand, rhizomatous grasses introduced with the livestock have evolved with these animals and thrive in their presence. Similarly, Eurasian weeds spread with the expansion of farming and are adapted to it, while native plants suffer. Ultimately the entire landscape changes—plants, soil, and hydrology.

On a lesser scale, coevolved groups of species can successfully invade together and effect greater change than each would alone. The mosquito, *Culex quinquefasciatus*, was introduced to Hawaii in 1826. In the early part of this century, after the introduction of many Asian bird species, the mosquito was able to transmit avian malaria to the native birds, leading to a wave of range reduction and extinction (van Riper et al. 1986). The introduced vine, *Passiflora mollisima*, affects patterns of mineral cycling, and thus the entire plant community, in mountainous Hawaiian rainforests (Scowcroft 1986). Feral pigs aid the vine's spread by consuming and dispersing its seeds. Further, they deposit these seeds in mounds of organic fertilizer—the pigs' feces—in "seedbeds" cleared by their own rooting activity (Ramakrishnan and Vitousek 1989).

1.7 CONCLUSIONS

Earlier I noted that the idiosyncrasies of individual ecological communities limit the predictions one might make about the effects of any perturbations, including an introduction. Nevertheless, the above examples, though they do not suggest precise ways of forecasting how an introduction will change its target community, at least point toward classes of introductions that might be especially likely to have large effects.

First, any plant species that might produce a physically substantial forest in a previously treeless habitat, such as the mangrove and salt-cedar described in Section 1.3 or *Casuarina* in southern coastal regions of Florida, will almost certainly affect the previously existing native plants, provide novel habitats for insects, and have further physical effects (such as on soil or sediment) that will lead to an entirely new community.

Second, a plant can be a keystone species even without providing substantial structure if it modifies the physical environment to make it inimical to the existing plant species, especially the dominant ones. Marram grass, water-hyacinth, ice plant, yellow bush lupine, *Myrica*, and the fire-facilitators fall into this category. Animals (like the *Cactoblastis* moth in Australia, pig in Hawaii and North America, coypu in Britain, and periwinkle in northeastern North America) or pathogens (like the chestnut blight in North America) can similarly act as keystone species by controlling the dominant plants that provide physical structure to the community.

Third, a species can be a keystone by virtue of depressing or removing an entire taxon, an effect that will probably be propagated in myriad ways and ultimately drastically modify the community. The Argentine ant in South Africa and Hawaii and *Wasmannia* in the Galapagos will almost certainly have this effect by direct and indirect effects on many other animals and plants.

There is also another form of major community change wrought by introduced species that I have not mentioned yet because it is so well known even in the earliest systematic introduction literature (e.g., Elton 1958) and popular works (e.g., King 1984). This is the massive destruction of native bird faunas by predators like rats and mustelids introduced to remote islands that lacked similar indigenous species. The birds' nesting habitats thus made them easy prey for the invaders. The tree snake on Guam is the first such reptilian keystone species. The Nile perch exemplifies a similar common problem–introduction of predator fishes into freshwater communities whose native fishes had evolved in their absence (Moyle et al. 1986; Payne 1987).

Finally, complexes of coevolved species, such as the European grazers and rhizomatous grasses or introduced birds and their diseases, might be particular cause for concern, even if no single keystone species exists.

All of the samples I have cited above are essentially post facto rationalizations for observed effects, though many of these hypotheses have been supported by elegant manipulative experiments (such as controlled pig and periwinkle exclusion), intensive observation, and thoughtful historical reconstruction entailing innovative uses of data such as those on power outages or hunt records. One might ask if it is possible to do better–to predict the effects of an introduction *before* it happens or at least in the very earliest stages. Elsewhere (Simberloff 1985) I have suggested that, with careful experimental study of the potential invader and key elements of the target community, prediction should often be possible on a case-by-case basis with the investment of about one doctoral dissertations' worth of time and effort. As an example, J.R. Pickavance performed just such doctoral research at the University of Liverpool in 1968 on *Dugesia tigrina*, an introduced North American flatworm, and accurately predicted its effects on native flatworms (Reynoldson 1985). With the burgeoning interest in biological introductions, summarized by Drake et al. (1989), there is every reason to think that such efforts would be at least as successful in the future.

REFERENCES

Allison, A.C. (1982) in *Population Biology of Infectious Diseases* (Anderson, R.M., and May, R.M., eds.), pp. 245–267, Springer-Verlag, Berlin.

Ashton, P.J., and Mitchell, D.S. (1989) in *Biological Invasions: A Global Perspective* (Drake, J.A., Mooney, H.A., diCastri, F., et al., eds.), pp. 111–154, Wiley, Chichester.

Barbour, M.G., DeJong, T.M., and Johnson, A.F. (1976) *J. Biogeogr.* 3, 55–69.

Barbour, M.G., and Johnson, A.F. (1988) in *Terrestrial Vegetation of California*, new expanded edn. (Barbour, M.G., and Major, J., eds.), pp. 223–261, California Native Plant Society, Sacramento.

Bertness, M.D. (1984) *Ecology* 65, 370–381.
Bond, W., and Slingsby, P. (1984) *Ecology* 65, 1031–1037.
Bratton, S.P. (1975) *Ecology* 56, 1356–1366.
Brenchley, G.A., and Carlton, J.T. (1983) *Biol. Bull.* 165, 543–558.
Carey, J. (1982) *Internat. Wildl.* 12(5), 19–28.
Chanin, P.R.F., and Jefferies, D.J. (1978) *Biol. J. Linn. Soc.* 10, 305–328.
Crosby, A.W. (1986) *Ecological Imperialism: The Biological Expansion of Europe, 900–1900.* Cambridge University Press, Cambridge.
Crowell, K.L. (1968) *Ecology* 49, 551–555.
Dean, C. (1988) "Tiny snail is credited as a force shaping the coast," *New York Times*, 23 August 1988, pp. 15,19.
Diamond, J. (1984) *Nature* 310, 452.
Diamond, J. (1986) in *Community Ecology* (Diamond, J., and Case, T.J., eds.), pp. 3–22, Harper & Row, New York.
Diamond, J., and Case, T.J. (1986) in *Community Ecology* (Diamond, J., and Case, T.J., eds.), pp. 65–79, Harper & Row, New York.
Drake, J.A., Mooney, H.A., diCastri, F., et al., eds. (1989) *Biological Invasions: A Global Perspective*, Wiley, Chichester.
Ebenhard, T. (1988) *Swedish Wildlife Research (Viltrevy)* 13(4), 1–107.
Ehrenfeld, D.W. (1970) *Biological Conservation*, Holt, Rinehart and Winston, New York.
Ehrlich, P.R. (1986) in *Ecology of Biological Invasions of North America and Hawaii* (Mooney, H.A., and Drake, J.A., eds.), pp. 79–95, Springer-Verlag, New York.
Ehrlich, P.R. (1989) in *Biological Invasions: A Global Perspective* (Drake, J.A., Mooney, H.A., diCastri, F., et al., eds.), pp. 315–328, Wiley, Chichester.
Elton, C.S. (1958) *The Ecology of Invasions by Animals and Plants*, Methuen, London.
Erlinge, S. (1972) *Oikos* 23, 327–335.
Ewald, P.W. (1983) *Annu. Rev. Ecol. Syst.* 14, 465–485.
Ewel, J.J. (1986) in *Ecology of Biological Invasions of North America and Hawaii* (Mooney, H.A., and Drake, J.A., eds.), pp. 214–230, Springer-Verlag, New York.
Fenner, F., and Ratcliffe, F.N. (1965) *Myxomatosis*, Cambridge University Press, Cambridge.
Floyd, R.B., and Easteal, S. (1986) in *Ecology of Biological Invasions* (Groves, R.H., and Burdon, J.J., eds.), pp. 151–157, Cambridge University Press, Cambridge.
Ford, D. (1988) *New Yorker*, 25 July 1988, pp. 34–63.
Fritts, T.H., Scott, N.J., Jr., and Savidge, J.A. (1987) *The Snake* 19, 51–58.
Funasaki, G., Lai, P-Y., Nakahara, L.M., Beardsley, J.W., and Ota, A.K. (1988) *Proc. Haw. Entomol. Soc.* 28, 105–160.
Gilbert, L.E. (1980) in *Conservation Biology: An Evolutionary-Ecological Perspective* (Soule, M.E., and Wilcox, B.A., eds.), pp. 11–33, Sinauer, Sunderland, MA.
Gosling, M. (1989) *New Scientist*, March 4, pp. 44–49.
Grue, C.E. (1985) *Nature* 316, 301.
Herbold, B., and Moyle, P.B. (1986) *Amer. Natur.* 128, 751–760.
Holdridge, L.R. (1940) *Caribb. For.* 1, 19–29.
Howe, H.F., and Westley, L.C. (1988) *Ecological Relationships of Plants and Animals*, Oxford, New York.
Hughes, N.F. (1986) *J. Fish. Biol.* 29, 541–548.
Jenkins, J.M. (1983) *Ornithological Monogr.* 31, 1–61.

Johnston, R.F., and Selander, R.K. (1964) *Science* 144, 548–550.
Johnston, R.F., and Selander, R.K. (1971) *Evolution* 25, 1–28.
Kerfoot, W.C., and Sih, A., eds. (1987) *Direct and Indirect Impacts on Aquatic Communities*, University Press of New England, Hanover, NH.
King, C. (1984) *Immigrant Killers: Introduced Predators and the Conservation of Birds in New Zealand*, Oxford, Auckland.
Kitching, J.A., and Ebling, F.J. (1967) *Adv. Ecol. Res.* 4, 197–291.
Knopf, F.L. (1986) *Wildl. Soc. Bull.* 14, 132–142.
Knopf, F.L., and Olson, T.E. (1984) *Wildl. Soc. Bull.* 12, 289–298.
Knutson, L., Sailer, R.I., Murphy, W.L., and Dogger, J.R. (1986) *Western Hemisphere Immigrant Arthropod Database*, Biosystematics and Beneficial Insects Institute, Agricultural Research Service, U.S. Department of Agriculture, Washington, D.C.
Krebs, C.J. (1985) *Ecology: The Experimental Analysis of Distribution and Abundance*, 3rd ed., Harper & Row, New York.
Lawton, J.H. (1984) in *Ecological Communities: Conceptual Issues and the Evidence* (Strong, D.R., Simberloff, D., Abele, L.G., and Thistle, A.B., eds.), pp. 67–100, Princeton University Press, Princeton, NJ.
Lever, C. (1977) *The Naturalised Animals of the British Isles*, Granada, London.
Levin, B.R., Allison, A.C., Bremermann, H.J., et al. (1982) in *Population Biology of Infectious Diseases* (Anderson, R.M., and May, R.M., eds.), pp. 213–243, Springer-Verlag, Berlin.
Long, J.L. (1981) *Introduced Birds of the World*, Universe Books, New York.
Loope, L.L., and Scowcroft, P.G. (1985) in *Hawai'i's Terrestrial Ecosystems: Preservation and Management* (Stone, C.P., and Scott, J.M., eds.), pp. 377–402, Cooperative National Park Resources Studies Unit, University of Hawaii, Honolulu.
Lubchenco, J. (1983) *Ecology* 64, 1116–1123.
Lubchenco, J., and Menge, B.A. (1978) *Ecol. Monogr.* 48, 67–94.
Lubin, Y.D. (1984) *Biol. J. Linn. Soc.* 21, 229–242.
MacArthur, R.H., and Wilson, E.O. (1967) *The Theory of Island Biogeography*, Princeton University Press, Princeton, NJ.
Macdonald, I.A.W., Loope, L.L., Usher, M.B., and Hamann, O. (1989) in *Biological Invasions: A Global Perspective* (Drake, J.A., Mooney, H.A., diCastri, F., et al., eds.), pp. 215–255, Wiley, Chichester.
Mack, R.N. (1989) in *Biological Invasions: A Global Perspective* (Drake, J.A., Mooney, H.A., diCastri, F., et al., eds.), pp. 155–179, Wiley, Chichester.
McCormick, J.F., and Platt, R.B. (1980) *Amer. Midl. Natur.* 104, 264–273.
Medeiros, A.C., Loope, L.L., and Cole, F.R. (1986) in *Proceedings of Sixth Conference in Natural Sciences, Hawaii Volcanoes National Park* (Smith, C.W., and Stone, C.P., eds.), pp. 39–51, Cooperative National Park Resources Studies Unit, University of Hawaii, Honolulu.
Miller, L. (1988) *Fremontia* 16(3), 6–7.
Monro, J. (1967) *J. Anim. Ecol.* 36, 531–547.
Mooney, H.A., Hamburg, S.P., and Drake, J.A. (1986) in *Ecology of Biological Invasions of North America and Hawaii* (Mooney, H.A., and Drake, J.A., eds.), pp. 250–272, Springer-Verlag, New York.
Moyle, P.B., Li, H.W., and Barton, B.A. (1986) in *Fish Culture in Fisheries Management* (Stroud, R.H., ed.), pp. 415–426, American Fisheries Society, Bethesda, MD.

Myers, J.H., Monro, J., and Murray, N. (1981) *Oecologia* 51, 7–13.
Myers, R.L. (1984) in *Cypress Swamps* (Ewel, K.C., and Odum, H.T., eds.), pp. 358–364, University of Florida Press, Gainesville.
Olson, T.E., and Knopf, F.L. (1986) *Wildl. Soc. Bull.* 14, 492–493.
Opler, P.A. (1978) in *Proceedings American Chestnut Symposium*, pp. 83–85, University of West Virginia, Morgantown.
Osmond, C.B., and Monro, J. (1981) in *Plants and Man in Australia* (Carr, D.J., and Carr, S.G.M., eds.), pp. 194–222, Academic Press, New York.
Paine, R.T. (1966) *Amer. Natur.* 100, 65–75.
Paine, R.T. (1969) *Amer. Natur.* 103, 91–93.
Paine, R.T. (1988) *Ecology* 69, 1648–1654.
Pankakoski, E., and Nurmi, K. (1986) *Ann. Zool. Fennici* 23, 1–32.
Payne, I. (1987) *New Scientist* Aug. 27, pp. 50–54.
Peters, R.H. (1988) *Ecology* 69, 1673–1676.
Petraitis, P.S. (1983) *Ecology* 64, 522–533.
Pietsch, M. (1970) *Zeitschr. für Säugetierkunde* 35, 257–288.
Porter, S.D., Van Eimeran, B., and Gilbert, L.E. (1988) *Annals Entomol. Soc. Amer.* 81, 913–918.
Price, P.W. (1984) in *Ecological Communities: Conceptual Issues and the Evidence* (Strong, D.R., Simberloff, D., Abele, L.G., and Thistle, A.B., eds.), pp. 510–524, Princeton University Press, Princeton, NJ.
Quimby, P.C. (1982) in *Biological Control of Weeds with Plant Pathogens* (Charudattan, R., and Walker, H.L., eds.), pp. 47–60, Wiley, New York.
Race, M.S. (1982) *Oecologia* 54, 337–347.
Ralph, C.J., and Sakai, H.F. (1979) *Elepaio* 40, 20–26.
Ramakrishnan, P.S., and Vitousek, P.M. (1989) in *Biological Invasions: A Global Perspective* (Drake, J.A., Mooney, H.A., diCastri, F., et al., eds.), pp. 281–300, Wiley, Chichester.
Reynoldson, T.B. (1985) *Brit. Ecol. Soc. Bull.* 16, 80–86.
Savidge, J. (1984) *Biolog. Conserv.* 30, 305–317.
Savidge, J. (1985) *Nature* 316, 301.
Savidge, J. (1987) *Ecology* 68, 660–668.
Schardt, J.D. (1985) *1985 Florida Aquatic Plant Survey*, Bureau of Aquatic Plant Research and Control, Florida Department of Natural Resources, Tallahassee.
Scowcroft, P.G. (1986) *Proceedings of Sixth Conference in Natural Sciences, Hawaii Volcanoes National Park* (Smith, C.W., and Stone, C.P., eds.), Cooperative National Park Resources Study Unit, University of Hawaii, Honolulu.
Selander, R.K., and Johnston, R.F. (1967) *Condor* 99, 217–258.
Simberloff, D. (1981) in *Biotic Crises in Ecological and Evolutionary Time* (Nitecki, M.H., ed.), pp. 53–81, Academic Press, New York.
Simberloff, D. (1985) in *Engineered Organisms in the Environment: Scientific Issues* (Halvorson, H.O., Pramer, D., and Rogul, M., eds.), pp. 152–161, American Society for Microbiology, Washington, D.C.
Simberloff, D. (1986) in *Ecology of Biological Invasions of North America and Hawaii* (Mooney, H.A., and Drake, J.A., eds.), pp. 3–26, Springer-Verlag, New York.
Singer, F.J., Swank, W.T., and Clebsch, E.E.C. (1984) *J. Wildl. Management* 48, 464–473.
Slobodchikoff, C.N., and Doyen, J.T. (1977) *Ecology* 58, 1171–1175.

Stone, C.P. (1985) in *Hawai'i's Terrestrial Ecosystems: Preservation and Management* (Stone, C.P., and Scott, J.M., eds.), pp. 251–297, Cooperative National Park Resources Studies Unit, University of Hawaii, Honolulu.
Taylor, A.D. (1990) *Ecology* 71, 429–436.
Usher, M.B. (1986) *Phil. Trans. R. Soc. Lond. B* 314, 695–710.
Usher, M.B. (1989) in *Biological Invasions: A Global Perspective* (Drake, J.A., Mooney, H.A., diCastri, F., et al., eds.), pp. 463–489, Wiley, Chichester.
van Riper, C., van Riper, S.G., Goff, M.L., and Laird, M. (1986) *Ecol. Monogr.* 56, 327–344.
Vitousek, P.M. (1986) in *Ecology of Biological Invasions of North America and Hawaii* (Mooney, H.A., and Drake, J.A., eds.), pp. 163–176, Springer-Verlag, New York.
Vivrette, N.J., and Muller, C.H. (1977) *Ecol. Monogr.* 47, 301–318.
von Broembsen, S.L. (1989) in *Biological Invasions: A Global Perspective* (Drake, J.A., Mooney, H.A., diCastri, F., et al., eds.), pp. 77–83, Wiley, Chichester.
Walsh, G.E. (1967) in *Estuaries* (Lauff, E.H., ed.), pp. 420–431, A.A.A.S., Washington, D.C.
Waterhouse, D.F. (1974) *Scientific Amer.* 230(4), 100–109.
Williamson, M.H., and Brown, K.C. (1986) *Phil. Trans. R. Soc. Lond. B* 314, 505–522.
Yamada, S.B., and Mansour, R.A. (1987) *J. Exp. Mar. Biol. Ecol.* 105, 187–196.

CHAPTER
2

Planned Introductions in Biological Control

L.E. Ehler

The potential environmental impact of the planned release of organisms modified through biotechnology continues to generate discussion and debate (Halvorson et al. 1985; National Academy of Sciences 1987; Hodgson and Sugden 1988; Marois and Bruening 1990). One of the most recent developments comes in the form of a perspective on the release of genetically engineered organisms published by the Ecological Society of America. In their report on behalf of the society, Tiedje et al. (1989) note that most engineered organisms will probably pose a minimal environmental risk; nevertheless, there remains considerable concern over potential undesirable effects. The authors classified such effects into seven categories: (1) creation of new pests, (2) enhancement of effects of existing pests, (3) harm to nontarget species, (4) disruptive effects on biotic communities, (5) adverse effects on ecosystem processes, (6) incomplete degradation of hazardous chemicals, and (7) squandering of valuable biological resources. These categories are necessarily speculative because of the very limited experience thus far with release of genetically engineered organisms into the environment. In view of the small number of empirical examples involving transgenic organisms, we must resort to examining other kinds of planned introductions in order to develop an ecological perspective. The purpose of the present chapter is to consider the role and effect of planned introductions of biological-control agents. The environmental impacts of these introductions relate largely to categories (3) and (4) of Tiedje et al. (1989).

2.1 CLASSICAL BIOLOGICAL CONTROL

Biological control can be operationally defined as the impact of a natural enemy (or enemies) that maintains a population (often a pest) at levels lower than would occur in the absence of the enemy (or enemies). Such biological control can be broken down into different categories, as summarized in Table 2–1. The major concern in this chapter is *classical* biological control, with particular emphasis on those planned introductions directed against arthropod pests and weeds.

2.1.1 Theory and Practice

Classical biological control has been practiced on an organized basis for about 100 years. In fact, the year 1989 marks the centennial observance of the first major success—the complete biological control of cottony-cushion scale (*Icerya purchasi* Maskell) in California by two imported natural enemies from Australia (vedalia beetle, *Rodolia cardinalis* [Mulsant] and the parasitic fly, *Cryptochaetum iceryae* [Williston]). Most of the effort in the

TABLE 2–1 The Major Categories of Modern Biological Control

Category	*Description*
Natural biological control	Biological control brought about by native (coevolved) natural enemies in the native home (area of origin) of a given species
Applied biological control	Biological control brought about through human intervention of one sort or another
Classical biological control	Biological control brought about through intentional introduction of exotic enemies, usually from the native home of the target pest
Augmentative biological control	Biological control brought about either through manipulation of natural enemies or modification of the environment; applies to both native and exotic pest species

last 100 years has been directed against phytophagous insects and weeds. As indicated in Table 2–2, there is an ecological relationship between control of phytophagous insects and weeds, and in this context, classical biological control of weeds can be thought of as the "inverse" of biological control of phytophagous insects. In the former, as shown in the lower part of the table, a phytophagous insect is intentionally introduced without its natural enemies, whereas in the latter, as shown in the upper part of the table, the natural enemies of the phytophagous insect are intentionally introduced.

In practice, classical biological control is generally carried out in the following sequence:

1. Foreign exploration for candidate natural enemies, usually in the native home of the target pest;
2. Quarantine or containment of imported material, designed to destroy unwanted or harmful species and to propagate candidates for release;
3. Mass culture of those biological-control agents cleared for release;
4. Colonization or field release of appropriate natural enemies; and
5. Evaluation of the release (if successful), including evaluation of ecological, economic, and environmental impacts.

TABLE 2–2 The Relationship between Classical Biological Control of: (Upper Part) Phytophagous Insects Utilizing Natural Enemies and (Lower Part) Weeds Utilizing Phytophagous Insects

Area of Origin	*Area Where Introduced*	*Comments*
Plants	Crops	Intentionally introduced in most cases
Phytophagous insects	Pests	Accidentally introduced without their natural enemies and reach outbreak levels
Natural enemies	Biological-control agents	Intentionally introduced to bring about classical biological control of target pests
Plants	Weeds	Either intentionally or unintentionally introduced
Phytophagous insects	Biological-control agents	Highly host-specific phytophagous insects intentionally introduced without their natural enemies
Natural enemies	—	Natural enemies intentionally excluded so as to permit "outbreaks" of phytophagous biological-control agent, designed to bring about classical biological control of target weed

The practical aspects of these procedures are discussed in DeBach (1964a) and Huffaker and Messenger (1976).

In many ways, the practice of classical biological control has preceded, and is more developed than, the attendant theory (Ehler 1990a). In this context, two of the most fundamental theoretical problems remain to be solved. First, there is an inadequate theoretical framework for devising realistic introduction strategies. By introduction strategy, I mean the choice of a species or combination of species to release for control of a target pest in a given ecological situation. Secondly, we lack a generally acceptable theory to account for the operation of successful natural enemies in the field. Turning biological control into a predictive science should thus be an important goal for both theoreticians and practitioners. Improved predictability with respect to potential environmental impact of candidate natural enemies would also be helpful.

2.1.2 Kinds of Biological-Control Agents

For a given insect pest, we can envision a "natural-enemy pool," from which candidate species can be drawn for possible introduction. This pool could contain natural enemies from a variety of taxa, including the following: vertebrates, such as entomophagous fish, mammals, reptiles, amphibians, and birds; arthropods, primarily predaceous and parasitic insects, predaceous mites, and spiders; other invertebrates, such as entomophathogenic nematodes; and pathogenic microorganisms, such as viruses, bacteria, protozoa, and fungi. For weeds, a similar "herbivore pool" exists and would include herbivorous vertebrates, phytophagous arthropods (chiefly insects and mites), other herbivorous invertebrates, and plant pathogens. However, in classical biological control of weeds, only host-specific agents are generally introduced and this greatly reduces the number of available candidates.

In the past 100 years, most introductions have involved parasitic and predaceous insects (versus insect pests) and phytophagous insects (versus weeds). Therefore, much of our knowledge and understanding of environmental impact of introduced biological-control agents comes from projects involving these insects. Insectan natural enemies of insect pests and weeds will continue to be important in biological control; however, additional kinds of enemies will likely play an increased role in the future. These include entomogenous nematodes that vector pathogenic bacteria (families Steinernematidae and Heterorhabditidae), predaceous mites in the family Phytoseiidae, and pathogenic microorganisms altered through biotechnology.

2.1.3 Empirical Record

Since 1960, a number of reviews of classical biological-control projects have been published. The major ones are listed in chronological order in Table 2-3. Most of these reviews are well known among biological-control spe-

TABLE 2–3 Major Reviews Summarizing Classical Biological Control of Arthropod Pests and Weeds throughout the World (Since 1960)

Reference	*Coverage*	*Comments*
Wilson (1960)	Insect pests and weeds in Australia and Australian New Guinea	Some analysis of data by Wilson (pp. 69–79)
McLeod (1962)	Pests of crops, fruit trees, ornamentals, and weeds in Canada up to 1959	Some analysis of data by McLeod (pp. 21–26); further analysis of data by Beirne (1975)
McGugan and Coppel (1962)	Forest insect pests in Canada from 1910–1958	Analysis of data by Beirne (1975)
DeBach (1964b)	Successes against insects; global coverage	
CIBC (1971)	Insect pests and weeds in Canada from 1959–1968	Analysis of data by E.G. Munroe (pp. 213–255)
Rao et al. (1971)	Insects and other pests in southeast Asia and Pacific region	
Greathead (1971)	Insect pests and weeds in the Ethiopian region	Analysis of data by Greathead (pp. 95–102).
Greathead (1976)	Primarily insect pests in western and southern Europe	
Laing and Hamai (1976)	Insect pests and weeds; global treatment of successful cases	Some analysis of data by Laing and Hamai
Clausen (1978)	Arthropod pests and weeds; global treatment of all attempts	Analysis of data by DeBach (1971b), Hall and Ehler (1979), Hall et al. (1980), Ehler and Hall (1982), Hokkanen and Pimentel (1984), Hokkanen (1985)
Luck (1981)	Parasitic insects introduced to control arthropod pests; global treatment	Based on Clausen (1978) and additional papers; data analysis by Stiling (1990)
Julien (1982)	Global treatment of introductions against weeds	Data analysis by Julien et al. (1984)
Kelleher and Hulme (1984)	Insect pests and weeds in Canada from 1969–1980	

Continued

TABLE 2–3 Continued Major Reviews Summarizing Classical Biological Control of Arthropod Pests and Weeds throughout the World (Since 1960)

Reference	*Coverage*	*Comments*
Cock (1985)	Insect pests and weeds in the Commonwealth Caribbean and Bermuda up to 1982	
Julien (1987)	Global treatment of introductions against weeds	Revised edition of Julien (1982)
Waterhouse and Norris (1987)	Selected pests and weeds in the southwest Pacific	
Cameron et al. (1987)	Invertebrates imported against invertebrate pests and weeds in New Zealand	
Funasaki et al. (1988)	Biological-control agents purposely introduced into Hawaii from 1890–1985	Analysis of data by the authors

cialists, so the list is meant primarily for the benefit of those in other disciplines who are concerned with biology of invasions. Many of the reviews contain a wealth of detailed information on target pests, natural enemies introduced, outcomes of introductions, and reference to the original literature. As such, they are a valuable source of empirical data for testing hypotheses concerning invasions and should be explored by ecologists and evolutionary biologists. Although some analyses of portions of the data (including hypothesis testing) have been carried out (see Comments column in Table 2–3), the fact remains that this massive data set has not been fully exploited.

Probably the best data sets available (global coverage) are being assembled in England and as yet are unpublished. The Silwood Project on the Biological Control of Weeds augments the summaries by Julien (1982, 1987) with first-hand, unpublished accounts (Moran 1985; Crawley 1989). The CAB International Institute of Biological Control (CIBC) database covers introductions against arthropod pests and has already proven valuable in testing hypotheses concerning introduction strategies (Waage and Greathead 1988; Waage 1990).

2.2 ENVIRONMENTAL IMPACT OF INTRODUCED BIOLOGICAL-CONTROL AGENTS

The interest in environmental impact of introduced biological-control agents is relatively recent. There are at least two reasons for this: First, practitioners of classical biological control have traditionally regarded their method as environmentally safe (Ehler 1990b), and this view was seldom if ever questioned. Howarth (1983) was one of the first to seriously challenge the assumption that biological control was without environmental risks, and he went on to call for more rigorous protocols for clearing candidate natural enemies for importation (see also Howarth 1985; Gagne and Howarth 1985). As a result, the environmental impact of planned introductions of natural enemies has suddenly been added to the research agenda in biological control (Harris 1990). Secondly, in recent years we have witnessed a growing concern over the environmental effects of introduced species in general (Wilson and Graham 1983; Groves and Burdon 1986; Macdonald et al. 1986; Mooney and Drake 1986; Kornberg and Williamson 1987), and genetically engineered organisms in particular (Halvorson et al. 1985; National Academy of Sciences 1987; Hodgson and Sugden 1988; Tiedje et al. 1989; Marois and Bruening 1990). As there is only a limited amount of information on potential environmental effects of genetically engineered organisms, it can be instructive to consider results from planned introductions in biological control when developing an ecological framework for assessing the environmental impact of the introduction of transgenic species.

Since 1986, there have been at least four reviews by biological-control workers that address various aspects of environmental impact of introduced natural enemies. Legner (1986b) noted that environmental risk relates not only to nontarget organisms, but also to the risks of making introductions that may preclude or adversely affect biological control at a later date. Legner also ranked various biological-control agents according to their potential risk, such that the amount of pre-introductory assessment of a candidate species would be directly related to its potential environmental impact. Coulson and Soper (1989) reviewed the current federal legislation and protocols for the introduction of biological-control agents in the United States; some examples of environmental impact were also given. Ehler (1990b) suggested that the environmental impact of a biological-control agent was a function of the attributes of the introduced species, the nature of the target zone, and the introduction strategy employed. He also reviewed several cases involving effects of introduced natural enemies on nontarget species, and pointed out some implications for the planned release of genetically engineered organisms. Harris (1990) reviewed several pertinent case histories, with particular emphasis on biological control of weeds. Collectively, these reviews cover most of the important issues and many of the relevant case histories. Therefore, the present discussion is designed to complement what has already been stated. I begin with an ecological perspective.

2.2.1 Ecological Perspective

In an ecological context, environmental impact can be defined as any effect of an introduced species on a non-target organism. Thus, every introduction of an exotic species can be expected to have an environmental impact of one sort or another. These impacts may or may not be predictable, and their effects can vary according to a number of factors, including the following:

1. Outcome (negative or positive),
2. Magnitude (from very minor to major),
3. Duration (from temporary or short-term to long-term),
4. Nature of interaction (direct versus indirect), and
5. Timing (immediate versus delayed).

Clearly, environmental "risk" and environmental "impact" are different matters, and I suggest that all kinds of impacts (as opposed to just major negative ones) be assessed in a given release program. Every release of a biological-control agent or a transgenic species constitutes a perturbation experiment, and all attendant ecological effects should be documented so as to enhance our understanding of the structure and function of the target system.

There are a number of projects in classical biological control that provide examples of the kinds of effects described above (Ehler 1990b) and many of the relevant examples have been discussed elsewhere (Legner 1986b; Coulson and Soper 1989; Ehler 1990b; Harris 1990). In the present discussion, examples of environmental impact of introduced biological-control agents will be used to illustrate ecological effects due to the following: lack of host specificity, reduction in abundance of target pest, competitive interactions among natural enemies, evolutionary change, agronomic change, and hyperparasites.

2.2.2 Lack of Host Specificity

Many of the negative impacts associated with introduced insects can be attributed to the lack of host specificity. For example, *Cactoblastis cactorum* (Berg), the well-known biological-control agent for *Opuntia* spp., was also found to feed on melon and green tomato in Australia; however, this was evidently a temporary event, and there have been no reports of damage to nontarget species since 1931 (Dodd 1940; Waterhouse and Norris 1987; Harris 1990). In East Africa, the lace bug *Teleonemia scrupulosa* (Stal), which was introduced for control of lantana (*Lantana camara* L.), was observed feeding on sesame (*Sesamum indicum* L.) following a population explosion on the target weed (Davies and Greathead 1967; Greathead 1971). The lace bug evidently could not successfully develop on this crop, however, and according to Harris (1990), the problem "... seems to have disappeared." For both *Cactoblastis* and *Teleonemia*, exploitation of nontarget hosts followed massive defoliation of the target weed and was presumably the result of starving insects seeking a suitable host—as opposed to being a true host shift. The case of *Chrysolina quadrigemina* (Suffr.) in northern

California is of greater concern (see Section 2.2.5). The final example from biological control of weeds is *Rhinocyllus conicus* (Froelich), a flowerhead weevil introduced into North America from Europe for control of weedy thistles (Schroeder 1980). This weevil is not strictly host specific (Zwolfer and Harris 1984), and in recent years, it has been reared from a number of native *Cirsium* thistles in California (Turner et al. 1987). The latter authors noted that the extension of the host range of *Rhinocyllus* was perhaps predictable, as European *Cirsium* spp. were known to be suitable hosts. The magnitude of the problem in California is not clear at present.

Negative impacts can also attend the introduction of natural enemies for insect control. This is especially critical if the nontarget species happens to be a beneficial one. Goeden and Louda (1976) describe several instances where an imported predator or parasite of an insect pest also exploited a phytophagous insect imported for weed control. For example, the mealybug destroyer, *Cryptolaemus montrouzieri* Mulsant, was introduced into South Africa for control of citrophilus mealybug, *Pseudococcus calceolariae* (Maskell); however, it also preyed on *Dactylopius opuntiae* (Cockerell), a cochineal insect imported for control of *Opuntia* cactus. The impact on *Dactylopius* was considerable and evidently precluded effective biological control of *Opuntia megacantha* Salm-Dyck in certain areas. In Hawaii, the nearctic egg parasite *Trichogramma minutum* Riley also parasitized eggs of *Bactra truculenta* Meyrick, an introduced biological-control agent for purple nutsedge, *Cyperus rotundus* L. This parasitism was evidently a factor in preventing *Bactra* from achieving the levels of biological control desired.

The environmental impact of imported natural enemies that are not strictly host specific can actually be positive in those cases where the nontarget species is also a pest (Ehler 1990b). When this is unexpected and the impact is significant, it is termed "fortuitous biological control" (cf. DeBach 1971a). There are many examples of fortuitous biological control of insect pests—as a result of both planned and unplanned introductions (Wilson 1965; DeBach 1971a; Clausen 1978).

Examples of environmental impact due to lack of host specificity are not restricted to insectan natural enemies. Vertebrate predators have been introduced in a number of cases for control of either invertebrates, other vertebrates, or weeds (Davis et al. 1976; Wood 1985; Coulson and Soper 1989; Harris 1990). These introductions have not been without negative impacts, and as a result, vertebrate predators have received some "bad publicity" (Coulson and Soper, 1989). Good examples include the Indian mongoose (*Herpestes auropunctatus* Hodg.) in Puerto Rico and Hawaii, and the giant toad (*Bufo marinus* [L.]) in a number of locations. Because of these experiences, there has been little enthusiasm in recent years for importation of vertebrate agents. A major exception to this involves fish—for either insect or weed control. One of the best examples is the mosquito fish, *Gambusia affinis* (Baird and Girard). This predator is native to the southeastern United States and has been introduced throughout the world for mosquito control. Although *Gambusia* is often an effective biological-control agent, it can also have a considerable effect on nontarget organisms, ultimately resulting in a restructuring of the target system (Hurlbert et al. 1972; Walters and Legner 1980; Miura et al. 1984). Similar effects could presumably result from introduction of herbivorous fish for control of weeds.

Introduced mollusks have generated considerable controversy in recent years. In Hawaii, the rosy snail (*Euglandina rosea* [Ferussac]) was imported from Florida for control of the giant African snail (*Achatina fulica* Bowdich), an agricultural pest. This predator also feeds on other mollusks and has been implicated in the extinction of many endemic tree snails (Howarth and Medeiros 1989). This same carnivorous snail was introduced into the island of Moorea (French Polynesia) for control of *A. fulica*, where it has apparently led to the extinction of at least one endemic snail and seriously threatens many more (Clarke et al. 1984). To make matters worse, *E. rosea* is not considered to be an effective biological-control agent in either place (Howarth and Medeiros 1989; Clarke et al. 1984), leading one to seriously question the continued introduction of this predator.

2.2.3 Reduction of Target Pest

When imported natural enemies result in a substantial reduction in abundance of the target pest, we can expect concomitant effects on nontarget species that have interacted with the target pest prior to introduction. For example, Neuenschwander et al. (1987) reported that, following the biological control of cassava mealybug (*Phenacoccus manihoti* Matile-Ferrero) in Africa, densities of native predaceous insects that previously exploited the exotic mealybug declined sharply. This was attributed to interspecific competition for hosts with the newly introduced parasite *Epidinocarsis lopezi* (DeSantis). These kinds of results are important and should be reported; however, they are primarily of intellectual value and should not be construed as a criticism of an introduction which results in successful control of an exotic pest. The situation for native target pests might be different however, depending on the circumstances. This is especially true in the case of native weeds (see below).

Apart from affecting associated natural enemies, the reduction in abundance of a target pest can also have an indirect effect on nontarget pest species. Here I have in mind "pest replacement," which occurs when a nontarget pest increases in abundance following the biological control of a target pest. This can be expected when the nontarget species has also been suppressed by interspecific competition with the target species and (or) by the chemical-control measures required for suppression of the target pest. Ehler (1990b) describes two examples of this phenomenon in California: (1) increases in dusky-veined walnut aphid (*Callaphis juglandis* [Goeze]) following the biological control of walnut aphid (*Chromaphis juglandicola*

[Kalt.]), and (2) increases in black scale (*Saissetia oleae* [Oliv.]) following biological control of olive scale (*Parlatoria oleae* [Colvee]). In both cases, the replacement pests were also exotic and soon became targets for classical biological control. Both projects continue at present. Pest replacement among exotic insects can probably be predicted in many situations; this capability would be especially helpful in those cases where the predicted replacement species is likely to cause major problems. The situation with native insect pests is complicated by the presence of native natural enemies, which may or may not prevent a potential native replacement species from reaching outbreak levels.

Biological control of weeds has been particularly successful against introduced plants in rangeland ecosystems. The associated plant communities are evidently "competitive" because other plant species normally replace the plant that was controlled. Of course, replacement plants need not be weeds; Waterhouse and Norris (1987) and Harris (1990) cite examples in which another exotic weed replaced the target exotic weed following successful biological control of the latter. In many cases, such replacement may also be predictable and should of course be taken into account in planning the project. The environmental impact associated with the biological control of a native weed is another matter. A great reduction in abundance of a native plant could have far-reaching effects in an ecosystem, including changes in structure of the associated plant community and various effects on endemic wildlife. Biological control of native weeds is a rather contentious issue at present and the reader is referred to reviews by Andres (1981), DeLoach (1981), Johnson (1985), Turner (1985), and Harris (1988) for further information.

2.2.4 Competitive Interactions

Many projects in classical biological control involve introduction of two or more natural-enemy species. Such multiple-species introductions can be carried out either simultaneously or sequentially; in either case, competition among the species released can result in an environmental impact. Two phenomena are of practical concern in biological control: competitive coexistence and competitive exclusion. Competitive coexistence occurs when two or more introduced species establish and compete for resources (e.g., the host). As the target pest population is reduced, interspecific competition among the natural enemies can be expected to intensify. However, this competition should not necessarily be viewed as a major negative impact, because it may very well result in increased levels of pest suppression. Although definitive empirical evidence is hard to come by, current theory suggests that the more natural-enemy species that are packed into a host-enemy system, the greater the impact on the host population will be. Competitive exclusion is perhaps of greater concern; this occurs when an introduced species fails to establish because of competition from one or more species released simultaneously or established previously. Several biological-control workers have suggested that competitive exclusion prevented establishment of certain introduced species; these include Flanders (1940), DeBach (1965), Oatman and Platner (1974), Cock (1986), and Legner (1986b). In a global analysis of the hypothesis, Ehler and Hall (1982) showed that the rates of establishment of natural enemies introduced against insect pests in the orders Homoptera, Lepidoptera, and Coleoptera were inversely related to (1) the number of species released simultaneously and (2) the number of exotic incumbent species.

Competitive exclusion remains controversial in biological control. This is due in large measure to the lack of definitive "proof" that it has occurred in a given situation. Although the available evidence is correlational in nature, there is good reason to believe that competitive exclusion has indeed occurred and thereby contributed to the rather low rate of establishment (about 34%) of introduced parasitic and predaceous insects. In this regard, one of the most critical issues involves the competitive exclusion of an introduced natural enemy that is capable of either controlling the target pest or at least contributing significantly to such control. Unfortunately, there is a lack of appropriate evidence to either confirm or reject this conjecture for a given species, and it is hoped that in future projects, the necessary experiments can be conducted during the introduction phase. In summary, where competitive exclusion does occur, it represents a kind of environmental impact that is "strategy dependent" (Ehler 1990b); as such, it can be either immediate or delayed, depending on the circumstances.

2.2.5 Evolutionary Change

Introduced natural enemies that establish and persist must make the genetic adjustments necessary for survival in the new environment. In the context of environmental impact, a major concern is the evolution of new races that can exploit new host species. Such genetically based host shifts are especially worrisome in biological control of weeds, because a new host race of an introduced agent could easily become a pest. The case of *Chrysolina quadrigemina* (Suffr.) in northern California may be a case in point. This well-known and successful biological-control agent for Klamath weed (*Hypericum perforatum* L.) has in recent years become somewhat of a pest on a congeneric plant, *H. calycinum* L., an exotic plant used extensively as a ground cover in northern California (C.S. Koehler, personal communication). Observations by Andres (1985) suggest that this is the result of the evolution of a new host race of *Chrysolina.* If so, this would represent a kind of "delayed" environmental impact of an introduced species. Fortunately, the problem on *H. calycinum* is not considered to be major; however, this could change as the presumed host-race formation continues in *Chrysolina.* This situation should be monitored closely, if for no other reason than to learn more about post-colonization evolutionary change in a biological-control agent.

2.2.6 Agronomic Change

Annecke et al. (1976) describe a rather unusual example of a delayed environmental impact involving biological control of *Opuntia* cactus in South Africa. Introduced natural enemies were relatively successful in controlling *Opuntia ficus-indica* (L.) Miller (Annecke and Moran 1978); these enemies included *Cactoblastis* and *Dactylopius* discussed earlier. However, this success in biological control of weeds also brought about problems when spineless varieties of *Opuntia* were developed later for livestock feed. This agronomic change in the status of *Opuntia* (i.e., from weed to crop) was primarily in response to recurring drought. As expected, exotic phytophagous insects, including *Cactoblastis* and *Dactylopius*, became pests in spineless-*Opuntia* plantations. The fact that phytophagous insects introduced in weed control are imported without their own natural enemies serves to emphasize that these biological-control agents are the ecological equivalents of phytophagous insect pests imported without their natural enemies. Thus, it behooves us to consider both the potential short-term and long-term environmental impacts of introduced natural enemies, particularly in biological control of weeds. Clearly, what is a biological-control agent today could become a pest at some point in the future.

2.2.7 Hyperparasites

A hyperparasite is a species which parasitizes another parasite. This condition can be either obligate or facultative, depending on the species in question. Obligate hyperparasites, which exploit primary parasites used in biological control, are generally considered to be detrimental and are carefully screened out during the quarantine phase of the project. (A primary parasite is a parasite of a nonparasitic host.) However, there have been instances in which a hyperparasite was actually released and eventually established. This would presumably result in a negative environment impact, although the necessary empirical evidence is often lacking. One of the putative examples is *Quaylea whittieri* (Girault), a hyperparasite that was introduced into California in the mistaken belief that it was a primary parasite of black scale (*S. oleae*). It was later discovered that *Quaylea* actually parasitizes *Metaphycus lounsburyi* (Howard), one of the more important exotic primary parasites of black scale in California. The environmental impact of *Quaylea* is evidently minor because black scale is considered to be under good biological control on citrus (DeBach 1964b; Laing and Hamai 1976). Black scale remains a serious pest on olive in many parts of northern California; however, *Quaylea* is evidently not present in these areas (L.E. Ehler, unpublished observations; Kennett 1986).

2.2.8 Future Research

In future biological-control programs, the environmental impact of an introduced natural enemy that establishes should be assessed whenever possible—if for no other reason than to learn more about the ecological structure of the target system. For imported predators and parasites of insect pests and for phytophagous insects imported in weed control, there is now a sufficient amount of data on environmental impact to serve as a basis for future investigations. With respect to other kinds of natural enemies, however, comparatively little information is available. Pathogenic microorganisms used in insect control are of particular concern. Whereas some data are available on their toxicity to nontarget beneficial species (Podgwaite 1986; Flexner et al. 1986), we know very little about the impact of these microbial insecticides at the level of the ecological community. Pathogens that are not host-specific could have a considerable impact, depending on the nature of the target system. For example, Miller (1990) has demonstrated that *Bacillus thuringiensis* Berliner var. *kurstaki*, which was applied for control of gypsy moth (*Lymantria dispar* [L.]) in southern Oregon, altered the structure of a complex of oak-feeding Lepidoptera. This is one of the few studies of community-level effects of an introduced pathogen and should serve as a guide for future investigations involving both *B. thuringiensis* and other pathogens with similar host ranges.

Introduced dung beetles should also be on the research agenda. In recent years, there has been considerable interest in the use of dung-burying scarabs for both pasture improvement and suppression of symbovine flies, which breed in cattle dung (Waterhouse, 1974; Legner, 1986a). In some cases, the imported dung beetles may be added to an existing complex of predaceous insects, and perhaps even a guild of parasites. Such a situation apparently exists in southern California, and this has led Legner (1983, 1986a) to question whether or not imported dung beetles are indeed compatible with the resident complex of natural enemies (chiefly predators) of the pests in question. Because of the ecological complexity of the target system, Legner (1986a) suggests that dung reduction and (or) dispersion will not necessarily lead to diminished fly abundance in the target zone. Similar projects are underway in Australia and Texas, and these would seem to be excellent opportunities for assessing the environmental impact of exotic dung beetles. The impact on nontarget species should not be restricted to economically important species, but should include endemic dung beetles and other members of the dung–insect community as well.

Finally, the environmental impact of inundative releases of predators and parasites should not be overlooked. In these programs, large numbers of insects (or mites) from commercial insectaries are released so as to "inundate" the target system and thereby achieve almost-immediate pest suppression. Inundative releases could also be employed as experimental manipulations designed specifically to assess community-level effects of introduced natural enemies.

2.3 CONCLUDING REMARKS

As the result of recent advances in recombinant DNA technology, it is readily apparent that we are entering a new and exciting era in both basic and applied biology. The development of transgenic organisms holds considerable promise in agriculture and medicine, and the planned release of such organisms poses a great challenge for biotechnologists, ecologists, and biological-control workers. In the context of environmental impact, predictability will be of paramount importance.

Biotechnologists and molecular biologists must recognize that planned introductions of genetically engineered species that result in permanent establishment will likely have an impact on nontarget species also. The examples from classical biological control described in this chapter should amply illustrate the variety of effects on nontarget species that can occur. From an ecological standpoint, it might be helpful for the biotechnology industry to develop transgenic species that, when released into an agricultural environment, would have the desired impact on the target problem, but would not permanently establish. This would seem to be a good corporate strategy also, as the same product could be offered again the following growing season. However, if permanent establishment of a transgenic organism is the desired goal, then careful consideration of the potential environmental impact will be required. It is hoped that ecologists and biotechnologists can work together in order to ensure an ecologically sound and economically rewarding release.

The planned release of transgenic species poses a special challenge to ecologists who will be expected to predict the ecological effects of the species in question. This will also provide a good test of the predictive power of ecological science. At this point, it is difficult to be very optimistic about the contributions of ecological theory. In recent years, ecology has been overly influenced by mathematical models that are often oversimplified and of questionable validity in the real world. At the same time, too much attention has been devoted to the solution of theoretical problems, as opposed to real ones, including those of practical importance. The planned release of genetically engineered organisms is welcome in the sense that it will help force ecologists to address those real-world, environmental problems whose solutions rightfully belong to ecological science.

For biological-control workers, environmental impact of introduced natural enemies will continue to be of concern. The development of a predictive theory of biological control will require increased attention. In the past 20 years, mathematical models of host-enemy dynamics have played a major role in the development of biological-control theory. However, these models have been used largely either to explain what we have already observed (e.g., successful biological control) or to justify conventional practices (e.g., multiple-species introductions). Clearly, the time has come for a more predictive theory of biological control, so that for a given planned introduction, the potential ecological impact on both the target population and nontarget populations can be assessed in advance. As transgenic biological-control agents become available, the need for predictive theory will become even greater. At the same time, practitioners of biological control should make every effort to document the environmental impact of the species they introduce, for such cases will continue to provide some of the best examples of environmental impact attendant to planned introduction of exotic beneficial species.

At recent meetings of the International Astronomical Union, concern was expressed over the degradation of the astronomical environment (Waldrop 1988). The problem of orbital debris, or "space junk," such as functioning and derelict satellites, spent rocket casings, and shrapnel from various explosions, is of particular significance. For example, the North American Aerospace Defense Command currently tracks about 6,000 orbital objects that are at least 10 cm in diameter. A collision between a piece of space junk and a surveillance satellite could of course have serious consequences, and hence the growing concern. This astronomical problem should remind us that an analogous problem can occur in the earthbound environment as the number of introduced species in a habitat increases through human activity, including that of biotechnologists and biological-control workers. In other words, we can think of a "space-junk hypothesis," which predicts that, as the number of introduced organisms (including ineffective ones) in the habitat increases, the greater the chance of incompatibility, particularly among those agents released simultaneously or sequentially against a common target. In contrast to the apparent lack of adequate concern for the astronomical environment at the beginning of the space age, it behooves biologists to demonstrate their concern for the earthbound environment by implementing sensible, well-planned introductions of exotic species, including genetically engineered ones.

REFERENCES

Andres, L.A. (1981) in *Biological Control in Crop Production* (Papavizas, G.C., ed.), pp. 341–349, Allanheld & Osmun, Totawa, NJ.

Andres, L.A. (1985) in *Proceedings VI International Symposium on Biological Control of Weeds* (Delfosse, E.S., ed.), pp. 235–239, Agriculture Canada, Ottawa.

Annecke, D.P., Burger, W.A., and Coetzee, H. (1976) *J. Entomol. Soc. Sth. Afr.* 39, 111–116.

Annecke, D.P., and Moran, V.C. (1978) *J. Entomol. Soc. Sth. Afr.* 41, 161–188.

Beirne, B.P. (1975) *Can. Entomol.* 107, 225–236.

Cameron, P.J., Hill, R.L., Valentine, E.W., and Thomas, W.P. (1987) DSIR Bull. no. 242, Wellington, NZ.

Commonwealth Institute of Biological Control (CIBC) (1971) *Biological Control Programmes Against Insects and Weeds in Canada, 1959-1968*. CIBC Tech. Comm. 4, 1–226.

Clarke, B., Murray, J., and Johnson, M.S. (1984) *Pac. Sci.* 38, 97–104.
Clausen, C.P., ed. (1978) *Introduced Parasites and Predators of Arthropod Pests and Weeds: A World Review*, USDA Agric. Hdbk. 480, 1–545.
Cock, M.J.W., ed. (1985) *A Review of Biological Control of Pests in the Commonwealth Caribbean and Bermuda up to 1982*, CIBC Tech. Comm. 9, 1–218.
Cock, M.J.W. (1986) *Biocontr. News Info.* 7, 7–16.
Coulson, J.R., and Soper, R.S. (1989) in *Plant Protection and Quarantine* (Kahn, R.P., ed.), vol. 3, pp. 1–35, CRC, Boca Raton, FL.
Crawley, M.J. (1989) *Annu. Rev. Entomol.* 34, 531–564.
Davies, J.C., and Greathead, D.J. (1967) *Nature* 213, 102–103.
Davis, D.E., Myers, K., and Hoy, J.B. (1976) in *Theory and Practice of Biological Control* (Huffaker, C.B., and Messenger, P.S., eds.), pp. 501–519, Academic Press, New York.
DeBach, P., ed. (1964a) *Biological Control of Insect Pests and Weeds*, van Nostrand-Reinhold, Princeton, NJ.
DeBach, P. (1964b) in *Biological Control of Insect Pests and Weeds* (DeBach, P., ed.), pp. 673–713, van Nostrand-Reinhold, Princeton, NJ.
DeBach, P. (1965) in *Genetics of Colonizing Species* (Baker, H.G., and Stebbins, G.L., eds.), pp. 287–306, Academic Press, New York.
DeBach, P. (1971a) in *Entomological Essays to Commemorate the Retirement of Professor K. Yasumatsu*, pp. 293–307, Hokuryukan, Tokyo.
DeBach, P. (1971b) *Proc. Tall Timb. Conf. Ecol. Anim. Contr. Hab. Mgt.* 3, 211–233.
DeLoach, C.J. (1981) in *Proceedings V. International Symposium Biological Control of Weeds* (Delfosse, E.S., ed.), pp. 175–199, Commonwealth Scientific and Industrial Research Organization, Melbourne.
Dodd, A.P. (1940) *The Biological Campaign Against Prickly Pear*, Commonwealth Prickly Pear Board, Brisbane.
Ehler, L.E. (1990a) in *Critical Issues in Biological Control* (Mackauer, M., Ehler, L.E., and Roland, J., eds.), pp. 111–134, Intercept, Andover, U.K.
Ehler, L.E. (1990b) in *Risk Assessment in Agricultural Biotechnology: Proceedings of the International Conference* (Marois, J.J., and Bruening, G., eds.), University California Division of Agriculture and Natural Resources, Publication No. 1928 (in press).
Ehler, L.E., and Hall, R.W. (1982) *Environ. Entomol.* 11, 1–4.
Flanders, S.E. (1940) *Ann. Entomol. Soc. Am.* 33, 245–253.
Flexner, J.L., Lighthart, B., and Croft, B.A. (1986) *Agric. Ecosyst. Environ.* 16, 203–254.
Funasaki, G.Y., Lai, Po-Yung, Nakahara, L.M., Beardsley, J.W., and Ota, A.K. (1988) *Proc. Haw. Entomol. Soc.* 28, 105–160.
Gagne, W.C., and Howarth, F.G. (1985) in *Proceedings Third Congress European Lepidopterology Cambridge* (1982), pp. 74–84.
Goeden R.D., and Louda, S.M. (1976) *Annu. Rev. Entomol.* 21, 325–342.
Greathead, D.J. (1971) *A Review of Biological Control in the Ethiopian Region*, CIBC Tech. Comm. 5, 1–162.
Greathead, D.J., ed. (1976) *A Review of Biological Control in Western and Southern Europe*, CIBC Tech. Comm. 7, 1–182.
Groves, R.H., and Burdon, J.J., eds. (1986) *The Ecology of Biological Invasions: An Australian Perspective*, Cambridge University Press, Cambridge.

Hall, R.W., and Ehler, L.E. (1979) *Bull. Entomol. Soc. Am.* 25, 280–282.
Hall, R.W., Ehler, L.E., and Bisabri-Ershadi, B. (1980) *Bull. Entomol. Soc. Am.* 26, 111–114.
Halvorson, H.O., Pramer, D., and Rogul, M., eds. (1985) *Engineered Organisms in the Environment: Scientific Issues*, American Society for Microbiology, Washington, D.C.
Harris, P. (1988) *Bioscience* 38, 542–548.
Harris, P. (1990) in *Critical Issues in Biological Control* (Mackauer, M., Ehler, L.E., and Roland, J., eds.), pp. 289–300, Intercept, Andover, U.K.
Hodgson, J., and Sugden, A.M., eds. (1988) *Planned Release of Genetically Engineered Organisms, Trends in Biotechnology/Trends in Ecology and Evolution*, Special Publication, Elsevier, Cambridge.
Hokkanen, H., and Pimentel, D. (1984) *Can. Entomol.* 116, 1109–1121.
Hokkanen, H.M.T. (1985) *CRC Crit. Rev. Plant Sci.* 3(1), 35–72.
Howarth, F.G. (1983) *Proc. Haw. Entomol. Soc.* 24, 239–244.
Howarth, F.G. (1985) in *Hawai'i's Terrestrial Ecosystems: Preservation and Management* (Stone, C.P., and Scott, J.M., eds.), pp. 149–178, University of Hawaii, Honolulu.
Howarth, F.G., and Medeiros, A.C. (1989) in *Conservation Biology in Hawai'i* (Stone, C.P., and Stone, D.B., eds), pp. 82–87, University of Hawaii, Honolulu.
Huffaker, C.B., and Messenger, P.S., eds. (1976) *Theory and Practice of Biological Control*, Academic Press, New York.
Hurlbert, S.H., Zedler, J., and Fairbanks, D. (1972) *Science* 175, 639–641.
Johnson, H.B. (1985) in *Proceedings VI International Symposium on Biological Control of Weeds* (Delfosse, E.S., ed.), pp. 27–56, Agriculture Canada, Ottawa.
Julien, M.H. (1982) *Biological Control of Weeds. A World Catalogue of Agents and Their Target Weeds*, Commonwealth Agricultural Bureaux, Farnham Royal, U.K.
Julien, M.H., ed. (1987) *Biological Control of Weeds. A World Catalogue of Agents and Their Target Weeds*, 2nd ed., CAB International, Wallingford, U.K.
Julien, M.H., Kerr, J.D., and Chan, R.R. (1984) *Prot. Ecol.* 7, 3–25.
Kelleher, J.S., and Hulme, M.A., eds. (1984) *Biological Control Programmes Against Insects and Weeds in Canada 1969-1980*, CIBC Tech. Comm. 8, 1–410.
Kennett, C.E. (1986) *Pan-Pac. Entomol.* 62, 363–369.
Kornberg, H., and Williamson, M.H., eds. (1987) *Quantitative Aspects of the Ecology of Biological Invasions*, Royal Society, London.
Laing, J.E., and Hamai, J. (1976) in *Theory and Practice of Biological Control* (Huffaker, C.B, and Messenger, P.S., eds.), pp. 685–743, Academic Press, New York.
Legner, E.F. (1983) *Proceedings and Papers 51st Annual Conference California Mosquito and Vector Control Association*, pp. 99–101, California Mosquito and Vector Control Association, Sacramento.
Legner, E.F. (1986a) *Entomol. Soc. Am. Misc. Publ.* 61, 120–131.
Legner, E.F. (1986b) *Fortschritte der Zoologie* 32, 19–30.
Luck, R.F. (1981) in *CRC Handbook of Pest Management in Agriculture* (Pimentel, D., ed.), vol. 2, pp. 125–284, CRC, Boca Raton, FL.
Macdonald, I.A.W., Kruger, F.J., and Ferrar, A.A., eds. (1986) *The Ecology and Management of Biological Invasions in Southern Africa*, Oxford, Capetown.
Marois, J.J., and Bruening, G., eds. (1990) *Risk Assessment in Agricultural Biotechnology: Proceedings of the International Conference*, University California, Division of Agriculture and Natural Resources, Publication No. 1928 (in press).

McGugan, B.M., and Coppel, H.C. (1962) *A Review of the Biological Control Attempts Against Insects and Weeds in Canada. Part II. Biological Control of Forest Insects, 1910-1958*. CIBC Tech. Comm. 2, 35–216.
McLeod, J.H. (1962) *A Review of the Biological Control Attempts Against Insects and Weeds in Canada. Part I. Biological Control of Pests of Crops, Fruit Trees, Ornamentals, and Weeds in Canada up to 1959*, CIBC Tech. Comm. 2, 1–34.
Miller, J.C. (1990) *Am. Entomol.* (in press).
Miura, T., Takahashi, R.M., and Wilder, W.H. (1984) *Mosq. News* 44, 510–517.
Mooney, H.A., and Drake, J.A., eds. (1986) *Ecology of Biological Invasions of North America and Hawaii*, Springer-Verlag, New York.
Moran, V.C. (1985) in *Proceedings VI International Symposium on Biological Control of Weeds* (Delfosse, E.S., ed.), pp. 65–68, Agriculture Canada, Ottawa.
National Academy of Sciences. (1987) *Introduction of Recombinant DNA-Engineered Organisms into the Environment: Key Issues*, National Academy of Sciences, Washington, D.C.
Neuenschwander, P., Hammond, W.N.O., and Hennessey, R.D. (1987) *Insect Sci. Applic.* 8(4-6), 893–898.
Oatman, E.R., and Platner, G.R. (1974) *Environ. Entomol.* 3, 262–264.
Podgwaite, J.D. (1986) *Fortschritte der Zoologie* 32, 279–287.
Rao, V.P., Ghani, M.A., Sankaran, T., and Mathur, K.C. (1971) *A Review of the Biological Control of Insects and Other Pests in Southeast Asia and the Pacific Region*, CIBC Tech. Comm. 6, 1–149.
Schroder, D. (1980) *Biocont. News Info.* 1, 9–26.
Stiling, P. (1990) *Am. Entomol.* (in press).
Tiedje, J.M., Colwell, R.K., and Grossman, Y.L., et al. (1989) *Ecology* 70, 298–315.
Turner, C.E. (1985) in *Proceedings VI International Symposium on Biological Control of Weeds* (Delfosse, E.S., ed.), pp. 203–225, Agriculture Canada, Ottawa.
Turner, C.E., Pemberton, R.W., and Rosenthal, S.S. (1987) *Environ. Entomol.* 16, 111–115.
Waage, J.K. (1990) in *Critical Issues in Biological Control* (Mackauer, M., Ehler, L.E., and Roland, J., eds.), pp. 135–157, Intercept, Andover, U.K.
Waage, J.K., and Greathead, D.J. (1988) *Phil. Trans. R. Soc. Lond.* B 318, 111–128.
Waldrop, M.M. (1988) *Science* 241, 1288–1289.
Walters, L.L., and Legner, E.F. (1980) *Hilgardia* 48(3), 1–18.
Waterhouse, D.F. (1974) *Sci. Am.* 230, 100–109.
Waterhouse, D.F., and Norris, K.R. (1987) *Biological Control—Pacific Prospects,*, Inkata, Melbourne.
Wilson, C.L., and Graham, C.L., eds. (1983) *Exotic Plant Pests and North American Agriculture*, Academic Press, New York.
Wilson, F. (1960) *A Review of the Biological Control of Insects and Weeds in Australia and Australian New Guinea*, CIBC Tech. Comm. 1, 1–102.
Wilson, F. (1965) in *Genetics of Colonizing Species* (Baker, H.G., and Stebbins, G.L., eds.), pp. 307–329, Academic Press, New York.
Wood, B.J. (1985) *J. Pl. Prot. Tropics* 2, 67–79.
Zwolfer, H., and Harris, P. (1984) *Z. Ang. Entomol.* 97, 36–62.

PART II

Ecology and Genetics of Microbial Populations

CHAPTER 3

Surface Transport of Microorganisms by Water

James A. Moore

There are several ways in which microorganisms can be transported from one location to another. The two major vehicles are water and air. Other important methods include insects, birds, humans, and field equipment used to work the soil and crops or to apply a specific microbial population. This chapter will only cover transport by water over the surface of the earth. Chapter 5 discusses the mechanisms of the aerial transport of microorganisms.

3.1 ROLE OF THE APPLICATION FORM

It is difficult to look into the future and draw sound conclusions on the application technology that will be used for introducing new organisms. Presumably, however, they may be packaged in one of three forms. (1) They may be applied as a dry powder or dust. (2) They could be in a liquid solution that would be sprayed onto the crop or receiver system. (3) They could be incorporated into pellets, or granules. These and other new techniques are important, as they influence where the organisms will be placed within the soil/plant interface. These application differences will alter their availability and the rate of their movement from the site.

A dusting with a powder on a thick canopy or heavy vegetation crop will place most of the organisms on the crop and very little on the soil surface. This location will allow opportunity for resuspension from a wind and loss from the target site. In this location the organisms would be influenced by moisture from dew during certain climatic conditions. Dew may wet and then attach organisms to the plant surface with subsequent drying. This type of application would provide a greater transportation opportunity for the organisms than the other two forms when a rainfall or irrigation event occurs.

The application of organisms in a liquid form will move into and perhaps through any crop canopy to or toward the soil surface. In the case of no vegetative cover and an unsaturated soil, the application will carry the organisms into the soil profile. Under these conditions, the organisms will be placed in contact with the soil and likely experience some movement into the soil particles and even attach during the drying process. This location will increase the difficulty of detaching and suspending the organisms in runoff.

The final location of organisms applied to a field encapsulated in pellets will depend in part on the size and density of the pellet. Heavier pellets will move further into and through a crop toward the soil surface. These organisms will also be delayed in their movement from the site with runoff. It will take some time to dissolve or melt the pellet and expose the organisms.

The application mode, therefore, is very important as it affects the location and availability of introduced organisms to water transport. There is no evidence that any research has been done to evaluate the differences between the release and transport characteristics of different shapes, sizes, or kinds of microorganisms.

3.2 RUNOFF EFFECTS

Many microbiologists have observed that the runoff from agricultural lands carries microorganisms into adjacent streams. Doran and Linn (1979) summarize some of the earlier work and report on runoff quality from pastureland. Their work compares the bacteriological quality of runoff water from grazed and ungrazed areas over a 3-year period. There was little difference in total coliform (TC) counts, while the grazed area showed 5 to 10 times more fecal coliforms (FC) than the ungrazed area. Wildlife droppings on the ungrazed area are suggested as the reason the fecal streptococci (FS) counts were higher than for the grazed area.

The differences were not as clear in another 3-year study of summer-grazed and ungrazed areas (Jawson et al. 1982). Numbers of TC from runoff of both watersheds routinely exceeded 10,000/100 ml. The absence of grazing animals did not seem to affect numbers of TC and FS in the runoff. Other researchers, who have reported on the bacterial populations in runoff, confirmed that many factors influence the results collected. In a recent review of the literature on bacterial quality in runoff from agricultural lands, Baxter-Potter and Gilliland (1988) report that while stream discharge during storm events influences the bacterial density, the relationship is not simple. A brief review of the runoff-rainfall relationship will provide some understanding of the hydrology which transports microorganisms from agricultural lands. Much research has been done in the field of hydrology. Many texts are available for the serious reader, and the *Handbook of Applied Hydrology* by Chow (1964) is an excellent single reference.

A runoff event can be caused by rainfall, snowmelt, or an excess irrigation application. However, it is unlikely that an introduced organism would be applied during the season that snow would be expected and proper management can limit excess irrigation applications. Rainfall, therefore, will provide most of the transportation opportunity for microorganism movement. Much work has been directed to the relationship between rainfall and runoff. Unfortunately, most of this has been focused on a much-larger scale than is needed for this application. The peak hydrography, which describes the *maximum* flow from a watershed, is required to size a culvert or build a dam, but not to describe the flow and transport characteristics of runoff.

A very simple rainfall-runoff relationship is described in Figure 3–1. For those extreme conditions where the soil is impervious, all rainfall will immediately become runoff as is shown by line a in Figure 3–1. A more typical relationship is shown by line b, where some infiltration into the profile will reduce the fraction of rainfall going into runoff. The runoff under these conditions will be delayed until the soil surface is wet and the small surface depressions are filled. The quantity of rainfall required to cause a runoff event will depend upon the rainfall intensity and the infiltration rate

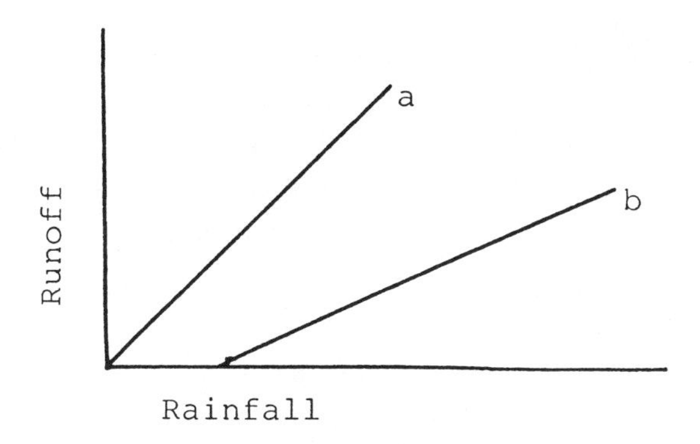

FIGURE 3–1 Sample rainfall-runoff relationship for (a) impervious soils and (b) typical soils.

of the soil. As an extreme example, a light rain on a sandy soil may produce no runoff. In many situations, at least a centimeter of rain must fall before runoff will begin.

There have been few modeling efforts that take into account some of the many parameters influencing the rainfall-runoff relationship. In a publication entitled *Hydrologic Modeling of Small Watersheds,* just one of the 13 chapters addresses "Modeling the Quality of Water from Agricultural Land" (Haan et al. 1982). Unfortunately, only one paragraph is devoted to microorganisms, and it notes that no modeling has been attempted. However, this chapter does have some items addressing movement of chemical constituents that the interested reader may find has application to microorganism movement. An international symposium on "Modeling Agricultural, Forest, and Rangeland Hydrology" produced a proceeding which devotes six papers to water quality, but again no papers on microorganisms are included (American Society of Agricultural Engineers, 1988). However, these proceedings contain a section on erosion, and the 11 papers on rainfall-runoff have some coverage of topics that indirectly relate to the movement of microorganisms.

As discussed earlier, some of the microorganisms may be transported as part of a soil particle. For that fraction, the research on erosion of soils is quite relevant. The work and publications in this area are quite lengthy and of benefit to an understanding of the processes of detachment, transport, and deposition of soils. One of the common methods for predicting soil loss is the use of the universal soil loss equation (equation 1). This equation takes into account the total rainfall energy for a specific area rather than the total rainfall amount (Chow 1964) and is written as follows:

$$A = RKLSCP \tag{1}$$

where A is the average annual loss in tons per acre, R is the rainfall factor, K is a soil-erodibility factor, L is a slope length factor, S is a steepness factor, C is a cropping and management factor, and P is a conservation practice factor. While this equation is not immediately transferable to predicting the transport of microorganisms, this approach does address many of the salient points. With sufficient data, a microorganism-release coefficient could be developed for an equation similar to the universal soil loss equation.

3.3 MICROORGANISM-RELEASE CHARACTERISTICS

There is no literature on the release rate to runoff of introduced organisms. However, there is some literature on the release rate of organisms from organic solids as they receive a rain or irrigation event. Most of the studies have investigated the movement of organisms from livestock manures. Almost all of the studies have monitored fecal coliform (FC), with fecal streptococci (FS) and total coliform (TC) the next-most-common organisms studied. There are a few reports that have evaluated the organism movement from sewage sludge that has been spread on agricultural lands. Typically, the same indicator organisms have been studied.

The best research work that evaluated the release of organisms under a simulated rain event was conducted at the Watershed Science Unit of Utah State University. Springer et al. (1983) reported that standard cowpies were made and then rained on at ages of 2 through 100 days. While the young, less-than-5-day-old, cowpies released millions of organisms per 100 ml of water, the number declined to 4,200 FC per 100 ml of water with 100-day-old material. Peak counts in the runoff were reduced when deposits were rained on more than once.

In another paper from the same study, Thelin and Gifford (1983) reported that the cowpies were rained on for 15 min at a rate of 6.1 cm/m. The runoff water sampled at 10 and 15 min gave almost identical results. They indicated there was a lag in organism numbers in the 5-min-runoff samples resulting from an apparent required wetting period for the dry cowpies. This may have application for the movement of organisms from soils that are relatively dry when they experience their first runoff-producing rain.

In a third paper from the study, Kress and Gifford (1984) report that rainfall intensity had no significant effect on peak FC release from fecal deposits that were 2–10 days old. However on cowpies that were 20 days old, the effect of intensity was significant. The highest intensity gave the lowest peak counts and the lowest intensity gave the highest peak counts. With a fixed number of microorganisms and different rainfall volume, this appears to be due to dilution. This may be transferable to applications of organisms to bare soils, but, unfortunately, no studies to date have evaluated this possibility.

There is another factor that is difficult to separate when one tries to determine the transport characteristics of organisms in any data involving a time lag. This is the die-off rate of organisms, which occurs naturally with time. It is difficult to calculate the release-rate coefficient with time without knowing the number of actual organisms still viable at the source. A partial list of environmental factors that influence the die-off rate would include: temperature, moisture content (humidity and precipitation), sunlight, available nutrients, and soil pH. While these factors influence the die-off rate and indirectly affect the measured transport of organisms in water, a summary of the literature on microorganism die-off is beyond the scope of this chapter. Reviews of this topic have been written by Burge and Marsh (1978), Ellis and McCalla (1978), Gerba et al. (1975), Lance (1976), Menzies (1977), Mitchell and Starzyk (1975), Morrison and Martin (1977), and Reddy et al. (1981).

One method to separate the die-off of organism from the release rate in transport is to model the expected die-off rate. This would allow an investigator to predict the organism population each day at the site after application and before a rainfall or irrigation event. Again, complete coverage of this topic is beyond the scope of this chapter, but an article that may be helpful to interested readers is entitled "Modeling Enteric Bacterial Die-Off: A Review" (Crane and Moore 1986).

3.4 CASE STUDIES

Research completed on bacterial pollution within the environment has generally centered on the various parameters involved in bacterial die-off and transport. It must be recognized that actual quantities of organisms transported from land-application sites are dependent on the complex interaction of many factors. Investigation of this problem by several researchers on a macro-scale has been attempted by determining a "bacterial numbers balance" between the total organisms applied to a site and their subsequent removal with rainfall or irrigation water. This approach tends to ignore the individual processes involved (i.e., die-off, adsorption, infiltration, etc.) but is useful in determining or estimating the quantity of bacteria that might leave the disposal site and enter the water resources of an area. This approach is also utilized to estimate the relative effectiveness of various management practices, such as the use of buffer areas or different application methods, for reducing bacterial losses.

Robbins et al. (1971), studying various livestock operations in North Carolina, determined that 2–23% of the FC deposited on fields by manure application or defecated directly by animals were lost in runoff from these areas on an annual basis. The actual number was dependent on the individual livestock operation presumably because of differences in or lack of manure management techniques. These percentages losses seem very high in comparison to other research results and are probably the reason why samples from waterways draining these watersheds contained bacterial densities substantially above water quality standards.

McCaskey et al. (1971) investigated the quality of runoff from dairy application sites where manure was applied frequently in liquid, semiliquid, or solid form at annual application rates averaging from 20–300 t (metric tons) of dry matter per year. Using their data, it was determined that the maximum annual removal of applied TC, FC, and FS was 0.06%, 0.007%, and 0.008%, respectively, of those applied, by analyzing runoff from these areas. Bacterial losses were highest for the solid-spreading technique and lowest for the liquid-manure-application method (0.0005–0.00012%). This study was performed on a minimally sloped, sandy loam soil with Bermuda grass cover. The movement was minimal but it is not clear exactly why. The slope, grass cover, and soil type appear to be very influential.

In a pilot study by Kunkle (1979), under summer conditions (25–30°C) utilizing surface-spread liquid dairy manure at 8 t (wet weight) per acre on a Cabot silt loam soil with grass cover, it was observed that only FC populations in runoff declined with time after manure application. Total coliforms, fecal streptococci, and enterococci runoff population densities remained constant throughout the study at levels similar to those prior to application. This was attributed to extremely high background contamination levels for these organisms. Total losses of FC in runoff from simulated rainfall totaled 6.73% of those applied during the 23-day period after application. However, most of this loss was recorded in the first irrigation event, which was initiated several hours following manure application. After the initial irrigation, all subsequent events removed only 0.061% of the applied fecal coliforms. Emphasis is placed on how critical the first runoff event seems to be in removal of bacterial by surface runoff. A similar finding was expressed by Dunigan and Dick (1980) when sewage sludge was land-applied at excessive rates. They reported that high FC counts were found in runoff until after a dry-weather period during which the sludge became thoroughly dried.

In a study by Crane et al. (1978), applying liquid swine wastes to pasture plots, it was shown that residence time of manure on the surface seemed to be a controlling factor in the number of bacteria that were transported in runoff. If runoff occurred on the day of application, 58–90% of the FC and 20–32% of the FS applied with the manure were removed with the runoff. If manure residence time was increased to 3 days, the percentage removal was dramatically reduced (0.10–0.22% and 0.14–0.32%, respectively). This decline was not due to die-off because bacterial counts in the surface soil revealed that a constant population of these bacteria were present during this 3-day period.

These observations indicate that time-dependent processes are involved in the transfer of bacteria from soil to liquid runoff. Marshall (1971 and 1980) and Daniels (1980) have presented comprehensive reviews of research on bacterial-solid interactions and adsorption processes. These reviews suggest that the increased residence time allows greater contact between soil materials and applied microorganisms. This increases adsorption and fixation by ion exchange, surface-charge attractive forces, and polymer bridging between solids and bacterial surfaces. Since the kinetics of these mechanisms are controlled by diffusional processes, the above results seem feasible. In general, there has been little study of the mechanisms involved in bacterial transfer from the solid (soil) to liquid phase prior to transport in runoff and to the factors controlling these processes. This information is essential for estimating the potential for bacterial movement via surface runoff.

3.5 BUFFER AREAS AND VEGETATIVE FILTERS

The use of pasture or forest buffer areas that act as overland flow treatment systems below waste application sites has proven very effective in reducing levels of nutrients and sediment in runoff (Vanderholm and Dickey 1978;

Bingham et al. 1978; Thomas 1974; Thomas et al. 1974; Johnson and Moore 1978; Dickey and Vanderholm 1981; Doran et al. 1981; Vanderholm et al. 1979). Many of the mechanisms involved in reducing the loss of nutrients and sediment will also apply to reducing the transport and loss of introduced organisms from an application site. The important mechanisms involved in this pollutant concentration reduction were shown by Johnson and Moore (1978) to be:

1. A reduction in volume of runoff from increased infiltration;
2. A decrease in runoff velocity caused by the vegetative cover with a resultant increase in sedimentation of pollutants that are adsorbed to particulate matter; and
3. Increased adsorption of pollutants by soil particles under a lower ionic concentration regime than that found on the waste application site.

While the infiltration rate will vary with different soils, it will vary over an even wider range depending on the conditions at the soil surface. Vegetation on the surface will protect the soil surface from the pounding of rain drops. By intercepting the rain and dissipating the energy, the soil structure is not broken down and the infiltrative capacity is maintained. As water infiltrates it will carry microorganisms into the soil which will remove them from potential loss in runoff.

Any vegetative cover, such as grass, will reduce the runoff velocity and will allow increased settling of organisms. This reduction in velocity and improved settling will affect both organisms that are suspended as individual cells and all those attached to a larger particle.

There is a body of literature that has measured the settling and concentration of organisms in stream sediments. Interest in this subject grew from the tremendous increases in microorganism concentrations recorded in stream water during rainfall events. It was measured during base flow and hydrographic events in England (McDonald and Kay 1981). These same researchers also evaluated bacterial water quality increases during nonrain storm events, such as releases from an upstream reservoir (McDonald et al. 1982). They concluded that large quantities of microorganisms accumulate in the bottom sediment and are resuspended with increased flow. These conclusions were supported with data reported by Sherer et al. (1988).

Increases in bacterial concentrations in streams in the high mountains of Wyoming were reported to be 1.7-fold for FC and 2.7-fold for FS, when the bottom sediment was disturbed (Gary and Adams 1985). In a similar study, increases in *Escherichia coli* concentrations of up to 760-fold greater than the overlying water were measured after disturbing bottom sediments of rangeland streams (Stephenson and Rychert 1982). In a study that raked stream sediment at different locations relative to cattle access, microorganism concentrations reflected cattle access and numbers (Sherer et al. 1988). Directly below a feedlot, the numbers of FC increased in the stream from 1.8×10^6 organisms/100 ml to 760×10^6 organisms/100 ml when 1 square meter was raked. All of these studies reflect the settling characteristics of microorganisms in streams. They support the use of buffer strips to reduce runoff velocities and enhance settling.

Research on the use of buffer strips and vegetative filters for bacterial removal has brought about conflicting results. Jenkins et al. (1978), using an overland flow system for treatment of primary and secondary wastewater effluents, found that 96–99% of FC in the effluents were removed in the summer. This was reduced, however, to less than 65% during the winter and was thought to occur because of decreased infiltration into the frozen soil. Opposite results were shown in an investigation by Peters and Lee (1978) utilizing overland flow on a reed-canary grass-fescue cover treating municipal wastewater. Fecal coliform concentrations were increased after overland treatment during summer months and were attributed to bacterial regrowth at warmer temperatures. Maximum removal during the winter period was only 60% on a concentration basis. Hunt et al. (1979) reported from this same investigation that coliforms and FS concentrations in effluent after overland flow treatment were always higher than the initial wastewater applied, regardless of season.

In all these studies (Jenkins et al. 1978; Peters and Lee 1978; Hunter et al. 1979), the investigators suggest that removal of chemical constituents from the effluents was unrelated to bacterial removal, being much less efficient as well as unpredictable. This conclusion is substantiated by Reese et al. (1980) and Barker and Sewell (1973) who found high concentrations of total and fecal coliforms in runoff from pasture areas where the chemical makeup of the runoff was similar to that from uncontaminated "background" sites. Summarizing the work of Johnson and Moore (1978), it would seem that vegetative filters are only reliably effective in removing bacteria at high concentrations (more than 10^5 organisms/100 ml). The bacterial populations in runoff from these buffer areas seem to equilibrate at about 10^4 to 10^5 organisms/100 ml, regardless of experimental conditions. The work of Doyle et al. (1975) and Young et al. (1980) seems to contradict the above conclusions, but under closer examination it was found that these differences could be explained.

Doyle et al. (1975) applied fresh dairy waste to pasture plots and used a forested strip as a buffer area. They found fecal coliforms and fecal streptococci were effectively removed from the runoff (99%) within the first 4 m from the edge of the application site. Bacterial concentrations on the order of 10^4/100 ml, however, were still found in the runoff from the buffer area in most samples. Young et al. (1980), investigating the effectiveness of vegetated buffer strips in controlling pollution from feedlot runoff, predicted that total coliform removal from this effluent followed the following statistical relationship:

$$\text{Total coliforms} \times 10^{-6}/100\ \text{ml} = 67.34 - 1.90L \qquad (2)$$

where L = length of buffer strip (flow distance) in meters ($r^2 = 0.77$). They suggested from this relationship that a 36-m-length of buffer area would be sufficient to reduce bacterial concentrations in the runoff to below 1,000 organisms/100 ml. The buffer lengths used in this study were only 27 m and at this distance total and fecal coliform concentrations were still on the order of 10^5 to 10^6/100 ml. Therefore, the extrapolation of these results to much greater removal efficiencies is highly questionable.

In a study of three Utah watersheds, Glenne (1984) found that the percent reduction of bacteria could be correlated to the buffer strip length and slope. As expected, flatter slopes and longer lengths are more effective in removing bacteria. However, at steep slopes (15-20%) the effect of length is almost eliminated. Working with Glenne's data, Moore et al. (1988) developed the following equation:

$$PR = 11.77 + 4.26 \times S \tag{3}$$

where: PR is the percent removal of bacteria, and S is the buffer width in feet ÷ slope in %. The following limits apply to equation 3: $0 < PR < 75\%$; $0\% < \text{slope} < 15\%$; and $10 \text{ ft} < \text{buffer strip width}$.

In summary, it would seem that buffer areas can be used to advantage in removing bacteria from wastes or effluents containing more than 10^5 organisms/100 ml. At this time it appears unlikely that most applications of introduced microorganisms will provide concentrations in this magnitude. Reductions below 10^4 organisms/100 ml seem to depend on season, soil infiltration rates, and other factors that need investigation. One would expect the same results for an organism present in low numbers, but no data is available to support that statement.

3.6 MODELING MICROORGANISM MOVEMENT

While the lengthy list above of parameters influencing the movement of microorganisms does create a seemingly impossible task for researchers, the technological age has brought a new tool. The personal computer offers the opportunity for researchers to analyze data and play "what if" games that were not possible several years ago. Perhaps the best example is the analysis and utilization of daily weather data in evaluating the effect of rainfall on runoff volume and water quality. Researchers can now go back for many years for several stations and reconstruct real weather events before and during water-sampling periods to determine the influence of rainfall on water quality. The time demand made this task impossible before the computer.

Unfortunately, these opportunities are just being processed, and the current literature does not contain many results of research effort to model watersheds where water-quality data exist. However, some work is being done to identify those areas where data are lacking, and this is critical to providing a complete understanding to some questions of microorganism transportation.

One of the first researchers to present a simple model to predict bacterial densities in runoff from feedlots was McElroy (1976). His relationship includes terms to input the direct runoff, the concentration of bacteria in the runoff, a delivery ratio term to reflect the distance between the source and the stream, and the area of the livestock facility. While including several critical factors, it does not include hydrologic factors such as rainfall intensity or runoff rate. There are no terms to cover the stocking factor of the feedlot, the season, or the daily temperature.

A nonpoint-source conceptual model was developed for sediment and manure transport that considers various transportation mechanisms (Khaleel et al. 1979). This model includes four different particle size fractions. While it does not address the problem of microorganisms, it contains many of the relevant components needed to describe their transport in runoff. A few years later, part of this same group published an excellent review of the behavior and transport of microorganisms from land-spread organic wastes (Reddy et al. 1981). The major transport processes were reviewed, and a simple conceptual model of bacterial retention on soil particles was developed and presented.

A new model that has potential for understanding surface transport of microorganisms is called WASP3; it is an aquatic fate model (Ambrose et al. 1986). It was developed to evaluate the movement of chemicals in large lakes and estuaries. It could also be useful in tracking organism movement in water bodies influenced by tides, discharge, wind, sedimentation, and both dispersion and advection. Currently it does not have a bacterial segment, but that could be added.

There is also a model called MICROBE-SCREEN currently under development (Reichenbach et al. 1987) in order to assess the movement and fate of microorganisms in air, water, and soil. However, it assumes that organisms are passive, rather than active, and it does not yet have the release and transport coefficients necessary to characterize the transport of microorganisms.

A model that requires limited input, but is currently available to calculate microorganism transport is called MWASTE (Moore et al. 1988). It was developed using field and laboratory data to track bacterial movement in dairy cattle manure from the cow, through a variety of manure management systems and land spreading to transport in runoff to a stream (Moore et al. 1983). It can be used to calculate organism population daily and uses the Soil Conservation Service Model CREAMS (Knisel 1980) to generate the hydrology or runoff portion of the model. MWASTE can be used to predict the number of bacteria in the runoff. The equation that calculates the number of bacteria removed in runoff uses a percent reduction method. The equation is shown below:

$$F = FO \times (1 - P)^r \tag{4}$$

where: F = number of bacteria remaining on soil; FO = original number of bacteria on soil; P = percent reduction factor of runoff and infiltration; and r = runoff or infiltration water depth.

The following conditions exist for values of the coefficients:

Waste Type	Spreading	Infiltration	Runoff
Solid	—	0.05	0.40
Liquid	Day 1	0.20	1.00
Liquid	7 day 1	0.05	0.40

The equations needed to address other factors that affect die-off rate are covered in earlier parts of the MWASTE program.

3.7 SUMMARY

Several studies have reported on the quality of runoff from land that has received either an application of livestock waste or been utilized as a pasture for livestock. Unfortunately, these studies have not directed their efforts to understanding and developing the relationships among several of the important parameters that influence runoff quality. One of the reasons for this deficiency is that the list of influencing parameters is quite long.

Nevertheless, it is important to identify the parameters and their probable impact on movement of organisms in water. The microbiological aspects are influenced by the fate of organisms in the environment. Radiant energy (sunlight), temperature, available nutrients, presence of toxic materials, available moisture (precipitation and humidity), and soil pH all influence the death/growth rate of the organisms in question. Site characteristics, such as slope, vegetative cover, antecedent moisture content, soil type, organic matter content, infiltration rate, and surface condition of the soil, all influence microorganism movement.

Hydrologic factors, such as frequency, duration, and intensity of rainfall, are very critical in determining the characteristics of runoff events that provide the transportation to move introduced organisms from their application site.

There are very few models today that can be used to calculate the microorganism population in runoff. While many of the influencing parameters have been identified, there has been little research on the surface transport of microorganisms.

REFERENCES

Ambrose, R.B., Jr., Vandergrift, S.B., and Wool, T.A. (1986) EPA 600/3-86/034, U.S. EPA Environmental Research Laboratory, Washington, DC.

American Society of Agricultural Engineers (ASAE) (1988) in *Modeling Agricultural, Forest, and Rangeland Hydrology, International Symposium*, ASAE publ. no. 07-88, ASAE, St. Joseph, MI.

Barker, J.C., and Sewell, J. (1973) *Transactions of the ASAE* 16(4), 804–807.
Baxter-Potter, W.R., and Gilliland, M.W. (1988) *J. of Environmental Quality* 17(1), 27–34.
Bingham, S.C., Overcash, M.R., and Westermann, P.W. (1978) *ASAE*, paper no. 78-2571, ASAE, St. Joseph, MI.
Burge, W.D., and Marsh, P.B. (1978) *J. of Environmental Quality* 7(1), 1–9.
Chow, V.T., ed. (1964) in *Handbook of Applied Hydrology*, 17-7, McGraw-Hill, New York.
Crane, S.R., and Moore, J.A. (1986) *Water, Air, and Soil Pollution* 27, 411–439.
Crane, S.R., Overcash, M.R., and Westerman, P.W. (1978) Unpublished paper no. 15, Agricultural Engineering Department, North Carolina State University, Raleigh.
Daniels, S.L. (1980) in *Adsorption of Microorganisms to Surfaces* (Bitton, G., and Marshall, K.C., eds.), Wiley, New York.
Dickey, E.C., and Vanderholm, H. (1981) *J. of Environmental Quality* 10(3), 279–284.
Doran, J.W., and Linn, D.M. (1979) *Appl. Environmental Microbiol.* 37, 985–991.
Doran, J.W., Schepers, J.S., and Swanson, N.P. (1981) *J. of Soil and Water Conservation* 36(3), 166–171.
Doyle, R.C., Wolfe, D.C., and Bezdicek, D.V. (1975) *ASAE*, pub. proc. no. 275, pp. 299–302, ASAE, St. Joseph, MI.
Dunigan, E.P., and Dick, R.P. (1980) *J. of Environmental Quality* 9(2), 243–250.
Ellis, J.R., and McCalla, T.M. (1978) *Transactions of ASAE* 21(2), 307–313.
Gary, H.L., and Adams, J.C. (1985) *Water, Air, and Soil Pollution* 25, 133–144.
Gary, H.L., Johnson, S.R., and Ponce, S.L. (1983) *J. of Soil and Water Conservation* March–April 124–128.
Gerba, C., Wallis, C., and Mellnick, J. (1975) *J. of Irrigation and Drainage Division*, American Society of Civil Engineering, 101, 157–174.
Glenne, B. (1984) *Water Resources Bulletin* 20, 211–217.
Haan, C.T., Johnson, H.P., and Brakensiek, D.L. (1982) *Hydrologic Modeling of Small Watersheds*, ASAE, St. Joseph, MI.
Hunt, P.G., Peters, R.E., Sturgis, T.C., and Lee, C.R. (1979) *J. of Environmental Quality* 8(3), 301–304.
Jawson, M.D., Elliott, L.F., Saxton, K.E., and Fortier, D.H. (1982) *J. of Environmental Quality* 11(4), 621–627.
Jenkins, T.F., Martel, C.J., Gaskin, D.A., Fisk, D.J., and McKim, M.L. (1978) in *Land Treatment of Wastewater*, pp. 61–77, International Symposium, U.S. Army Corps of Engineers, Cold Regions Research and Engineering Laboratory, Hanover, NH.
Johnson, G.D., and Moore, L.A. (1978) *The Effects of Conservation Practices on Nutrient Loss*, Dept. of Agricultural Engineering, Special Report, University of Minnesota, Minneapolis.
Khaleel, R., Foster, G.R., Reddy, K.R., Overcash, M.R., and Westerman, P.W. (1979) *Transactions of ASAE* 22(6), 1353–1361.
Knisel, W.G., ed. (1980) *USDA Conservation Research*, report no. 26.
Kress, M., and Gifford, G.F. (1984) *Water Resources Bulletin, American Water Resources Assoc.* 20(1), 61–66.
Kunkle, S.H. (1979) *Using Bacteria to Monitor the Influences of Cattle Wastes on Water Quality*, USDA-Science and Education Administration—Agricultural Research Results, ARR-NE-3.

Lance, J.C. (1976) *Fate of Bacteria and Viruses in Sewage Applied to Soil, ASAE,* paper no. 76-2558, ASAE, St. Joseph, MI.

Marshall, K.C. (1971) in *Soil Biochemistry*, vol. 2 (McLaren, A.D., and Skujins, J.J., eds.), pp. 527–538, Marcel Dekker, New York.

Marshall, K.C. (1980) in *Adsorption of Microorganisms to Surfaces* (Bitton, G., and Marshall, K.C., eds.), pp. 439–448, Wiley, New York.

McCaskey, T.A., Rollins, G.H., and Little, J.A. (1971) in *Livestock Waste Management and Pollution Abatement, Proceedings 2nd International Symposium on Livestock Wastes*, pp. 239–242, ASAE, St. Joseph, MI.

McDonald, A., and Kay, D. (1981) *Water Research* 15, 961–968.

McDonald, A., Kay, D., and Jenkins, A. (1982) *App. and Environmental Microbiology* 44(2), 292–300.

McElroy, A.D. (1976) *EPA* publ. no. EPA/600/2-76/515, Su. Doc. no. EP 1.23/2:600/2-76-151, Washington, D.C.

Menzies, J.D. (1977) in *Soils For Management of Organic Wastes and Wastewaters*, pp. 574–578, Soil Science Society of America, Madison, WI.

Mitchell, D.O., and Starzyk, M.J. (1975) *Can. J. Microbiol.* 21, 1420.

Moore, J.A., Grismer, M.E., Crane, S.R., and Miner, J.R. (1983) *Transactions of the ASAE* 20(4):1194–1200.

Moore, J.A., Smyth, J., Baker, S., and Miner, J.R. (1988) *Evaluating Coliform Concentrations in Runoff from Various Animal Waste Management Systems, Special Report* no. 817, pp. 1–80, Agricultural Experiment Stations, Oregon State University, Corvallis, OR.

Morrison, S.M., and Martin, K.L. (1977) in *Land as a Waste Management Alternative* (Loehr, R.C., ed.) pp. 371–389, Proceedings 1976 Cornell Agricultural Waste Management, Cornell University, Ithaca, NY.

Peters, R.E., and Lee, C.R. (1978) in *Land Treatment of Wastewater*, pp. 45–60, International Symposium, U.S. Army Corps of Engineers, Cold Regions Research and Engineering Laboratory, Hanover, NH.

Reddy, K.R., Khaleel, R., and Overcash, M.R. (1981) *J. of Environmental Quality* 10(3), 255–266.

Reese, L.E., Hegg, R.O., and Gantt, R.E. (1980) *ASAE* paper no. 80-2536, St. Joseph, MI.

Reichenbach, N., Wickramanayake, G., Lordo, R., and Hetrick, D. (1987) Battele, Columbia Division—Washington Operations to U.S. EPA.

Robbins, J.W., Kriz, G.J., and Howells, D.H. (1971) in *Proceedings 2nd International Symposium on Livestock Wastes*, ASAE pub. proc. 271, pp. 166–169, ASAE, St. Joseph, MI.

Sherer, B.M., Miner, J.R., Moore, J.A., and Buckhouse, J.C. (1988) *Transactions of ASAE* 31(4), 1217–1222.

Springer, E.P., Gifford, G.F., Windham, M.P., Thelin, R., and Kress, M. (1983) Hydraulics and Hydrology Series UWRL/H-83/02, pp. 1–108, Utah Water Research Laboratory, Utah State University, Logan, UT.

Stephenson, G.R., and Rychert, R.C. (1982) *J. of Range Management* 35(1), 119–123.

Thelin, R., and Gifford, G.F. (1983) *J. of Environmental Quality* 12(1), 57–62.

Thomas, R.E. (1974) *Feasibility of Overland-Flow Treatment of Feedlot Runoff*, U.S. EPA Report EPA-660/2-74-062, Corvallis, OR.

Thomas, R.E., Jackson, K., and Penrod, L. (1974) *Feasibility of Overland Flow for Treatment of Raw Domestic Wastewater*, U.S. EPA Report EPA-660/2-74-087, Corvallis, OR.

Vanderholm, D.H., and Dickey, E.C. (1978) *ASAE* paper no. 78-2570, ASAE, St. Joseph, MI.

Vanderholm, D.H., Dickey, E.C., Jackobs, J.A., Elmore, R.W., and Spahr, S.L. (1979) *Livestock Feedlot Runoff Control by Vegetative Filters*, EPA-600/2-79-143, Aela, OK.

Young, R.A., Huntrods, T., and Anderson, W. (1980) *J. of Environmental Quality* 9(3), 483–487.

CHAPTER 4

Soil and Groundwater Transport of Microorganisms

Duane F. Berry
Charles Hagedorn

During the 1970s, the controversy over recombinant DNA research focused on the risks presumed to arise from the possible escape of genetically engineered microorganisms (GEMs) from research laboratories. In the 1980s, concern shifted to the nature of the long-term environmental impact of GEMs that are intentionally released. This change in emphasis was brought about primarily by the existing potential for the commercial use of GEMs.

At present, GEMs are being developed for commercial use to accomplish various tasks such as:

- Pollutant degradation,
- Pesticide control,
- Wastewater treatment,
- Polymer production,
- Bioleaching of ores,
- Biomonitoring,
- Fuel desulfurization,
- Biomass conversion, and
- Crop enhancement.

Table 4–1 lists the number of patent applications for GEMs from 1980–1986 on the basis of the genus involved and the purpose of the patent.

For many GEMs to perform their designated tasks, they must be deliberately released into the natural environment. In order for these GEMs to be successful they must be competitive, or at the very least persistent, within the habitat into which they are released. Presumably, a successful GEM is one that will not only survive, but multiply as well. Unfortunately, these characteristics also may allow GEMs to move out of their proposed release areas.

The physical containment of a GEM within a release area is highly improbable. Clearly, strategies for the containment of GEMs are not numerous. One of the most intriguing methods involves the creation of GEMs that are potentially suicidal due to the incorporation of conditionally lethal genes (Bej et al. 1988). Other possible strategies include constructing GEMs that are dependent upon the presence of a specific plant, are temperature

TABLE 4–1 Number of GEM Patent Applications from 1980–1986 by Genus and Use

	Number of Patents by Proposed Use[2]										
Genus Involved[1]	*A*	*B*	*C*	*E*	*G*	*M*	*N*	*O*	*P*	*R*	*W*
Bacillus	17	97	53	18	30	1	1	3	307	8	83
Clostridium	0	269	204	1	42	1	0	4	14	3	49
Pseudomonas	3	9	44	1	16	14	0	5	57	7	34
Escherichia	2	48	45	10	13	8	5	4	134	1	28
Saccharomyces	0	96	75	1	51	6	0	5	16	0	31
Candida	0	74	58	0	29	2	0	2	35	2	57
Aspergillus	0	13	20	3	2	2	0	0	132	1	72
Klebsiella	0	75	21	0	14	0	5	7	4	0	24
Alcaligenes	0	0	31	0	6	0	3	8	23	1	55
Aerobacter	0	49	30	0	9	0	0	0	1	0	3
Zymomonas	0	33	25	0	23	0	0	1	0	0	5
Streptomyces	0	15	14	0	1	2	0	0	20	0	33
Kluyveromyces	0	16	15	0	13	0	0	0	10	0	20
Methylosinus	0	1	49	0	0	0	0	1	4	0	8
Thiobacillus	0	0	0	0	0	42	1	10	5	1	4
Penicillium	0	0	0	2	0	2	0	6	23	1	26
Proteus	0	14	8	0	18	0	0	0	2	0	2

[1] The genus of the microorganisms most likely to be used is indicated.

[2] Abbreviations for proposed use: A, agricultural chemicals; B, conversion of biomass; C, industrial chemical production; E, monitoring/measurement/biosensor; G, energy; M, mining/metal recovery; N, nitrogen fixation; O, other; P, polymer/macromolecule production; R, enhanced oil recovery; W, waste/pollutant degradation.

Source: *Report of the Biotechnology Science Advisory Committee of the U.S. Environmental Protection Agency*, 1987.

sensitive, or require a substrate that must be added to maintain viability. Theoretically, such GEMs would survive only as long as the specific situation permitted.

Although a reasonable approach, the development of conditional suicide systems for the containment of GEMs has not been perfected. As a result, researchers must concern themselves with the scenario that deliberately released GEMs will likely be mobile and take up residence in nontarget habitats. Questions pertaining to the fate of GEMs center around mechanisms and rates of transport, as well as the possible transfer of genetic material to indigenous bacteria. Information on the fate of GEMs is preemptive to the development of adequate guidelines that address issues such as risk assessment and potential ecological impact. The purpose of this paper is to review scientific information on the transport of microorganisms in soil and groundwater environments as well as the factors affecting their transport.

4.1 OVERVIEW OF ISSUES RELATING TO THE SURVIVAL OF GEMs

The rationale for concern over the survival and colonization of GEMs in the environment is threefold:

1. The framework for biotechnology risk assessment requires knowledge of the existing potential for GEMs to induce an undesired impact on nontarget species and important ecological processes, such as nutrient cycling.
2. The commercial use of GEMs may require repetitive applications of amended organisms that only persist over a discrete period of time.
3. The specific use category of a GEM may require long-term or infinite survival to achieve desired results.

Information on the factors that exert either positive or negative effects on the survival and colonization of GEMs in the environment is necessary before these issues can be adequately addressed. Concern over the fate of GEMs in the environment must be focused both on their ability to survive and on the fate of the recombinant DNA sequences they contain.

Much of the current understanding of the fate of GEMs in the environment has been inferred from data on the survival and colonization of nongenetically engineered microbes. However, results from recent field trials involving genetically altered "ice-minus" *Pseudomonas syringae* and nitrogen-fixing *Rhizobium meliloti* indicated that these particular GEMs had a relatively short survival period once released into the environment (*Chemical & Engineering News* 1989). On the other hand, the competitive abilities of GEMs may not necessarily be reduced, simply as a result of carrying additional genes (Devanas et al. 1986; Devanas and Stotzky 1986). It should

TABLE 4–2 Abiotic Factors Affecting Survival of Bacteria in the Soil Environment

Factor	*Comment*	*Reference*
Temperature	An increase of 10°C in temperature increases die-off at about a twofold rate between 5–30°C	Reddy et al. 1981
Soil texture	Organisms may be better protected in finer-textured, clay-type soils	van Veen et al. 1985; van Elsas et al. 1986
pH	Optimal pH range of 5–8 with shorter survival times in acid soils (pH 3–5)	Bitton and Gerba 1984
Nutrient availability	Heterotrophic organisms require a carbon source for their energy-yielding, oxidation-reduction reactions	
Electron acceptors	Oxygen (or nitrate, in some instances) is required by heterotrophic microbes for respiration purposes	
Water availability	Most bacteria require water activities (a_w) of 0.95–0.99 for growth	Brock 1974

also be pointed out that many nonindigenous soil organisms can demonstrate significant, long-term persistence under conditions of physiological stress in a foreign environment. Under such conditions, microorganisms (including GEMs) may require "resuscitation" to overcome physiological stress effects. Thus, difficulty may arise in enumerating these organisms by commonly used microbiological techniques.

It appears that critical factors which contribute to the successful establishment, persistence, and survival of GEMs in the environment are related to: (1) the physiological fitness of the GEM relative to the physicochemical conditions to which it will be exposed, and (2) specific properties of the ecosystem, such as availability of microhabitats, competition by indigenous microbes, temporal and spatial variability of microbial population density, and minimum population size required for establishment (inoculum effect). While some of these factors are rather intangible, experimental research approaches must be developed to evaluate these effects, and to elucidate fundamental (and, hopefully, unifying) principles that can be extrapolated to many GEM-release scenarios. The critical need for field investigation and

TABLE 4–3 Factors Influencing Virus Survival in Soil

Factor	*Comment*	*Reference*
Temperature	Virus deactivation rate increases as temperature increases	Yeager and O'Brien 1979; Hurst et al. 1980; Yates et al. 1985
Desiccation	Virus survival rate decreases as soil moisture content decreases	Alexander 1977; Yeager and O'Brien 1979
Adhesion	Virus survival may be enhanced as a result of adhesion to soil surfaces	Hurst et al. 1980
Biological degradation	Some viruses are susceptible to proteolytic enzymes	Cliver and Herrmann 1972

validation should become self-evident as information is obtained through the monitoring of actual release situations.

Survival times of bacteria vary greatly with species and are difficult to assess other than individually. Table 4–2 summarizes the major abiotic factors that affect the survival and persistence of bacteria in the soil environment.

Survival times of viruses in soil vary quite dramatically depending upon viral type and soil conditions. Some viruses, especially those associated with plant disease, can persist in soil for periods of up to one year or more (Alexander 1977). Two of the more important factors which influence virus survival in soil are temperature and moisture (Alexander 1977; Yeager and O'Brien 1979; Hurst et al. 1980). Table 4–3 briefly summarizes the major factors known to affect the survival of viruses in soils.

4.2 TRANSPORT OF MICROORGANISMS IN SOIL AND SUBSURFACE ENVIRONMENTS

Microbes are transported through the soil matrix primarily as a result of water movement. Laboratory investigations involving undisturbed versus disturbed "packed" soil columns and field studies have demonstrated that water movement in soil is either through small pores or through larger pores or channels, termed macropores (Elrick and French 1966; Wild 1972; Ritchie et al. 1972; McMahon and Thomas 1974). Evidence suggests that macropore water flow is a significant factor in groundwater recharge (Thomas and Phillips 1979). This observation may have profound implications with regard to transport of microbes from the surface to subsurface environments.

Soil porosity is a function of the soil matrix structure and is influenced by climate, mineral composition, organic matter composition, and vegetative cover.

Hydrodynamic forces play an extremely important role in the transport processes of microorganisms. A thin water layer is held tightly to the solid surface by molecular forces. The thickness of this essentially immobile layer is dependent upon the magnitude of the fluid shear stress. Shearing forces, along with microbial and solid surface characteristics (i.e., hydrophobic versus hydrophilic properties), have been shown to be important in controlling rates of bacterial cell deposition from laminar flows onto solid surfaces in laboratory investigations (Powell and Slater 1982, 1983). Results of Powell and Slater's studies demonstrate very clearly that in laminar flow, rates of microbial transport increase with increasing fluid shear stress.

When water moves through soil primarily as a result of gravity, the movement is referred to as saturated flow. The high rates of water movement associated with saturated flow create optimal transport conditions for microorganisms. Unsaturated flow occurs when an air phase is introduced into the flow channel, resulting in a negative pressure, so that matrix forces become the predominant cause of flow. Decreased fluid shear stress and increased tortuosity are among the reasons for less-than-optimal transport under conditions of unsaturated flow. Water path tortuosity increases as water content decreases because water is forced to flow along surfaces and through smaller pores that still retain water at the prevailing matrix potential. An increase in tortuosity of the flow path leads to increases in the ability of soil to restrict transport of microorganisms. Also, as the water layer becomes thinner, the opportunity for a microbe to adhere to surfaces becomes greater as a result of increased frequency of collision.

The frequency of microbial adhesion can be significantly affected by surface roughness of the substratum. Characklis (1984) reported that rates of microbial deposition are greater on rougher surfaces. Two reasons were given in support of this contention: (1) greater surface area leads to an increase in adhesion sites, and (2) microbes attached in cavities will be shielded from fluid shear forces.

Reversible adhesion between bacteria or viruses (which can both be considered as biocolloidal particles [Marshall 1976]) and soil particulates occurs despite the net negative surface charge on these entities. Traditionally, this anomaly has been explained by the DLVO theory (developed by Derjaguin and Landau 1941 and Verwey and Overbeek 1948), which describes the interactions between solid surfaces in a liquid environment. The extent of interaction between two particles suspended in an electrolyte solution is dependent upon the relative strengths of van der Waals attractive forces and electrostatic double-layer repulsive forces. When two solid surfaces are in close proximity, short-range forces, such as hydrogen bonding and London dispersion interactions, come into play.

In an aqueous electrolyte suspension, negatively charged surfaces accumulate counter charge near the solid-water interface. The surface charge, along with compact and diffuse counterions, form an electrical double layer. The electrical double-layer potential is a long-range repulsive force that is approximately exponential in distance dependence (Israelachvili and McGuiggan 1988). The strength of the double-layer potential is determined by surface charge density, while the range of the potential is dependent upon electrolyte concentration and the nature of the counterion species (divalent ions are more effective screening agents than monovalent ions).

In their interaction with solid surfaces, biocolloids experience the double-layer repulsive force as a result of counterion atmosphere interference in dilute electrolytic suspensions. The range of the repulsive force is reduced considerably with an increase in electrolyte concentration or in the proportion of divalent or trivalent ions (Marshall et al. 1971; Abbott et al. 1983; Taylor et al. 1981; Kott 1988). A decrease in the repulsive double-layer force relative to the van der Waals attractive forces, results in an increased frequency of adhesion. Van der Waals attractive forces, which have an inverse power-law distance dependence (i.e., the force between two spheres is proportional to $-1/D$, where D is equal to the distance from surface), dominate between surfaces at close range (Israelachvili and McGuiggan 1988). Further, van der Waals energy of attraction is independent both of valence and concentration of counterions (Singh and Uehara 1986).

In soil and in the subsurface, attractive forces presumably operate between hydrophobic regions on biocolloids and the substratum. Recent investigations by Stenström (1989) and van Loosdrecht et al. (1987) have helped to elucidate the role of bacterial cell wall hydrophobicity in adhesion processes. Stenström has shown that bacterial adhesion to quartz, albite, feldspar, and magnetite increases with an increase in cell surface hydrophobicity as measured by hydrophobic chromatography. In a similar investigation, van Loosdrecht et al. (1987) recorded a positive correlation between the surface hydrophobicity of 16 bacterial strains (all differing in surface hydrophobicity) and a negatively charged, sulfated-polystyrene solid phase. Valeur et al. (1988) conducted adhesion studies with *Pseudomonas* species and discovered that bacterial cells adhering to hydrophobic surfaces had a higher ratio of saturated to unsaturated C_{16}-fatty acids than those adhering to hydrophilic surfaces. Research conducted by Marshall and Cruickshank (1973) has shown that certain bacteria possess specific hydrophobic regions located on the cell wall or membrane which tend to orient the cell during the adhesion process. For distances (D) of up to approximately 10 nm, hydrophobic forces are 10–100 times stronger than van der Waals attractive forces and decay exponentially with a force proportional to $\exp(-D/1.4)$ (Israelachvili and Pashley 1982; Pashley et al. 1985). It is reasonable, therefore, to assume that hydrophobic forces play a more substantial role than van der Waals attractive forces in the adhesion of biocolloids to soil particle surfaces.

The use of model systems to elucidate the nature of biocolloid-solid surface interactions has provided researchers with a useful "first approach" to explain microbial adhesion phenomena (Rutter and Vincent 1984). Application of the electrical double-layer model to elucidate surface interactions in soil-water systems has merit, but, as Tadros (1980) has pointed out, these models assume clean homogeneous surfaces. Because surfaces in soil are usually heterogeneous, and the characteristics of biocolloids are in a dynamic state of flux, caution must be exercised when attempting to interpret biocolloid-substratum interactions in soil and subsurface environments based on model systems.

4.2.1 Bacterial Movement and Irreversible Adhesion

Generally speaking, bacteria are 0.2–1.5 μm in diameter (Lamanna et al. 1973). Most bacteria are subject to Brownian motion since their effective radius is less than 1.0 μm. Even for a particle the size of a bacterial cell, Brownian motion can be quite significant. For example, the Brownian displacement Δ of a microorganism in water (20°C) with a radius of 0.1 μm would be approximately 1.23 mm/hr given by the equation

$$\Delta = 2Dt \qquad (1)$$

where t is time in seconds and D, the diffusion coefficient, is given by the following equation:

$$D = RT/6\pi\eta rN \qquad (2)$$

where R is the gas constant, T is temperature in degrees Kelvin, η is the viscosity of the liquid (water in this case), r is the radius of the particle, and N is Avogadro's number (Marshall 1985). Equation 1 can be used to calculate Brownian displacement for uncharged spheres. Many bacteria possess the ability to move under their own power by means of flagella. Vaituzis and Doetsch (1969) have recorded a velocity of 56 μm/sec for *Pseudomonas aeruginosa*. Brownian motion and propulsion by means of flagella probably do not play a large role in microbial transport by laminar flow. However, these forms of motion may be significant in terms of either decreasing or increasing the frequency of adhesion once microbes penetrate the immobile "viscous" water layer adjacent to solid surfaces. A motile bacterium may increase adhesion frequency by continually moving toward a surface or decrease the frequency by propelling itself away.

The rate and extent of bacterial transport in soils depends upon a dynamic relationship between cell adhesion (reversible or irreversible), rate of water movement through soil pores, soil water content, and soil filtration processes. Irreversible adhesion of a bacterium to a solid surface can be broken down into two phases. Phase one involves an extremely fast or instantaneous reversible process. The irreversible phase, or phase two, is time dependent and may require expenditure of metabolic energy by the organism (ZoBell 1943; Marshall et al. 1971). For example, Marshall et al. (1971) found that irreversible adhesion of *Pseudomonas* strain R3 to a glass surface was related to increased production of an extracellular polymer. Adhesion was stimulated by lowering the substrate (glucose) concentration of the culture medium from 70 mg/liter to 7 mg/liter.

4.2.2 Bacterial Transport

Studies concerned with contamination of groundwater by sewage-borne pathogenic microorganisms have provided most of the information available in the literature pertaining to the transport of bacteria in soil and subsurface environments. This information is germane to any discussion concerning the fate of GEMs in the soil environment.

Bacteria are vertically and laterally transported in soil as a result of water movement (Hagedorn et al. 1981; McGinnis and DeWalle 1983; Romero 1970). Water moves through soil in response to force gradients brought about primarily by gravity and capillarity. Several factors influence the rate of water movement in soil, such as soil texture (i.e., the percentage of sand, silt, and clay), structure, water content, temperature, and organic matter content. The influence of soil texture on rates of groundwater movement and microbial transport has been summarized by McGinnis and DeWalle (1983). They reported a positive correlation between velocity of groundwater flow and distance of bacterial transport in soil. Current research has indicated that the faster groundwater velocities and the greater distances of bacterial transport corresponded to the coarsest soil texture, while the slower velocities were associated with the more finely textured soils.

Under conditions of saturated flow, bacteria applied to a soil surface layer could be expected to move vertically and laterally through the soil profile. While studies designed to examine the transport of surface-applied bacteria into subsurface and groundwater environments are not numerous, results from those that have been conducted suggest that, while bacteria are normally restricted in transport to the top 60 cm of the surface layer, substantial numbers of surface-applied bacteria can, in fact, be found in groundwater at considerable distances from the point of dispersal (Schaub and Sorber 1977; University of California 1955; Romero 1970). An investigation by Schaub and Sorber of a rapid-infiltration wastewater application site revealed that surface-applied bacteria could be transported horizontally a distance of greater than 180 m from the point of application.

The potential for rapid lateral transport of bacteria in the soil surface layer is primarily dependent upon three important factors: (1) soil water content, (2) soil texture, and (3) water table gradient. Field studies conducted by Hagedorn et al. (1978) illustrate the importance of these factors. In that study, lateral transport of antibiotic-resistant indicator bacteria under conditions of saturated subsurface flow through an A horizon (0-36 cm in depth) and a B2t horizon (36-50 cm in depth) was primarily dependent on soil water content and texture. Chow (1964) has defined subsurface flow as essentially the portion of precipitation that infiltrates the surface soil and moves in a lateral direction through the upper-most horizons lying above and distinct from groundwater. In their evaluation of bacterial transport, Hagedorn and his coworkers discovered that the presence of indicator bacteria in test wells, strategically located down field from the point of dispersal, reached maxima during intervals that were closely associated with a rise in the water table following a major rainfall event. They also noted that, in general, lateral transport of indicator bacteria was faster in the A (silty loam) than the B2t (silty clay loam) horizon. The differences in transport rate between the two horizons were attributed largely to the difference in clay content. Similar observations have been made by several investigators (Stewart and Reneau 1981; Viraraghavan 1978; Reneau and Pettry 1975) who also found that increases in the lateral transport of fecal coliforms corresponded to periods when the water table was at or above the point of dispersal, which in these cases were sewage drainfields. Transport of the bacteria was always in the direction of the water table gradient.

In some cases, transport of bacteria in soil occurs at a rate faster than would be predicted by measurement of hydraulic conductivity (McCoy and Hagedorn 1980; Rahe et al. 1978). One reasonable explanation for this phenomenon is the presence of horizontal and vertical pathways, more commonly referred to as macropores, that allow rapid movement of water in natural soils. Laboratory and field studies suggest that bacteria transported through these macropores bypass the bulk of the soil matrix (Smith et al. 1985; McCoy and Hagedorn 1980; Brown et al. 1979).

Clearly, saturated flow in soil provides an opportunity for bacteria to be transported significant distances from the point of dispersal, but this is not the case for unsaturated flow. Transport of bacteria under conditions of unsaturated flow is extremely limited due to increases in adhesion and the filtering ability of soil. The tendency of natural soil to retain bacteria under largely unsaturated conditions has been demonstrated by Brown et al. (1979) who concluded from their studies that a distance of 120 cm, in either the vertical or horizontal direction from a sewage discharge point, was sufficient for complete retention of coliform bacteria. Madsen and Alexander (1982) observed transport of surface-applied *Rhizobium japonicum* and *Pseudomonas putida* only when water was allowed to percolate through a 10-cm soil column. Without this percolating water, the bacteria were restricted to the top 2.7 cm of soil.

Transport of bacteria from soil surface layers to underlying groundwater has been documented (Schaub and Sorber 1977; McGinnis and DeWalle 1983; DeWalle et al. 1980; Romero 1970; Butler et al. 1954). One consequence of this is that once bacteria reach the groundwater they can be transported substantial distances in a relatively short period of time. Allen and Morrison (1973) found that tracer bacteria were readily transported by a groundwater gradient into a downslope well. At one particular site, the

TABLE 4–4 Factors Influencing Bacterial Transport in Soil

Factor	*Comment*	*Reference*
Hydrodynamic condition	Transport under conditions of saturated flow can be substantial (180 m), compared to unsaturated flow (3 m)	
Soil texture	Retention usually is inversely proportional to soil particle size	Bitton and Gerba 1984
Soil pH	Low pH enhances bacterial retention	Bitton et al. 1974
Water table gradient	Rate of transport increases with the water table gradient	
Ionic strength of soil solution	Increase leads to an increase in adhesion frequency	

rate of transport was 28.6 m in less than 30 hours. Table 4–4 summarizes several important factors that influence bacterial transport in soil and subsurface environments.

4.2.3 Virus Movement

Viruses, including phages, range in size from 20 to 200 nm and carry a net negative charge at pH values near neutrality (Bitton 1975). Because of their small size, virus particles can experience significant Brownian motion. Based on calculations by Marshall (1985), a virus particle approximately 10 nm in size would experience a Brownian displacement of about 390 μm/hr. Brownian motion likely plays a significant role in transport once a virus particle has penetrated the immobile water layer associated with solid surfaces. Cookson (1970) described the removal of bacteriophage from packed beds of activated carbon by use of a mass transfer model. He found that the virus removal rate closely fits the model, which could be used to predict the effect of flow rate, bed height, pore volume, and matrix texture on removal efficiency.

4.2.4 Viral Transport

Laboratory studies involving batch and column experiments have been widely used to elucidate factors involved in the transport of viruses through soil. Studies involving soil columns have been criticized because of the lack of standardization in experimental conditions (Bitton et al. 1979). Despite these problems, laboratory investigations have provided valuable information about the factors that influence virus transport in soil (Table 4–5).

It is generally agreed that fine-textured, clay type soils retain viruses more effectively than coarse-textured, sandy type soils. Clay minerals display increased sorptive capacities towards viruses as a result of their high surface area and ion exchange capacity. Drewry and Eliassen (1968) were able to show, with a series of nine soils from Arkansas and California, that virus retention increased with increases in clay content and specific surface area.

Studies using soil columns have indicated that for the most part, surface-applied virus particles are retained within the top 25 cm of soil (Goyal and Gerba 1979; Lance et al. 1976). Other investigations have shown, however, that virus particles are capable of being transported well beyond the top few centimeters of soil. The importance of the velocity of water movement

TABLE 4–5 Factors Influencing Virus Transport in Soil

Factor	*Comment*	*Reference*
Flow rate	Transport in soil increases with increased rates of water movement	Lance and Gerba 1980; Duboise et al. 1976
Hydraulic condition	Transport is greater during conditions of saturated flow than during unsaturated flow	Lance and Gerba 1984
Ionic strength of soil solution	Increase leads to enhanced adhesion to soil surfaces	Lance et al. 1976; Gerba and Lance 1978
Soil texture	Finely-textured soils retain viruses to a greater extent than coarse-textured soils	Bitton et al. 1979; Sobsey et al. 1980
pH	Adhesion to soil usually increases as pH decreases	Goyal and Gerba 1979; Burge and Enkiri 1978; Taylor et al. 1981
Virus type	Adhesion to soils varies with virus type	Gerba et al. 1981; Goyal and Gerba 1979; Landry et al. 1979
Humic substances	Soil organic matter may block virus adhesion to soil surfaces	Burge and Enkiri 1978; Bixby and O'Brien 1979; Scheuerman et al. 1979

through soil on poliovirus transport has been examined by Lance and Gerba (1980), who used soil columns packed with coarse sand to examine the effect of flow rates on virus retention. Their studies showed that an increase in flow rate from 0.6 m/day to 1.2 m/day increases the depth of virus penetrating from 160 cm to 250 cm after 4 days of effluent infiltration. In a field study, velocity of water movement has been reported to be a significant factor in virus contamination of groundwater at a groundwater recharge installation receiving tertiary-treated effluent. Vaughn et al. (1981) found that high infiltration rates (75–100 cm/hr) resulted in the transport of substantial numbers of surface-applied poliovirus into the groundwater. Reduced flow rates (from approximately 100 cm/hr to 6 cm/hr) lead to significant reductions in the numbers of viruses reaching groundwater.

In a field situation, surface-applied viruses are capable of being transported substantial distances. Wellings et al. (1975) discovered that viruses could be transported horizontally at least 7 m, at a depth of approximately 3 m, outside the boundary limits of a cypress wetland site receiving wastewater discharge. Their observations were made over a 28-day period during heavy rains and at a time when no new effluent had been added to the site. Even more remarkable are the findings of Schaub and Sorber (1977) who observed lateral subsurface transport of surface-applied tracer virus some 180 m from the point of dispersal within 20 days of release. These studies were conducted at a rapid-infiltration land wastewater application site composed largely of unconsolidated silty sand and gravel.

Under the appropriate conditions, viruses can be transported for significant distances in subsurface environment. From the investigations noted above, it can be concluded that conditions leading to transport include high soil water content, rapid water flow, and coarse-textured soil (see Table 4–5).

4.3 SUMMARY

Releases of GEMs into the environment are expected to increase in the next few years, with the most dramatic increases resulting from the application of pest-control agents in agriculture and forestry. Of major significance in assessing the environmental risk impact of GEMs is an understanding of their survival and transport in soil and subsurface environments. While information on the transport and survival of microorganisms through soil is available, it is neither abundant nor extensive in terms of microbial types tested or soils examined.

Though the transport of microorganisms from an application site depends primarily upon passive mechanisms, broad generalizations pertaining to the transportability of a specific microorganism within a particular soil environment may not be possible. Indeed, to extrapolate from information about one microbe to another, or from one geographical location to another, may not be appropriate. What is clear, however, is that the broader the data base, the more powerful the argument for making reasoned judgement, and consequently the more satisfactory the results of the predictive process.

Several inherent difficulties exist in studying the transport behavior of GEMs in soil and subsurface environments. Detection of low microbial numbers or of stressed microbial populations is exceedingly difficult with traditional technology. In an effort to improve detection sensitivity, many improved methods of monitoring GEMs in the soil and subsurface are currently under development (Chaudhry et al. 1989).

Beyond the difficulties of making accurate measurements of microorganisms to determine their spatial and temporal situation in the soil and subsurface environment, lies the need to ascertain the dynamic relationships between indigenous populations of microorganisms and how they may interact with a released GEM. Also, research strategies have not adequately addressed methods to predict the potential interactions between GEMs and natural microorganisms. These issues must be addressed if environmental risk assessment is to be valuable. There is clearly a need for focused research on the survival and transport of GEMs in these environments.

REFERENCES

Abbott, A., Rutter, P.R., and Berkeley, R.C.W. (1983) *J. Gen. Microbiol.* 129, 439–445.

Alexander, M. (1977) *Introduction to Soil Microbiology*, pp. 103–112, Wiley, New York.

Allen, M.J., and Morrison, S.M. (1973) *Groundwater* 11, 6–10.

Bej, A.K., Perlin, M.H., and Atlas, R.M. (1988) *Appl. Environ. Microbiol.* 54, 2472–2477.

Bitton, G. (1975) *Water Res.* 9, 473–484.

Bitton, G., Davidson, J.M., and Farrah, S.R. (1979) *Water, Air, Soil Pollut.* 12, 449–457.

Bitton, G., and Gerba, C. (1984) *Groundwater Pollution Microbiology*, pp. 65–88, Wiley, New York.

Bitton, G., Lahav, N., and Henis, Y. (1974) *Plant Soil* 4, 373–380.

Bixby, R.L., and O'Brien, D.J. (1979) *Appl. Environ. Microbiol.* 38, 840–845.

Brock, T.D. (1974) *Biology of Microorganisms*, pp. 300–306, Prentice Hall, Inc., Englewood Cliffs, NJ.

Brown, K.W., Wolf, H.W., Donnelly, K.C., and Slowey, J.F. (1979) *J. Environ. Qual.* 8, 121–125.

Burge, W.D., and Enkiri, N.K. (1978) *J. Environ. Qual.* 7, 73–76.

Butler, R.G., Orlob, G.T., and McGauhey, P.H. (1954) *U.S. J. Am. Water Works Assoc.* 46, 97–111.

Characklis, W.G. (1984) in *Microbial Adhesion and Aggregation* (Marshall, K.C., ed), pp. 137–157, Springer-Verlag, New York.

Chaudhry, G.R., Toranzos, G.A., and Bhatti, A.R. (1989) *Appl. Environ. Microbiol.* 55, 1301–1304.

Chemical & Engineering News (1989) 67, 28–34.

Chow, V.T. (1964) in *Handbook of Applied Hydrology* (Chow, V.T., ed.), pp. 1–56, McGraw-Hill Book Co., New York.
Cliver, D.O., and Herrmann, J.E. (1972) *Water Res.* 6, 797–805.
Cookson, J.T., Jr. (1970) *Environ. Sci. & Technol.* 4, 128–134.
Derjaguin, B.V., and Landau, L. (1941) *Acta Physicochim.* USSR 14, 633–662.
Devanas, M.A., Rafaeli-Eshkol, D., and Stotzky, G. (1986) *Curr. Microbiol.* 13, 269–277.
Devanas, M.A., and Stotzky, G. (1986) *Curr. Microbiol.* 13, 279–283.
DeWalle, F.B., Schaff, R.M., and Hatlen, J.B. (1980) *U.S. J. Am. Water Works Assoc.* 72, 533–536.
Drewry, W.A., and Eliassen, R. (1968) *J. Water Pollut. Control Fed.* 40, R257–271.
Duboise, S.M., Moore, B.E., and Sagik, B.P. (1976) *Appl. Environ. Microbiol.* 31, 536–543.
Elrick, D.E., and French, L.K. (1966) *Soil Sci. Soc. Am. Proc.* 30, 153–156.
Gerba, C.P., Goyal, S.M., Cech, I., and Bogdan, G.F. (1981) *Environ. Sci. & Technol.* 15, 940–944.
Gerba, C.P., and Lance, J.C. (1978) *Appl. Environ. Microbiol.* 36, 247–251.
Goyal, S.M., and Gerba, C.P. (1979) *Appl. Environ. Microbiol.* 38, 241–247.
Hagedorn, C., Hansen, D.T., and Simonson, G.H. (1978) *J. Environ. Qual.* 7, 55–59.
Hagedorn, C., McCoy, E.L., and Rahe, T.M. (1981) *J. Environ. Qual.* 10, 1–8.
Hurst, C.J., Gerba, C.P., and Cech, I. (1980) *Appl. Environ. Microbiol.* 40, 1067–1079.
Israelachvili, J., and Pashley, R. (1982) *Nature* 300, 341–342.
Israelachvili, J.N., and McGuiggan, P.M. (1988) *Science* 241, 795–800.
Kott, Y. (1988) *Water Sci. Technol.* 20, 61–65.
Lamanna, C., Mallette, M.F., and Zimmerman, L. (1973) *Basic Bacteriology*, The Williams & Wilkins Co., Baltimore, MD.
Lance, J.C., and Gerba, C.P. (1980) *J. Environ. Qual.* 9, 31–34.
Lance, J.C., and Gerba, C.P. (1984) *Appl. Environ. Microbiol.* 47, 335–337.
Lance, J.C., Gerba, C.P., and Melnick, J.L. (1976) *Appl. Environ. Microbiol.* 32, 520–526.
Landry, E.F., Vaughn, J.M., Thomas, M.Z., and Beckwith, C.A. (1979) *Appl. Environ. Microbiol.* 38, 680–687.
Madsen, E.L., and Alexander, M. (1982) *Soil Sci. Soc. Am. J.* 46, 557–560.
Marshall, K.C. (1976) *Interfaces in Microbial Ecology*, Harvard University Press, Cambridge, MA.
Marshall, K.C. (1985) in *Bacterial Adhesion* (Savage, D.C. and Fletcher, M., eds.), pp. 133–161, Plenum Press, New York.
Marshall, K.C., and Cruickshank, R.H. (1973) *Arch. Mikrobiol.* 91, 29–40.
Marshall, K.C., Stout, R., and Mitchell, R. (1971) *J. Gen. Microbiol.* 68, 337–348.
McCoy, E.L., and Hagedorn, C. (1980) *J. Environ. Qual.* 9, 686–691.
McGinnis, J.A., and DeWalle, F. (1983) *U.S. J. Am. Water Works Assoc.* 75, 266–271.
McMahon, M.A., and Thomas, G.W. (1974) *Soil Sci. Soc. Am. Proc.* 38, 727–732.
Pashley, R.M., McGuiggan, P.M., Ninham, B.W., and Evans, D.F. (1985) *Science* 229, 1088–1089.
Powell, M.S., and Slater, N.K.H. (1982) *Biotech. Bioeng.* 24, 2527–2537.
Powell, M.S., and Slater, N.K.H. (1983) *Biotech. Bioeng.* 25, 891–900.

Rahe, T.M., Hagedorn, C., McCoy, E.L., and Kling, G.F. (1978) *J. Environ. Qual.* 7, 487–494.

Reddy, K.R., Khaleel, R., and Overcash, M.R. (1981) *J. Environ. Qual.* 10, 255–266.

Reneau, R.B., Jr., and Pettry, D.E. (1975) *J. Environ. Qual.* 4, 41–44.

Report of the Biotechnology Science Advisory Committee of the U.S. Environmental Protection Agency. (1987) Office of Toxic Substances, vol. 1, no. 1, Summer, 1987, EPA, Washington, D.C.

Ritchie, J.T., Kissel, D.E., and Burnett, E. (1972) *Soil Sci. Soc. Am. Proc.* 36, 874–879.

Romero, J.C. (1970) *Groundwater* 8, 37–48.

Rutter, P.R., and Vincent, B. (1984) in *Microbial Adhesion and Aggregation* (Marshall, K.C., ed.), pp. 21–38, Springer-Verlag, Berlin.

Schaub, S.A., and Sorber, C.A. (1977) *Appl. Environ. Microbiol.* 33, 609–619.

Scheuerman, P.R., Bitton, G., Overman, A.R., Asce, M., and Gifford, G.E. (1979) *J. Environ. Eng. Div.* 105, 629–641.

Singh, U., and Uehara, G. (1986) in *Soil Physical Chemistry* (Sparks, D.L., ed.), pp. 1–38, CRC Press, Boca Raton, FL.

Smith, M.S., Thomas, G.W., White, R.E., and Ritonga, D. (1985) *J. Environ. Qual.* 14, 87–91.

Sobsey, M.D., Dean, C.H., Knuckles, M.E., and Wagner, R.A. (1980) *Appl. Environ. Qual.* 40, 92–101.

Stenström, T.A. (1989) *Appl. Environ. Microbiol.* 55, 142–147.

Stewart, L.W., and Reneau, R.B., Jr. (1981) *J. Environ. Qual.* 10, 528–531.

Tadros, Th. F. (1980) in *Microbial Adhesion to Surfaces* (Berkeley, R.C.W., Lynch, J.M., Melling, J., Rutter, P.R., and Vincent, B., eds.), pp. 93–116, Horwood, Chichester, UK.

Taylor, D.H., Moore, R.S., and Sturman, L.S. (1981) *Appl. Environ. Microbiol.* 42, 976–984.

Thomas, G.W., and Phillips, R.E. (1979) *J. Environ. Qual.* 8, 149–152.

University of California. (1955) *Studies in Water Reclamation*, Tech. Bull. no. 13, Berkeley.

Vaituzis, Z., and Doetsch, R.N. (1969) *Appl. Microbiol.* 17, 584–588.

Valeur, A., Tunlid, A., and Odham, G. (1988) *Arch. Microbiol.* 149, 521–526.

van Elsas, J.D., Dijkstra, A.F., Govaert, J.M., and van Veen, J.A. (1986) *FEMS Microbiol. Ecol.* 38, 151–160.

van Loosdrecht, M.C.M., Lyklema, J., Norde, W., Schraa, G., and Zehnder, A.J.B. (1987) *Appl. Environ. Microbiol.* 53, 1893–1897.

van Veen, J.A., Ladd, J.N., and Amato, M. (1985) *Soil Bio. Biochem.* 17, 747–756.

Vaughn, J.M., Landry, E.F., Beckwith, C.A., and Thomas, M.Z. (1981) *Appl. Environ. Microbiol.* 41, 139–147.

Verwey, E.J.W., and Overbeek, J.Th.G. (1948) *Theory of the Stability of Lyophobic Colloids*, Elsevier, Amsterdam.

Viraraghavan, T. (1978) *Water, Air, Soil Pollut.* 9, 355–362.

Wellings, F.M., Lewis, A.L., Mountain, C.W., and Pierce, L.V. (1975) *Appl. Environ. Microbiol.* 29, 751–757.

Wild, A. (1972) *J. Soil Sci.* 23, 315–324.
Yates, M.V., Gerba, C.P., and Kelley, E.M. (1985) *Appl. Environ. Microbiol.* 49, 778–781.
Yeager, J.G., and O'Brien, R.T. (1979) *Appl. Environ. Microbiol.* 38, 694–701.
ZoBell, C.E. (1943) *J. Bact.* 46, 39–59.

CHAPTER

5

Aerial Dispersal of Bacteria

Christen D. Upper
Susan S. Hirano

The presence of microorganisms in the air has been recognized since the pioneering work of Pasteur in the mid-19th century (Gregory 1973). Since then, many reports have been published documenting the presence of airborne bacteria, fungi, viruses, and other microorganisms in enclosed facilities as well as in the atmosphere over land, lakes, and oceans (Gregory 1973; Hers and Winkler 1973; Edmonds 1979; Pedgley 1982; Cox 1987). The air that surrounds us is a pathway for dispersal of microorganisms. Although aerial dispersal of microorganisms is a common phenomenon, the process has recently generated concern in relation to the planned introductions of genetically engineered microorganisms into the environment. Once introduced, microbes have the potential to be dispersed away from the specific site of introduction and to become established and flourish at an unintended site (i.e., to spread). Thus, an aspect of assessing the potential environmental impact of these microorganisms requires estimating the extent to which their dispersal through the air may lead to their spread.

To assess the risk associated with the aerial dispersal of introduced microorganisms, mere detection of their presence in the air is insufficient.

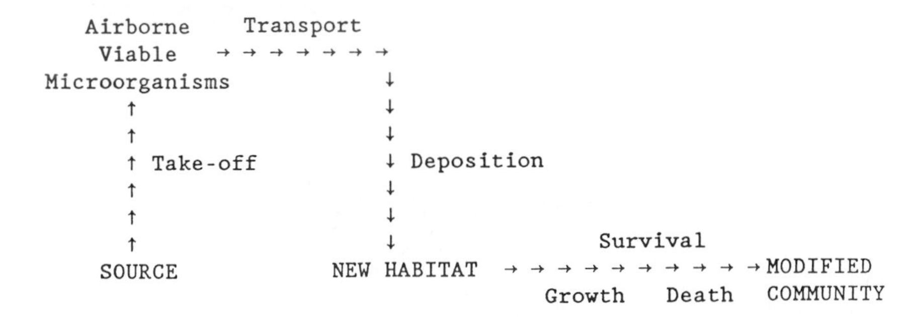

FIGURE 5–1 A diagram of the steps that are important to the risk assessment of the aerial dispersal and spread of microorganisms.

We need to consider quantitative aspects related to take-off, viability of the microorganisms once airborne, the likelihood that they may be transported a given distance, deposition, and, most importantly, survival and growth in the new habitat (Figure 5–1). If the likelihood of occurrence of any one step in this chain of events is zero, or nearly zero, then the ecological risk due to aerial dispersal (used in the broad sense to include all steps of the process) would also be zero. If the probability of occurrence of each step in the process is greater than zero, then it is important to consider how the environment influences the likelihood of each individual step in the process in order to assess the overall likelihood of aerial dispersal leading to spread of engineered microorganisms. Thus, we will discuss the factors (environmental and otherwise) that influence the likelihood that each step will occur, and the number of propagules likely to be involved at each step of the overall dispersal process.

Most of the released microorganisms for which assessment of environmental impact will be required will be expected to mimic at least some aspects of the normal function and behavior of indigenous environmental inhabitants. The most useful literature to examine with regard to these organisms is that dealing with the natural dispersal of microorganisms from managed and unmanaged ecosystems. Because of the growing interest in the use of genetically engineered microorganisms for enhanced crop production, we draw most of our examples from aerial dispersal of bacteria in agricultural ecosystems.

5.1 SOURCES AND TAKE-OFF

The number of bacteria becoming airborne from a given source per unit time (i.e., the source strength) will be a function of the concentration of bacteria in the source (source density), the area of the source, and the fraction of bacteria in the source that become airborne per unit time. The total number of bacteria dispersed from a particular source will be the integral of source strength over the duration of the lifetime of the source. In cases where bacteria actually grow in the source and take-off is dependent on ambient weather conditions, the concentration of bacteria in the source and the rate of take-off may be expected to be highly variable with time. In some cases, the area of the source can be expected to be relatively fixed (e.g., a soilborne bacterium applied to a given parcel of land). In others, the area of the source may change with time (a marine oil spill). Thus, source strength may be highly variable. The time scale with which source strength varies will itself depend on the time scale upon which the forces that influence the individual processes vary, ranging from seconds (variability in wind speed), to hours (bacterial doubling times), to days (variability in weather associated with frontal passage).

Sources of airborne bacteria may be grouped into three general (although not necessarily exclusive) categories: anthropogenic, aquatic, and terrestrial sources. Regardless of the source, unlike certain fungi with active discharge mechanisms, there are no known active take-off mechanisms for bacteria. Bacteria enter the air passively. In general, airborne bacteria are present in higher numbers in urban as compared to rural environments and higher over terrestrial as compared to aquatic systems (Bovallius et al. 1978a; Lighthart et al. 1979). For example, in a three-year investigation of concentrations of natural airborne bacteria, Bovallius et al. (1978a) observed average airborne bacterial densities of 850 bacteria/m^3 (range, 100–4,000) over a city street, 763 bacteria/m^3 (range, 100–2,500) over a city park, 99 bacteria/m^3 (range, 2–3,400) over an agricultural district, and 63 bacteria/m^3 (range, 9–560) over a coastal area.

Dispersal of bacteria into the air from anthropogenic sources has been most-often studied with regard to sewage treatment plants, cooling towers, irrigation of cropland with wastewater, and other sources of putative hazard to human health (Adams and Spendlove 1970; Katzenelson and Teltsch 1976; Sorber et al. 1976; Parker et al. 1977; Dondero et al. 1980; Bausum et al. 1983). Densities of up to 10^5 bacteria/m^3 have been detected downwind from a sewage treatment plant (Adams and Spendlove 1970). In rural environments, bacteria are dispersed into the air by human activities such as plowing, harvesting, and irrigation of agricultural cropland (Venette and Kennedy 1975; Pérombelon et al. 1979; Lighthart 1984; McInnes et al. 1988).

Dispersal from some of these anthropogenic sources may serve as useful models for dispersal during the actual release of microorganisms to the environment. For example, aerosols that are formed during spray application of a microorganism onto an application site (crop, oil spill) can be expected to be dispersed in a way that can be modeled as dispersal from a point source (Lighthart and Frisch 1976; Peterson and Lighthart 1977). Important considerations during this kind of application include the pressure and droplet size delivered by the sprayer, which influence the likelihood that bacteria will be in small droplets that will form a relatively stable aerosol, as opposed to larger droplets that will land on the intended target. The relative humidity and temperature of the water in which the bacteria are suspended will affect the rate of evaporation of water from the droplets and hence the final size and water content of the aerosols formed. This, in turn, will affect their stability and likelihood of downwind transport.

Bodies of water are an important source of airborne bacteria. Blanchard and coworkers (Blanchard and Syzdek 1970, Blanchard et al. 1981; Blanchard and Syzdek 1982) have studied the role of bursting bubbles in generating bacterial aerosols from aquatic sources. Some bacteria and viruses are concentrated on the surfaces of bubbles rising through a column of water (Blanchard and Syzdek 1970; Baylor et al. 1977). Thus, the concentration of bacteria at the bubble surface is higher than that in the bulk water. When bubbles burst at the water surface, the film that comprised the bubble surface is consolidated into a jet of water that is ejected vertically into the air. This jet forms "jet drops" whose size and number are determined very precisely by the size of the bubble from which they arose. Additionally, much smaller and more numerous droplets called film drops are often formed. Because the surface of the bubble concentrates bacteria as it rises and the jet drops arise from the bubble surface, the concentrations of microorganisms ejected into the air is often much greater than that in the bulk water from which the drops arise. Film drops can also carry microorganisms into the air. Bacterial aerosols can arise from bursting bubbles generated by waves breaking in the ocean, by bubbles generated during treatment of wastes subjected to aeration and agitation, or by any process that agitates the surface of a body of water. Knowledge of this kind of dispersal mechanism has been used to identify the putative source and mechanism of dispersal to susceptible individuals of *Mycobacterium intracellulare*, a pathogen of humans with otherwise-compromised respiratory or immune systems (Wendt et al. 1980).

The extent to which bacteria are concentrated on rising bubbles differs among bacterial species. In jet drops, *Serratia marinorubra* was enriched up to 30-fold more than *Escherichia coli* (Hejkal et al. 1980). The enrichment factor was greater for strains of *Mycobacterium intracellulare* than for *M. scrofulaceum* (Parker et al. 1983). Differential enrichment has also been observed among different strains within a species. Enrichment was greater for pigmented compared to nonpigmented *Serratia marcescens* (Blanchard and Syzdek 1978; Burger and Bennett 1985). It may be useful to assess the impact of the genetic alterations of engineered bacteria on the likelihood of take-off from the site of introduction.

Airborne bacteria from natural terrestrial sources may arise from the soil or from aerial parts of plants that cover the land masses. Wind and rain splash have been assumed to be the dominant natural forces involved in release of bacteria from the soil or vegetation. Aerosols of the plant-pathogenic bacterium, *Pseudomonas syringae* pv. *glycinea* were detected in an infected soybean plot during rainstorms and sprinkler irrigation (Venette and Kennedy 1975). Similar findings have been made with the soft-rot plant pathogen, *Erwinia carotovora* (Graham and Harrison 1975; Graham et al. 1977; Quinn et al. 1980; Fitt et al. 1983). Graham and Harrison (1975) demonstrated that when simulated raindrops fell onto potato stems infected with *E. carotovora*, aerosols were generated that contained viable cells of the pathogen. In the field, airborne *E. carotovora* was detected during rainfall but never when the weather was dry (Quinn et al. 1980). Development of gradients of plant diseases has frequently been attributed to the rain-splash dispersal of plant pathogens (Walker and Patel 1964; Ercolani et al. 1974). In these studies, rain-splash dispersal was suggested as a major mechanism of dispersal of phytopathogenic bacteria from diseased plants.

Lindemann et al. (1982) provided new insight into the importance of plant canopies as significant sources of airborne bacteria. Bacteria, including ice nucleation-active bacteria, were detected over corn, wheat, pea, bean, and alfalfa fields during dry weather conditions. Canopy level concentrations of airborne bacteria were as high as 6,500 viable cells/m^3 over a closed wheat canopy. Rain was not necessary for the generation of bacterial aerosols from plants. In addition, canopy-level bacterial concentration and upward flux were four times greater over alfalfa than over an upwind field of bare, dry soil despite greater bacterial numbers in the latter. Thus, the vegetation that covers the land masses constitutes a major source of airborne bacteria.

The effect of the physical environment on bacterial take-off from plant canopies was examined by Lindemann and Upper (1985). Although viable bacteria were present in the air at all times of day, significant take-off of bacteria from bean canopies, based on measurements of upward fluxes, were likely to occur only when atmospheric conditions were neutral or unstable, there was direct sunlight, windspeeds exceeded 1 m/s, and leaves were dry. Wind speed was positively correlated with airborne bacterial concentrations and upward flux measurements. Concentration and deposition were highly correlated. Thus, deposition and take-off were occurring simultaneously at the same location; a small fraction of the bacteria that take-off were almost immediately redeposited on nearby plants. Due to the strong interrelationships among temperature, humidity, and wind speed with time of day, rates of bacterial take-off were also related to time of day. A net introduction to the atmosphere of approximately 4.5×10^{10} particles bearing viable bacteria per hectare was estimated to occur daily. The mechanism by which leaf surface bacteria are dispersed from plant canopies remains to be elucidated.

The experiments of Lindemann and Upper (1985) illustrate several important considerations with regard to the role of the take-off stage in the overall impact of aerial dispersal in ecological risk assessments. Their experimental design allowed estimation of the relationship between a number of weather variables and several bacterial parameters (e.g., concentration, upward flux, and deposition). In addition, they studied the inherent variability in airborne bacterial concentrations and upward movement. No upward flux was detected on about half of the days during which samples were taken. During the other half, concentration, deposition, and upward flux were highly variable, approximating lognormal distributions. The very large sample-to-sample variability illustrates the need for frequent samples. Single estimates or designs that only sample once daily (Lindow et al. 1988) may result in substantially distorted estimates of take-off.

5.1.1 Summary

The likelihood of take-off occurring for a microorganism introduced into the environment must be considered in light of the size of the area to be treated, the way the microorganism is to be applied, the density of the microbe at the source following the introduction, and the environmental and other factors that affect the rate at which take-off is likely to occur. There are a number of biological considerations that may influence source strength, rate, and total numbers of bacteria entrained from a given site. Bacterial genotype may influence take-off, not only in terms of species but also in terms of genotypes within a species. The total number of bacteria available for dispersal is also an extremely important consideration. The larger the site and/or the greater the bacterial population size in or on the site, the greater the number of bacteria that are available for take-off. Small-scale field experiments such as that conducted by Lindow and Panopoulos (1988) should be expected to introduce orders-of-magnitude-fewer bacteria into the troposphere than the potential large-scale commercial use of successful genetically engineered microbial products.

5.2 AERIAL TRANSPORT

In the context of risk assessment, what we need to know about transport is the likelihood of a given cell reaching a given location in a viable condition. The input is an aerosol produced at take-off; the endpoint is deposition of a viable microorganism at a new site.

There is substantial literature on the atmospheric transport of particles. Atmospheric dispersal of inert particles has been frequently modeled, particularly as dispersal of pollutants from point sources. Dilution and loss from aerosols by deposition are the two major processes that decrease concentrations of these particles as they move away from the source. Microbial aerosols differ from other particulate aerosols in that their concentrations are also decreased by cell death. Indeed, the major decrease in concentrations of many microbial aerosols is probably decreased viability. The physical environment affects transport of microbial aerosols, then, by determining the mechanics of transport and dilution and by influencing bacterial viability.

Once take-off has occurred, the viability of the airborne bacteria plays a critical role in the likelihood that transport to a particular site will occur. Most studies on the viability of airborne bacteria have been conducted by varying one parameter at a time under controlled environmental conditions indoors. Most studies of survival outdoors under natural conditions have used the microthread technique of May and Druett (1968). In addition, most of these studies have focused on a single or relatively few selected species or strains within a bacterial species (cf. Bovallius et al. 1980; Cox 1987). Many interactive factors appear to affect the viability of airborne bacteria by as-yet-undetermined mechanisms. Relative humidity, temperature, radiation, chemical constituents in the air, growth conditions of the bacteria prior to take-off, the conditions by which bacterial aerosols are produced, the genotype/phenotype of the bacteria, and other factors affect viability of airborne bacteria. In addition, the methods used to collect airborne bacteria may influence estimates of viable airborne bacteria.

To add complexity to the issue, it has been suggested that viability of airborne bacteria differs in indoor versus outdoor air (Druett and May 1968; Cox 1987). The phenomenon has been referred to as the open air factor. Only a few bacterial species have been tested for their sensitivity to the open air factor. The open air effect has only been observed with air outside the laboratory; it vanishes when air is pumped into the laboratory. The effect has only been observed with bacteria attached to microthreads (spider webs), which subject bacteria to very different conditions than a bacterium suspended as an aerosol particle. Nonetheless, *Escherichia coli*, *Serratia marcescens*, *Micrococcus albus*, and the phytopathogenic bacteria, *Erwinia carotovora* pv. *carotovora*, *E. carotovora* pv. *atroseptica*, and *E. amylovora*, were sensitive to the open air factor whereas spores of *Bacillus subtilis* var *niger* and *B. anthracis* were not (Cox 1987). For example, survival of *E. amylovora* (Southey and Harper 1971) and *E. carotovora* (Graham et al. 1979) on microthreads was generally less than 10% and 1%, respectively, after two hours of exposure in open air. The open air factor should affect bacteria resident on plant surfaces to the same extent as it does those suspended on spider webs. However, the observed death of *E. amylovora* on spider webs is much too rapid to be consistent with the normal residence of this bacterium on the flowers of fruit trees, where it is fully exposed to the outdoor air (Miller and Schroth 1972). In light of the observed differences in viability of bacteria in indoor versus outdoor air, how useful are results from controlled environment studies in predicting the survival or death rates of bacteria that become airborne under natural conditions? Good, quantitative, careful research to compare bacterial survival on microthreads to survival in suspension in the same "air" in aerosols, and to evaluate the effect of the physiological condition of the bacteria on survival during transport is needed before this issue can be resolved.

Because bacteria traveling as single cells or on very small particles have small terminal sedimentation velocities, the potential for long-distance transport is not significantly affected by sedimentation. That is, concentrations of bacterial aerosols are not significantly diminished due to loss from the aerosol by sedimentation onto surfaces. Atmospheric turbulence and wind velocity are important meteorological parameters that affect the likelihood of movement away from a given source. The greater the turbulence, the more rapid the dilution of aerosols; the higher the windspeed, the more rapidly the particles travel. Studies have been done to estimate the distance that airborne microbes are likely to be detected downwind of what may be considered point sources, such as sewage treatment plants and spray applications of bacteria to plants in small-scale field plots. Airborne bacteria have been detected at least 1,290 m from a sewage treatment plant (Adams and Spendlove 1970). However, more frequently, bacteria will not be detected beyond a few hundred meters downwind of such plants or of sprinkler irrigation systems (cf. Bovallius et al. 1980). *Pseudomonas syringae* was detected on petri plates at 9.1 m and on oat plants up to 27 m from the source during the spray application of a marked strain of the bacterium onto oat plants in small-scale field plots (Lindow et al. 1988). After aerosols generated by the initial application of the bacteria to the plants had dissipated, the applied bacterium was not detected on deposition plates outside the plot area. It appears that bacteria released from sources of relatively small size are not likely to be detected at very long distances from the source. Caution is necessary in accepting this interpretation, however. The distance over which bacterial transport could be detected was limited by sample locations in some of these studies (Adams and Spendlove 1970; Lindow et al. 1988); bacteria were detected in the most distant samples taken. Further, lack of detection does not imply lack of transport beyond a given distance. What it does mean is that the combination of dilution and removal by death and deposition have diminished the concentration to the point that detection of the microorganism is unlikely.

The likelihood that bacteria will be detected at longer distances from the source increases as source strength increases. This is simply because greater distances will be required to dilute airborne concentrations to the level where they are no longer detectable. Evidence for the long-distance transport of bacterial spores was reported by Bovallius et al. (1978a, 1978b, 1980). Bacteria collected during a snowfall in central Sweden were reported to have originated from the Black Sea, 1,800 km away. Bacteria collected from the air were similar to that in the snow but different from the local microflora on the basis of several properties. An examination of the meteorological conditions prevailing at the time using a computer-based horizontal back-trajectory analysis indicated that the time required for the 1,800 km journey was about 36 hr. Bacteria collected over a soybean field during a rainfall were different from bacteria present on soybean leaves with respect to ice nucleation activity and pathogenicity to soybeans and snap beans (Constantinidou et al. 1990). On the basis of these results coupled with the work of Lindemann and Upper (1985), it has been suggested that ice-nucleation active and phytopathogenic bacteria released into the atmosphere during dry, sunny weather, may be carried intermediate to long distances before being returned to plant surfaces by rain (Constantinidou et al. 1990). Lindemann and Upper (1985) found peaks in airborne bacterial concentrations and deposition of bacteria several hours after dispersal from nearby plant canopies had ended for the day. These bacteria may have travelled for 40 to 65 km, after entrainment several hours earlier.

5.2.1 Summary

Insufficient empirical data preclude precise estimations of the probability that a given number of bacteria entering the air at a given place will be deposited at a given distance from the source. Models have been developed for dispersal of air pollutants and other aerosols (cf. Edmonds 1979). Modification of these models to better account for microbe viability may make them more useful tools for predicting the likelihood of aerial transport of bacteria. Lighthart and coworkers (Peterson and Lighthart 1977; Lighthart and Mohr 1987) have attempted to account for microbe viability in their model. These authors recognize that a severe limitation in the model, however, is that the inputs for the microbe viability parameter were obtained from laboratory estimates.

Thus, models have served one of the major functions of models in a developing field—to define limitations in knowledge and need for experimental data. The current models are useful for identifying weaknesses in estimates of the relative quantitative contributions that source size and strength, viability of airborne bacteria, and meteorological conditions have on the probability that a given number of bacteria entering the air are likely to be deposited at a given distance from the source. As areas needing more research are identified, and as results are obtained, improved models with substantial predictive power may eventually result. The interactions between atmospheric and biotic processes, however, are sufficiently complex that the amount of research that needs to be done before these models will be truly predictive is almost overwhelming.

5.3 DEPOSITION

Theoretical and experimental considerations of aerosol deposition have been amply reviewed in the literature, most frequently with regard to the deposition of fungal spores, pollen, and aerosol deposition during inhalation (Gregory 1973, Edmonds 1979; Pedgley 1982; Cox 1987). In general, airborne particles may be deposited onto surfaces by gravitational settling (sedimentation), impaction, turbulent deposition, or during precipitation due to rainout or washout. The mechanisms by which a particle is most likely to be delivered onto a surface is strongly dependent on the size of the particle. Sedimentation and impaction are likely to be important for relatively large particles, such as pollen and some large fungal spores, but much less important for particles the size of small fungal spores, individual bacterial cells, bacterial spores, or viruses. For these smaller particles, turbulent deposition or scrubbing by precipitation are the dominant factors in deposition. Similarly, rainout is important for extremely small particles carried by molecular diffusion onto cloud droplets or ice crystals, whereas, washout occurs when falling raindrops or snowflakes collide with and retain large aerosol particles (Gregory 1973). Although raindrops are too large to be efficient "collectors" of bacterial-sized particles, sufficient numbers of drops fall through a given column of air during a rainstorm of modest intensity to remove most of the bacteria from the air.

Under field conditions, deposition of airborne bacteria onto petri plates suspended within plant canopies was correlated with concentrations of bacteria in the air over the canopy (Lindemann and Upper 1985). Because airborne concentrations were largest during periods when take-off was greatest, deposition exhibited a diurnal pattern with peak numbers during midday. That is, deposition near the site of take-off followed the same diurnal pattern as did take-off. Approximately 3×10^3 viable bacteria were likely to be deposited per bean leaflet per day during dry sunny weather; less than 10% of the bacteria that became airborne were redeposited onto the plants. Large numbers of bacteria were deposited on bean plants (Lindemann and Upper 1985) even in the absence of detectable take-off. Bacteria deposited during these conditions apparently had taken off from other sites, often several hours earlier, and presumably had been transported many kilometers.

Risk assessment of the deposition stage of aerial dispersal of microorganisms should include estimates of the sizes of particles bearing viable propagules that are likely to be deposited, the geometry of potential sites where larger particles are likely to be deposited, and weather conditions.

5.4 FATE OF DEPOSITED BACTERIA

In many cases of planned introductions of microorganisms, it is reasonable to assume that the combined probabilities of take-off, aerial transport, and deposition are sufficiently large that some dispersal to a given nearby site is highly likely, particularly in an agroecosystem or in an aquatic ecosystem. Thus, the important question is not whether aerial dispersal will occur. Of importance, rather, are the questions related to frequencies and numbers of cells that are likely to be dispersed and reach suitable habitats. We will call this "the numbers game" for brevity. Because dispersal will almost certainly occur, the most important aspect of risk assessment of aerial dispersal will be to determine how the microorganism is likely to respond in the various communities on the various habitats in which it will arrive, and how "the numbers game" will influence this response. Will the introduced microorganism survive, multiply, and cause adverse effects in those new habitats that it reaches? What are some of the factors that are likely to affect the fate of airborne bacteria in their new habitats?

Different types of habitats support different types of bacterial species. Bacterial species commonly found in lakes and streams are often not abundant in soil or on vegetation (Starr et al. 1981). Even within these very general categories of habitats, there is ample evidence of site specialization. For example, both the phyllosphere and rhizosphere of plants normally harbor large numbers of bacteria (i.e., 10^7–10^8 bacteria per gram). The species of bacteria that colonize these plant-associated habitats differ (Stout 1960a,b). *Pseudomonas syringae* is a ubiquitous inhabitant of the phyllosphere (Lindow et al. 1978) but only rarely of the rhizosphere (Schroth et al. 1981; Lindow 1985; Lindow and Panopoulos 1988). Intensive monitoring of soil was conducted by Lindow and Panopoulos (1988) following the introduction of ice-minus (Ice^-) strains of *P. syringae* (bacteria containing an inactivated gene for ice nucleation) to field plots of potato plants. Although large numbers of Ice^- bacteria were introduced into the soil during the application, they were not detectable in soil from one week after spray inoculation to the termination of the experiment, approximately five months later.

There are, of course, species of bacteria that are regarded as less habitat-specific, at least in the sense that they have been recovered from different types of habitats. For many of these microorganisms, whether or not they are able to multiply (as compared to merely survive) in the various habitats where they have been found remains to be demonstrated. For example, although the soft rot phytopathogen *Erwinia carotovora* is commonly found in association with plants (e.g., potato tubers) (Pérombelon and Kelman 1980), the bacterium has also been detected in water from drains, ditches, streams, rivers, and lakes, and in seawater (McCarter-Zorner et al. 1984). *Pseudomonas fluorescens* has been isolated from leaves, roots, soil, and water (Stolp and Gadkari 1981). *Erwinia herbicola*, generally regarded as a common epiphyte (Lindow et al. 1978), has also been isolated from the ocean (Fall and Schnell 1985). Thus, the breadth of adaptation of the introduced organism to various habitats should be a consideration in risk assessment.

There have been numerous experiments in which large numbers of bacteria have been introduced into their natural habitat, and their survival, colonization, and growth have been followed over time. Experiments of this kind, although they are not usually performed with the evaluation of risk assessment of microbial introductions as a goal, can provide insight into the ease with which microorganisms can become established, even on (presumably) favorable habitats. For example, the fate of phyllosphere bacteria has been followed in the course of a number of experiments following spray application to leaves (Weller and Saettler 1980; Stadt and Saettler 1981;

Lindow 1982; Lindow et al. 1988; Lindow and Panopoulos 1988). Colonization of root systems by rhizosphere bacteria after application of bacteria to seed has been monitored by many scientists (Kloepper et al. 1980; Suslow and Schroth 1982; Weller 1983; Loper et al. 1985). The results of these experiments range from establishment of vigorous, multiplying, large bacterial populations to complete die-off of the introduced bacteria in a very short time. Although there are many reports in the literature of successful establishment of bacteria on plants, it is likely that many of the unsuccessful attempts have not been published (e.g., S.S. Hirano and C.D. Upper, unpublished observations). Thus, we are unable to determine the fraction of these experiments that have actually resulted in successful establishment of bacteria, but we suspect that it is very small. Indeed, the frequent inability to establish introduced microorganisms on favorable habitats is one of the factors limiting successful biological control (National Academy of Sciences 1989).

The nature of the microbial community that is already established on the habitat may also affect the fate of the new arrival. Intraspecific or interspecific competition may diminish the survival or ease of establishment of the new arrival. For example, Ice$^-$ strains of *P. syringae* are far more effective in excluding ice-nucleating (Ice$^+$) strains if they are applied to leaf surfaces prior to arrival and establishment of Ice$^+$ strains (Lindow 1986; Lindow 1987; Lindemann and Suslow 1987; Lindow and Panopoulos 1988). Lindow has proposed that saturation of favorable sites on leaf surfaces with the Ice$^-$ strain denies favorable sites to the Ice$^+$ strain, and hence diminishes its ability to colonize the leaf surface, a mechanism he has termed preemptive competitive exclusion. The specificity in competitive exclusion among strains within a species and among bacterial species remains to be elucidated (Lindow 1986; Lindow and Panopoulos 1988). It is distinctly possible that this specificity will be strongly influenced by climate and physical environment. The success of microbial introductions into the rhizosphere have often varied from highly successful to rather poor, and often dependent on the site where the experiment was conducted. One possible cause of this variability is the interaction of the introduced bacterium with the microbial communities already established at the various sites.

Even when experiments are successful in achieving domination of a habitat by a single microorganism, the domination is often short-lived. If the experiment is continued long enough, the microbial community on the habitat often returns to a level that resembles those on similar habitats where no introduction had occurred. For example, a marked strain of *P. syringae* applied to young bean plants was a dominant bacterium on bean leaflets for approximately 5 weeks (Hirano et al. 1981). At the time of pod-harvest, 7–8 weeks after its introduction, the marked strain was a minor component of the leaf-associated bacterial communities. Although still present on every leaf and nearly every pod sampled, it was largely replaced by naturally occurring conspecifics.

The physical environment is undoubtedly a major factor affecting establishment of bacterial populations on a given habitat. However, specific effects of the physical environment on bacterial survival and growth and the mechanisms that mediate these effects remain to be understood. Because many important environmental parameters differ substantially between regions with different climates, we expect that there will be regional differences in the likelihood that a given bacterium will successfully colonize a particular habitat. Regional differences in bacterial plant diseases probably reflect regional climatic differences. Although a bacterium may be deposited in a suitable habitat, the physical conditions may not be favorable for its survival or growth. Thus, the time of arrival in the new habitat may also be critical for the subsequent survival/growth of the immigrant bacteria.

The most important lesson to be learned from the above discussion may be that microbial communities appear to be quite robust, and domination of these communities by introduced organisms is possible, but often difficult to achieve. We expect that, given this level of difficulty in achieving domination of a habitat by deliberately applying large numbers of bacteria, inadvertent achievement of this feat by low numbers of bacteria that move aerially between sites will not be frequent. In Lindow's experiment with Ice^- strains of *P. syringae*, two sets of plants were evaluated for establishment of the Ice^- strains. One set of plants was young potato plants within the plot area, on which natural *P. syringae* population sizes were very small. Another set was established wheat plants outside the buffer zone, on which *P. syringae* populations were 10^4–10^5 cells per gram fresh weight. The numbers of Ice^- *P. syringae* deposited on the potato plants were very large and led to successful establishment of the strains. Although dispersal during the spray application was detected outside the plot area, the numbers deposited on the wheat were very small, and the Ice^- bacteria were not detected on the wheat. The lack of success of these bacteria on the wheat habitat may be related to the relatively small number of Ice^- bacteria dispersed to a habitat that was already occupied with relatively large natural populations of Ice^+ *P. syringae*.

Aerial dispersal is a normal part of the community dynamics of many bacteria. We do not know how important a continuous supply of immigrants to a given habitat is in determining the relative population sizes of bacterial species in some habitats. In one area, aerial dispersal apparently had a major effect on relative population sizes of *P. syringae* and on the relative abundances of different genotypes of *P. syringae* (Lindemann et al. 1984). We suspect that it may make the difference between survival and local extinction in at least some cases. Dispersal of introduced microorganisms may function in much the same way. The continuous, although relatively low-level introduction of a microorganism to a site may pose a greater threat of successful spread of that microorganism to that site than a single high-level introduction would. Frequent arrival of low numbers of bacteria via the air may favor subsequent establishment in the new habitat because of the increased likelihood of arriving when physical and biological conditions are suitable for survival/growth. Bacteria are known to grow rapidly in many natural habitats, for example, on leaves (Hirano and Upper 1989). Yet total numbers of bacteria fluctuate within a relatively small range (about three orders of magnitude on leaves or in soil). Thus, frequent death of large numbers of bacterial cells must also be occurring in these communities. Establishment of an immigrant may be better favored by arrival of small numbers at a time when conditions favor growth relative to death of that particular type, than a single large introduction at a random time.

In rare cases, an empty niche in a vulnerable habitat may be occupied by an introduced microorganism. In this case, introduction of only a few individuals may pose an environmental hazard. These rare cases should be relatively easily identified through risk assessment of the genotype and phenotype of the microorganism proposed for release.

5.4.1 Summary

Given that bacteria that inhabit or are applied to many different ecosystems are likely to be dispersed via the air route, the fate of bacteria deposited in a new habitat will depend on the type of habitat, the parameters of the physical environment, and the nature of the established indigenous microbial community. The frequency of arrival at a new habitat of bacteria aerially dispersed from a given site of introduction may have a significant effect on survival and growth in the new habitat. In general, we expect that low-level, infrequent introductions of microorganisms to the vast majority of habitats that are already occupied by active microbial communities will not result in successful immigration to those habitats. On the other hand, frequent and substantial introduction to these habitats probably will result in occasional establishment of the immigrant in the community. If the introduced microorganism has a property that possibly would be detrimental to the community or the habitat, then frequent, high-level introductions will certainly present a greater risk than infrequent, low-level introductions.

5.5 THE OVERALL PROCESS

Thus far we have established that the likelihood of survival/growth of an introduced microorganism in a new habitat away from the site of introduction is dependent, among other things, on the number of microorganisms deposited on the habitat. The numbers deposited are dependent on the likelihood of transport, and the likelihood of transport depends on the source strength and lifetime. Although each step is dependent on the preceding step, each can also be considered as an independent module in the overall process.

There is sufficient information on the steps involving take-off, transport, and deposition to provide a degree of predictability to these steps of the process. The factors that determine bacterial community dynamics and hence bacterial community composition are much less well understood for most outdoor microbial communities. Microbial ecology studies have usually focused on ecosystem issues such as nutrient cycling, rather than community or population dynamics. Thus, the data base for drawing conclusions about important aspects of risk assessment is rather thin.

Most models of dispersal are based on concentration, or numbers deposited per area per unit time, and not total number of particles. For some kinds of risk assessment, concentration is the measure of interest. For example, the inhalation volume of human lungs per unit time is quite well known. Thus, the probability that an individual will inhale a bacterium in a unit of time can be related to aerosol concentration, and this relationship can be used to evaluate risk. On the other hand, for estimation of other kinds of risks, the total number of bacteria transported to a particular site may be more important than either rate or concentration. For example, if a perceived risk is associated with the likelihood that a particular bacterium will reach a given habitat, then the criterion of interest is the gross number of bacteria transported. In this case, concentration and rate are important only as part of the calculation of total numbers of arrivals. For example, if a pathogen has a very high probability of causing disease if it lands on a susceptible host (as many fungal plant pathogens do), then the risk being assessed is whether or not a viable propagule is likely to land on a susceptible host throughout the entire period that the source remains active. Finally, if continuous immigration is necessary to sustain a population on a particular habitat until the population has achieved some minimal size, then rate, not total numbers or concentration, should be considered the criterion of interest in risk assessment. Thus, the kind of risk being considered or the mechanism by which an introduced microorganism may be expected to affect a given target should be considered in choosing the manner in which risk due to aerial dispersal should be evaluated.

The greatest risk comes from the possibility that a habitat exists where an introduced microorganism may flourish *and* cause harm. Opponents of planned introductions have often cited introductions of exotic species that caused such harm as examples of the kind of risk that introductions may pose. The amount of "exotic" genetic material that is introduced to a habitat by engineering a single gene (or a few genes) into an organism that is already present in that habitat is far smaller than the thousands of genes that would be introduced with an exotic species. Thus, the task of risk assessment of the phenotype and genotype of a well-characterized engineered microorganism should be substantially simpler.

We expect that, for the foreseeable future, risk assessment of planned introductions of microorganisms will include substantial experimentation. (We hope that it will be careful experimentation rather than some level of monitoring imposed by a regulatory agency upon a scientist who does not regard it as an important part of his/her research.) It can be particularly difficult to quantitate low-level dispersal of microorganisms from small experimental sites that may occur over a relatively long period. Low concentrations of airborne particles are difficult to measure quantitatively, and long durations of dispersal can require many days of continuous sampling, a process that is both expensive and tedious. Nonetheless, if the total number of bacteria arriving at a particular site is the measure of interest, some means of estimating it is essential.

For this purpose, consider a hypothetical field experiment. A bacterium is to be applied to a few thousand square meters of a particular crop plant. Application of the bacterium to the entire area may take only a few minutes. Because the sprayer will move steadily across the plot during application, the period during which any given location within the plot is likely to be a source of airborne bacteria is much less than one minute. Dispersal during such an event can be thoroughly documented by sampling continuously throughout application (cf. Lindow et al. 1988; Lindow and Panopoulos 1988). Thus, the first stage of dispersal is brief, intense, and relatively easy to monitor. After the initial application, the applied bacteria may grow on the plants and remain at high populations for a month or more. The bacteria on the plants will serve as a source of airborne bacteria throughout this period. What kind of sampling strategy will be necessary to compare total numbers of bacteria dispersed during application to that during the period when the bacteria are established on the plants? If we assume that the bacteria are aerially dispersed from the plants for about 7 hours a day (Lindemann and Upper 1985) for 30 days, the plot will be a source for a total of 12,600 minutes, or more than 12,600-fold greater than the period of time (i.e., less than one minute) when each point on the plot was a source during the initial application. Thus, to ask if more bacteria arrive at a given habitat downwind during application, or by dispersal of bacteria growing on the plants, we need to measure concentrations of bacteria dispersing from the plants that are about 1/12,600 of those during the initial application. Clearly, a careful experimental design may be required to approach the question in a useful way.

The distance over which bacteria are likely to disperse from a given introduction and the expected number of arrivals at a given site downwind will be highly dependent on the size of the introduction, the nature of the bacterium introduced, the site to which it is introduced, and a number of environmental factors. Thus, the number of sites exposed and the numbers of bacteria to which they are exposed, will increase as the size and number of introductions increase. To the extent that any given site poses no risk, increasing the number of these same sites will also pose no risk. Alternatively, where the risk to a given site is real but small, increasing the number and size of introductions will increase the overall risk of environmental harm. Thus, as planned introductions continue to occur, it will be very important to distinguish sites on which nothing at all happens, and no risk can be expected, from sites on which there is some modification by the introduced microorganism, but the risk of harm is small.

REFERENCES

Adams, A.P., and Spendlove, J.C. (1970) *Science* 169, 1218–1220.

Bausum, H.T., Schaub, S.A., Bates, R.E., et al. (1983) *J. Water Pollut. Control Fed.* 55, 65–75.

Baylor, E.R., Baylor, M.B., Blanchard, D.C., Syzdek, L.D., and Appel, C. (1977) *Science* 198, 575–580.

Blanchard, D.C., and Syzdek, L. (1970) *Science* 170, 626–628.

Blanchard, D.C., and Syzdek, L.D. (1978) *Limnol. Oceanogr.* 23, 389–400.

Blanchard, D.C., and Syzdek, L.D. (1982) *Appl. Environ. Microbiol.* 43, 1001–1005.

Blanchard, D.C., Syzdek, L.D., and Weber, M.E. (1981) *Limnol. Oceanogr.* 26, 961–964.

Bovallius, A., Bucht, B., Roffey, R., and Ånäs, P. (1978a) *Appl. Environ. Microbiol.* 35, 847–852.

Bovallius, A., Bucht, B., Roffey, R., and Ånäs, P. (1978b) *Appl. Environ. Microbiol.* 35, 1231–1232.

Bovallius, A., Roffey, R., and Henningson, E. (1980) *Ann. N.Y. Acad. Sci.* 353, 186–200.

Burger, S.R., and Bennett, J.W. (1985) *Appl. Environ. Microbiol.* 50, 487–490.

Constantinidou, H.A., Hirano, S.S., Baker, L.S., and Upper, C.D. (1990) *Phytopathology* (in press).

Cox, C.S. (1987) *The Aerobiological Pathway of Microorganisms*, Wiley, Chichester, U.K.

Dondero, T.J., Jr., Rendtorff, R.C., Mallison, G.F., et al. (1980) *New Eng. J. Med.* 302, 365–370.

Druett, H.A., and May, K.R. (1968) *Nature* 220, 395–396.

Edmonds, R.L., ed. (1979) *Aerobiology: The Ecological Systems Approach*, Dowden, Hutchinson & Ross, Stroudsburg, PA.

Ercolani, G.L., Hagedorn, D.J., Kelman, A., and Rand, R.E. (1974) *Phytopathology* 64, 1330–1339.

Fall, R., and Schnell, R.C. (1985) *J. Marine Res.* 43, 257–265.

Fitt, B.D.L., Lapwood, D.H., and Dance, S.J. (1983) *Potato Res.* 26, 123–131.

Graham, D.C., and Harrison, M.D. (1975) *Phytopathology* 65, 739–741.

Graham, D.C., Quinn, C.E., and Bradley, L.F. (1977) *J. Appl. Bacteriol.* 43, 413–424.

Graham, D.C., Quinn, C.E., Sells, I.A., and Harrison, M.D. (1979) *J. Appl. Bacteriol.* 46, 367–376.

Gregory, P.H. (1973) *The Microbiology of the Atmosphere*, 2nd ed., Leonard Hill, Plymouth, U.K.

Hejkal, T.W., LaRock, P.A., and Winchester, J.W. (1980) *Appl. Environ. Microbiol.* 39, 335–338.

Hers, J.F.P., and Winkler, eds. 1973. *Airborne Transmission and Airborne Infection*, Oosthoek, Utrecht, The Netherlands.

Hirano, S.S., Demars, S.J., and Morris, C.E. (1981) *Phytopathology* 71, 881.
Hirano, S.S., and Upper, C.D. (1989) *Appl. Environ. Microbiol.* 55, 623–630.
Katzenelson, E., and Teltsch, B. (1976) *J. Water Pollut. Control Fed.* 48, 710–716.
Kloepper, J.W., Schroth, M.N., and Miller, T.D. (1980) *Phytopathology* 70, 1078–1082.
Lighthart, B. (1984) *Appl. Environ. Microbiol.* 47, 430–432.
Lighthart, B., and Frisch, A.S. (1976) *Appl. Environ. Microbiol.* 31, 700–704.
Lighthart, B., and Mohr, A.J. (1987) *Appl. Environ. Microbiol.* 53, 1580–1583.
Lighthart, B., Spendlove, J.C., and Akers, T.G. (1979) in *Aerobiology: An Ecological Systems Approach* (Edmonds, R.L., ed.), pp. 11–22, Dowden, Hutchinson and Ross, Stroudsburg, PA.
Lindemann, J., Arny, D.C., and Upper, C.D. (1984) *Phytopathology* 74, 1329–1333.
Lindemann, J., Constantinidou, H.A., Barchet, W.R., and Upper, C.D. (1982) *Appl. Environ. Microbiol.* 44, 1059–1063.
Lindemann, J., and Suslow, T.V. (1987) *Phytopathology* 77, 882–886.
Lindemann, J., and Upper, C.D. (1985) *Appl. Environ. Microbiol.* 50, 1229–1232.
Lindow, S.E. (1982) in *Plant Cold Hardiness and Freezing Stress—Mechanisms and Crop Implications*, vol. II (Li, P.H., and Sakai, A., eds.), pp. 395–416, Academic Press, New York.
Lindow, S.E. (1985) in *Engineered Organisms in the Environment: Scientific Issues* (Halvorson, H.O., Pramer, D., and Rogul, M., eds.), pp. 23–35, American Society for Microbiology, Washington, D.C.
Lindow, S.E. (1986) in *Perspectives in Microbial Ecology* (Megušar, F., and Gantar, M., eds.), pp. 509–515, Slovene Society for Microbiology, Ljubljana, Yugoslavia.
Lindow, S.E. (1987) *Appl. Environ. Microbiol.* 53, 2520–2527.
Lindow, S.E., Arny, D.C., and Upper, C.D. (1978) *Appl. Environ. Microbiol.* 36, 831–838.
Lindow, S.E., Knudsen, G.R., Seidler, R.J., et al. (1988) *Appl. Environ. Microbiol.* 54, 1557–1563.
Lindow, S.E., and Panopoulos, N. (1988) in *The Release of Genetically-Engineered Micro-Organisms* (Sussman, M., Collins, C.H., Skinner, F.A., and Stewart-Tull, D.E., eds.), pp. 121–138, Academic Press, New York.
Loper, J.E., Haack, C., and Schroth, M.N. (1985) *Appl. Environ. Microbiol.* 49, 416–422.
May, K.R., and Druett, H.A. (1968) *J. Gen. Microbiol.* 51, 353–366.
McCarter-Zorner, N.J., Franc, G.D., Harrison, M.D., et al. (1984) *J. Appl. Bacteriol.* 57, 95–105.
McInnes, T.B., Gitaitis, R.D., McCarter, S.M., Jaworski, C.A., and Phatak, S.C. (1988) *Plant Dis.* 72, 575–579.
Miller, T.D., and Schroth, M.N. (1972) *Phytopathology* 62, 1175–1182.
National Academy of Sciences, National Research Council (1989) *The Ecology of Plant-Associated Microorganisms*, National Academy Press, Washington, D.C.
Parker, B.C., Ford, M.A., Gruft, H., and Falkinham, J.O., III (1983) *Am. Rev. Respir. Dis.* 128, 652–656.
Parker, D.T., Spendlove, J.C., Bondurant, J.A., and Smith, J.H. (1977) *J. Water Pollut. Control Fed.* 49, 2359–2365.
Pedgley, D. (1982) *Windborne Pests and Diseases: Meteorology of Airborne Organisms*, Ellis Horwood Limited, Chichester, U.K.
Pérombelon, M.C.M., Fox, R.A., and Lowe, R. (1979) *Phytopathol. Z.* 94, 249–260.

Pérombelon, M.C.M., and Kelman, A. (1980) *Annu. Rev. Phytopathol.* 18, 361–387.
Peterson, E.W., and Lighthart, B. (1977) *Microb. Ecol.* 4, 67–79.
Quinn, C.E., Sells, I.A., and Graham, D.C. (1980) *J. Appl. Bacteriol.* 49, 175–181.
Schroth, M.N., Hildebrand, D.C., and Starr, M.P. (1981) in *The Prokaryotes*, vol. I (Starr, M.P., Stolp, H., Truper, H.G., Balows, A., and Schlegel, H.G., eds.), pp. 701–718, Springer-Verlag, Berlin.
Sorber, C.A., Bausum, H.T., Schaub, S.A., and Small, M.J. (1976) *J. Water Pollut. Control Fed.* 48, 2367–2379.
Southey, R.F.W., and Harper, G.J. (1971) *J. Appl. Bacteriol.* 34, 547–556.
Stadt, S.J., and Saettler, A.W. (1981) *Phytopathology* 71, 1307–1310.
Starr, M.P., Stolp, H., Truper, H.G., Balows, A., and Schlegel, H.G., eds. (1981) *The Prokaryotes*, vols. I and II, Springer-Verlag, Berlin.
Stolp, H., and Gadkari, D. (1981) in *The Prokaryotes*, vol. I (Starr, M.P., Stolp, H., Truper, H.G., Balows, A., and Schlegel, H.G., eds.), pp. 719–741. Springer-Verlag, Berlin.
Stout, J.D. (1960a) *N. Z. J. Agric. Res.* 3, 214–223.
Stout, J.D. (1960b) *N. Z. J. Agric. Res.* 3, 413–430.
Suslow, T.V., and Schroth, M.N. (1982) *Phytopathology* 72, 111–115.
Venette, J.R., and Kennedy, B.W. (1975) *Phytopathology* 65, 737–738.
Walker, J.C., and Patel, P.N. (1964) *Phytopathology* 54, 140–141.
Weller, D.M. (1983) *Phytopathology* 73, 1548–1553.
Weller, D.M., and Saettler, A.W. (1980) *Phytopathology* 70, 500–506.
Wendt, S.L., George, K.L., Parker, B.C., Gruft, H., and Falkinham, J.O., III (1980) *Am. Rev. Respir. Dis.* 122, 259–263.

CHAPTER

6

Factors Affecting the Transfer of Genetic Information Among Microorganisms in Soil

Guenther Stotzky
Lawrence R. Zeph
Monica A. Devanas

The use of microorganisms as alternatives to traditional technologies is being explored in such areas as agriculture, pest control, and bioremediation of toxic wastes. These applications of biotechnology rely on the expression of useful phenotypic traits both in naturally occurring microorganisms and in microorganisms genetically modified by recombinant DNA techniques. In the latter case, it is the merging of the fields of molecular biology and ecology that is providing the most exciting alternative technologies (e.g., the substitution of biological pesticides for traditional chemical pesticides), as well as new uncertainties. These uncertainties are particularly associated with: (1) environmental uses of genetically engineered microorganisms (GEMs) capable of expressing traits not present in the unmodified parent

---

The preparation of this chapter and some of the studies discussed were supported, in part, by cooperative agreements CR812484, CR813431, and CR813650 between the U.S. Environmental Protection Agency—Corvallis Environmental Research Laboratory and New York University. The opinions expressed herein are not necessarily those of the Agency.

organism; (2) the probability of the transfer of these genetic traits to other organisms indigenous to the environment; and (3) the possibility of the new traits having a deleterious effect on the environment.

The potential risks to public health and other aspects of the environment from a deliberate or accidental release of GEMs to the environment are the most urgent concerns, both scientifically and with respect to public policy, associated with this aspect of biotechnology. Questions about the probabilities of survival, colonization, and activity of released GEMs and their novel DNA in established microbial communities in soil and other natural habitats and the ability to predict the consequences of their release will only be answered by applying the knowledge derived from the study of microbial ecology and molecular interactions in these habitats. Both biotic and abiotic environmental characteristics will affect the survival, perpetuation, efficacy, and potential risk associated with the release of GEMs to any natural habitat. In this chapter, the physicochemical and biological factors that can affect the survival and gene transfer of GEMs in soil and other natural habitats are discussed. Table 6–1 lists the most important of these factors. The survival of novel genetic information and its potential effect on the homeostasis of an ecosystem may be greater if the information is transferred to indigenous species that are more adapted to the specific habitat than the introduced GEMs.

The relative importance of individual environmental characteristics varies with the specific habitat (for example, electromagnetic radiation is probably relatively unimportant in soil, whereas it is extremely important in aquatic habitats), and their effect is usually greater on introduced than on indigenous microbes. Moreover, none of these characteristics exerts its

TABLE 6–1 Factors Affecting the Activity, Ecology, and Population Dynamics of Microorganisms in Natural Habitats

Factors
Carbon and energy sources
Mineral nutrients
Growth factors
Ionic composition
Available water
Temperature
Pressure
Atmospheric composition
Electromagnetic radiation
pH
Oxidation-reduction potential
Surfaces
Spatial relationships
Genetics of the microorganisms
Interactions among microorganisms

influence individually; it is always in concert with other characteristics, and although the influence of one or a few characteristics may predominate in a specific habitat, these influences have indirect, but cascading, effects on other characteristics. Consequently, an alteration in one environmental factor may result in simultaneous or subsequent changes in other characteristics and, ultimately, in the habitat and, therefore, in the ability of both introduced and indigenous microbes to survive, establish, grow, and transfer genetic information (Stotzky 1974 and 1986; Stotzky and Krasovsky 1981). Inasmuch as the possible permutations of interactions between these environmental characteristics are vast, the relative success of microbes containing new genetic information to transfer this information in soil cannot be easily predicted.

The survival, perpetuation, and efficacy of recombinant DNA in microbes in soil or in any natural environment depend on: (1) the nature of the microbial host and the vector (e.g., a plasmid or a bacteriophage), if involved, carrying the DNA; (2) the survival, establishment, and growth of the host-vector system; (3) the maintenance, replication, and segregation of the DNA; (4) the frequency of transfer of the DNA to other microbes; (5) the adaptability of the host-vector system, either the original or the subsequent new ones, to their ecological "niches"; and (6) the selective advantage or disadvantage conferred on the original or subsequent hosts by the DNA.

6.1 THE SOIL ECOSYSTEM

Soil is unique among microbial habitats, as it is a structured environment with a high ratio of solid to water. Inasmuch as all microbes require water, their metabolism in soil is restricted to those microhabitats where there is a continual supply of available water. Hence, their distribution is essentially restricted to microhabitats that contain clay minerals, as sand and silt do not retain water for long against gravitational pull. Clay minerals, because of their surface activity, retain water against this pull, as the water adjacent to these active surfaces and coordinated with charge-compensating ions on the clays becomes sufficiently ordered to form a quasi-crystalline structure (i.e., the strong attraction of water molecules to the negatively and positively charged surfaces of clay minerals and to their charge-compensating ions enhances the hydrogen bonding of adjacent water molecules). The ordering of this clay-associated water decreases with distance from the clay surface until a distance is reached at which water is no longer under the attraction of the clay and is susceptible to gravity.

Clay minerals do not exist free in soil but as coatings, or cutans, on larger sand and silt particles or as oriented clusters, or domains, between these particles. The clay-coated particles and domains cluster together, primarily as the result of electrostatic attraction between the net-negatively charged faces and the net-positively charged edges of clays, into microaggregates, which, in turn, cluster together to form aggregates that can range from 0.5 to 5 mm in diameter and that are stabilized by organic matter and precipitated inorganic materials. These aggregates retain water, the thickness and permanence of which depend on the type and amount of clay and organic matter within the aggregates. This water may form bridges with the water of closely neighboring aggregates. These aggregates or clusters of aggregates, with their adjacent water, comprise the microhabitats in soil wherein microbes function. The space between the microhabitats constitutes the pore space, which is filled with air and other gases and volatiles (e.g., Stotzky 1986).

As the result of the discreteness of microhabitats in soil, the probability of genetic exchange in soil is markedly less than the probability in a system wherein water is continuous. Except for periods when soil is saturated with water, such as after a heavy rain or snow melt or when irrigated, individual microhabitats are isolated by the surrounding pore space, and movement of bacteria, transducing bacteriophages, and transforming DNA between habitats is limited to areas where water bridges between microhabitats may occur. Even when the pore space is saturated, movement between microhabitats may be restricted, as the surface tension of the ordered water around aggregates may be too great to allow passive movement of bacteria or even active movement by flagellated cells. It should be noted that there is no convincing evidence that bacteria are flagellated in soil, even though they may have the genetic capacity to produce flagella when isolated from soil and cultured in liquid media or on agar. However, filamentous fungi appear to be able to bridge pore spaces between microhabitats, even when the pore spaces are not filled with water. These fungi grow apically from mycelia that have a food and water base in a microhabitat and, therefore, are independent of the nutrient and water conditions surrounding the growing mycelia. Moreover, the extending mycelia probably have surrounding water films in which bacteria, bacteriophages, and transforming DNA may be transported from one habitat to another.

Conditions within a microhabitat or a portion of it can also affect microbial activities, including the exchange of genetic information. For example, if a bacterium sticks on a clay cutan or domain, after its chemotropic attraction to adsorbed organic substances, it may be nutritionally deprived after it consumes the adsorbed substances, as it will be dependent on the rate of diffusion of new substances for its nutrition. If the adsorbed substances to which it was attracted are toxic, the bacterium will die and eventually release its DNA for possible transformation of other cells within the microhabitat. It must be emphasized that there is no proof that bacteria stick on clay cutans or domains, but there are theoretical reasons why they do not stick. In contrast, there is evidence for the adhesion of bacteriophages on clays and that such adhesion protects phages against inactivation. This adhesion would prolong the persistence of transducing phages in soil (e.g., Stotzky 1986).

The mechanisms by which organic molecules are bound on clays will determine the tenacity with which they are held and the ability of bacteria, usually with the aid of extracellular enzymes, to utilize these molecules as substrates. The presence of chaotropic ions (which decrease the structure of water and tend to disrupt hydrophobic interactions by increasing the accommodation of nonpolar compounds in aqueous solutions) and of antichaotropic ions (which increase the structure of water and thereby increase hydrophobic interactions by reducing the ability of aqueous solutions to accommodate nonpolar groups) will affect the nutritional status of the microhabitat.

Even though clay cutans and domains are relatively stable, some clay particles may become dislodged and attach to the surface of microbes in the microhabitat. This attachment not only reduces the effective surface of microbes for transmembrane transfer of nutrients and waste products, but it may also block sites for the attachment of pili, bacteriophages, or DNA, thereby reducing the potential transfer of genetic material (Stotzky 1986 and 1989).

These conditions in soil differ markedly from those in sediments of aquatic systems. Although clay minerals in sediments also occur as cutans on larger particles and as domains in aggregates, water-dependent microhabitats do not occur as they do in soil, as the water in sediments is essentially continuous from one aggregate to the next. Moreover, microbes appear to colonize primarily sand and silt particles, rather than clays, in sediments, as water surrounds these particles, and the need to overcome the electrokinetic repulsion between net-negatively charged clays and microbes is eliminated (Stotzky 1986). Hence, care must be exercised in extrapolating observations on gene transfer in sediments to soil and vice versa.

Unfortunately, knowledge of the structure, physicochemical composition, and nature of microbial events in microhabitats in soil, as well as in other environments, is sparse and mostly conjectural. What can be assumed with a high degree of certainty is that these microhabitats, and even sites within a single microhabitat, are heterogeneous and differ in the type and amount of clay minerals, organic and inorganic substances, and other physicochemical characteristics and, therefore, in microbial composition and activities, including the transfer of genetic information (Stotzky 1974 and 1986).

6.2 MICROBIAL INTERACTIONS

On the basis of numerous studies conducted in vitro, in vivo, and in situ, it is apparent that the survival, establishment, and growth of introduced microorganisms, whether they contain recombinant DNA or not, are usually reduced when other species of microorganisms are present. This is especially true in natural habitats, such as soil, where the indigenous microbial populations are usually not only better adapted to the specific habitat, but also can exert competitive, amensalistic, parasitic, and predatory pressures on the introduced organisms. For example, the survival and growth of bacteria that were not genetically engineered were significantly reduced when they were inoculated into nonsterile environments (such as soil, water, and sewage) in which they were not natural residents, whereas they survived, and even grew, in the same environments when the environments were sterilized; for example, the numbers of *Salmonella typhimurium*, *Agrobacterium tumefaciens*, and *Klebsiella pneumoniae* increased by one to two orders of magnitude in sterile soil, whereas there was a reduction in their numbers when they were inoculated into nonsterile soil (Liang et al. 1982). There are numerous other examples in the literature of the apparent lower survival and growth rate of introduced microbes in nonsterile than in sterile environments (e.g., Alexander 1971; Atlas and Bartha 1987; Stotzky and Babich 1986), and this relation appears intuitively to be valid. However, most of these studies did not consider the possibility of a "viable but nonculturable" phenomenon (e.g., Devanas and Stotzky 1988; Devanas et al. 1986; Stotzky et al. 1990). Both genetically engineered and nonengineered bacteria introduced into soil and other natural habitats sometimes become sufficiently debilitated or otherwise altered that they can not be recovered from these habitats, especially on selective media, even though they could actually be surviving and growing in their new environment. Because of the implications of such alteration to monitoring the fate of GEMs introduced into soil and other natural habitats, studies such as those conducted by Liang et al. (1982) should be viewed critically until it is clearly established that the apparent lack of survival and growth of the introduced microorganisms was not an artifact of the experimental procedures.

Various types of interactions occur between individual microbial species and even between individual cells in soil and other natural habitats. These interactions are in a constant state of change, and the result is a dynamic biological equilibrium among the microbes that shifts with changes in the physicochemical status of the environment (e.g., Alexander 1971 and 1977; Stotzky 1974).

Odum (1959) suggested seven relations that could exist among species in an environment, although the relevance of these relations to natural microbial habitats has not been unequivocally established. These are as follows:

1. Neutralism, in which species function independently and do not interact;
2. Mutualism (symbiosis), in which specific species rely on each other, usually in an obligatory manner, and both benefit by the relation;
3. Protocooperation, an association that is of mutual benefit to interacting species but is not required for their existence;

4. Commensalism, in which some species derive benefit from other species that are not affected by the relation;
5. Competition, in which there is a suppression of some species, and eventually of all interacting species, as species struggle for limited quantities of nutrients, oxygen, space, and other requirements for growth;
6. Amensalism, in which some species are suppressed by the activity or products of other species that are not affected; and
7. Parasitism and predation, which involve the direct attack of one species on another, either by continual feeding on or consumption of the host.

These relations may be either neutral (no. 1), beneficial (nos. 2 to 4), or deleterious (nos. 5 to 7), and all, with the exception of neutralism, have been demonstrated to some extent among microbes in soil and other natural habitats.

Among the beneficial relations are mutualism, protocooperation, and commensalism. The succession of microbial populations in soil is an example of commensalism; e.g., primary decomposers attack complex organic compounds and release simpler substances for secondary colonizers that lack the enzymes for the utilization of the complex compounds. Many soil bacteria require B-vitamins and amino acids for growth, even in pure culture, and the widespread occurrence of these fastidious bacteria in soil indicates that these compounds are provided by other microbes in soil. Other examples of commensalism include changes in the environment by some species to produce favorable conditions for others: for example, facultative bacteria reduce the E_h of microhabitats to levels that enable anaerobes to persist and grow, even in "apparently" aerobic soils; the hyphal extension of fungi provides a route along which bacteria can move between microhabitats in soil that the bacteria are unable to traverse, because of the intervening voids (the pore space) (e.g., Brock 1971; Stotzky 1974).

When organisms interact to their mutual benefit, but the relation is not obligatory for either (i.e., many organisms can provide the mutual benefit), the association is considered to be protocooperative. For example, Jensen and Swaby (1941) showed that cellulose could serve as an energy source for the fixation of atmospheric nitrogen by *Azotobacter* species if a cellulose degrader was present to convert the cellulose to simple sugars or organic acids that were then available as energy sources for the *Azotobacter* species. In turn, the fixed nitrogen was used by the cellulose degrader. Neither organism required this association for survival, but the growth of both organisms was enhanced.

When the relation is so specific as to be obligatory for both partners, it is considered to be mutualistic, or symbiotic. Mycorrhizae and nodule-forming nitrogen-fixing bacteria are examples of mutualistic associations with plants. Lichens are examples of a symbiosis between an alga and a fungus, in which the alga provides organic carbon compounds (via primary productivity) as energy sources and vitamins for the fungus, and the fungus provides the alga with minerals, water, and protection from desiccation and high light intensities. Although the mutualistic relations between nitrogen-fixing bacteria and their host plants are usually highly specific, the relations between mycorrhizal fungi and plants and between the fungal and algal symbionts in lichens are less specific.

The exchange of genetic information can also be considered to be a type of beneficial interaction, as it is a form of protocooperation between cells in a population that enables the dissemination of information necessary for adaptation to changing environmental conditions, e.g., resistance to antibiotics, heavy metals, and other antimicrobial agents; ability to utilize hitherto recalcitrant substrates (Atlas and Bartha 1987). Hence, gene transfer in soil and other natural habitats may increase the adaptability of the indigenous microbiota.

Examples of negative or deleterious interactions between microbes in soil are competition, amensalism, parasitism, and predation. Competition is an active demand by interacting species for materials or conditions, e.g., nutrients, water, oxygen, and, perhaps, space, that the species require but that are in short supply in soil (e.g., Clark 1965; Clements and Shelford 1939; Stotzky 1974). Nutrient conditions in soil are often inadequate to support the multiplication of microbial cells, and thus, competing microbial populations tend to exclude one another, a phenomenon that has been called competitive exclusion (e.g., Fredrickson and Stephanopoulis 1981). Competition in natural habitats is difficult to demonstrate, as competition results in the selection of species better fit to the prevailing environmental conditions and the elimination of those less fit. However, the effect of competition can be inferred from studies that show the utilization of an added substrate by the indigenous microbiota and an increase in the numbers of specific members of the microbiota, presumably those that have the enzymatic capacity to utilize the substrate (e.g., Alexander 1971).

Competition between microbes for carbon and energy sources, inorganic nutrients, and growth factors has been demonstrated in chemostat experiments with mixed cultures (e.g., Veldkamp et al. 1984). Competition has also been observed by varying natural environmental conditions; for example, *Chromatium vinosum* outcompetes and excludes *Chromatium weissei* under conditions of continuous light when sulfide is the growth-limiting substrate. The two sulfur bacteria coexist under conditions of intermittent light when both strains grow, and most of the sulfide is oxidized by *C. vinosum*; during the dark, however, sulfide accumulates, and with the return of the light part of the cycle, a greater amount of sulfide is oxidized by *C. weissei*. As long as there is a dark/light cycle, these two species, which compete for the same substrate, can coexist (van Gemerden 1974).

Competition for oxygen exists between filamentous fungi and facultative bacteria in soil when the supply of oxygen is limited, as in flooded soils (Griffin 1969). The competitive advantage when oxygen is limiting belongs to facultative bacteria.

A competitive advantage is provided to microorganisms if they have the ability to survive and multiply with changing abiotic conditions in soil. Gradations in pH, E_h, temperature, water activity, salinity, light, and oxygen concentrations are just some of the environmental conditions that selectively support the growth of different groups of microorganisms in natural habitats. Substrates and energy sources are probably the primary factors for which microbes compete (exploitative competition) (Richards 1987), and competition for space is usually of lesser significance (Stotzky 1974). Even though competition for energy sources may be significant in determining the level of microbial activity in soil, competition is not always the governing factor for survival in soil, because in many situations, soil microbes may be protocooperative (Richards 1987).

Amensalism has been defined as another form of competition (interference competition) (Richards 1987) in which the action of one species results in the inhibition of other species. Although microbes excrete metabolites that can be stimulatory and can be used as energy sources, nutrients, or growth factors by other microbes, some metabolites can also be inhibitory or amensalistic, even at very low concentrations, e.g., antibiotics. Although the antagonistic effects of antibiotics are often considered to be among the most important causes of amensalism in soil, there is limited, if any, evidence for the production of detectable amounts of antibiotics in soil. However, soil and other natural habitats contain many substances of microbial origin that are toxic to microbes when these substances are present in sufficient quantities; these include organic acids that inhibit fungi (especially in acid soils), ammonia, nitrite, carbon dioxide, and decomposition products of plant materials, such as resins, tannins, and phenolic compounds (e.g., Alexander 1977; Stotzky 1974). For example, *Thiobacillus thiooxidans*, which can increase the acidity to pH 1 to 2 while oxidizing sulfur, gains a competitive advantage in mine drainage by the exclusion of acid-sensitive species (Higgins and Burns 1975).

Several interactions between microbes can occur simultaneously, thereby confounding an already complex situation. For example, mycostasis represents a situation where both amensalism and competition can be operative, as there are at least two possible explanations why the germination of fungal spores is inhibited in soil (Grant and Long 1981). One possibility is the presence of microbe-produced inhibitors: e.g., ammonia appears to be a mycostatic agent in alkaline soils, whereas ethylene has been shown to inhibit the outgrowth from sclerotia of *Sclerotium rolfsii* in acidic soils in Australia. Another possibility is competition for nutrients released from spores after their hydration, especially in soils that are low in nutrients and with fungal species that have a requirement for a threshold concentration of nutrients for germination.

A variety of bacteria, protozoa, and fungi function as predators in natural environments, and their importance in microbial ecology is beginning to be elucidated. Among the first bacterial predators to be described were members of the myxobacteria (Wireman and Dworkin 1975). These bacteria produce extracellular proteolytic enzymes that lyse the cell walls of other bacteria, resulting in the release of nutrients. Numerous studies in pure culture have confirmed the ability of myxobacteria to lyse a broad range of bacteria. However, the activity of myxobacteria in situ is not well understood. Nonetheless, predatory myxobacteria are commonly found in soil, and isolates capable of even lysing algae have been characterized (Burnham et al. 1988).

Several other Gram-negative bacterial predators, as well as predatory actinomycetes, have been isolated from soil (Casida 1988; Zeph and Casida 1986), suggesting that predation by bacteria, other than myxobacteria, has a larger role in the microbial ecology of soil than previously thought. These soil bacteria are not obligate predators, but they were capable of lysing a variety of species of prey bacteria in microscopic studies that enabled the lytic events in soil to be followed visually. Several types of Gram-negative predatory bacteria were especially active in soil, based on their ability to attain high population levels and on the large number of prey species to which they responded. Sequential multiplication of predator bacteria and, at times, of protozoa was observed, suggesting that a complex microbial food chain exists in soil, that may be associated with the mineral requirements of the predator bacteria. However, some of these predator bacteria were resistant to attack by soil protozoa (Casida 1989).

Nevertheless, protozoa appear to be active predators of bacteria in terrestrial and aquatic habitats, based on the results of studies conducted in laboratory microcosms (e.g., Alexander 1977). For example, the addition of high concentrations of the plant-pathogenic bacterium *Xanthomonas campestris* and the dinitrogen-fixing bacterium *Rhizobium meliloti* caused significant increases in the numbers of protozoa in soil (Danso et al. 1975; Habte and Alexander 1978). Increases in protozoan populations required high numbers of prey cells, indicating that protozoa are primarily active in soil microenvironments where prey bacteria are actively multiplying and, thus, stimulating protozoan predation. Moreover, predator protozoa will probably not multiply if more energy is required to locate prey cells than the energy that is derived from the prey (Alexander 1981). Models that explain cycles in populations of protozoan predators and bacterial prey have been developed from chemostat studies (e.g., Williams 1980). These models incorporate such parameters as saturation kinetics and prey refuge, that can result from the physical isolation of the prey in natural environments.

Another example of predation by microorganisms is found in the nematode-trapping fungi (Barron 1977). These fungi have specialized hyphae that allow them to capture prey nematodes. For example, when a soil-dwelling nematode comes into contact with *Arthrobotrys*, the prey is entrapped in rings produced on the fungal hyphae. Other hyphae (haustoria) then penetrate the nematode, break down its tissues, and utilize the nutrients. Other types of nematode-trapping fungi (e.g., *Monacrosporium*) produce other specialized structures, such as sticky knobs, with which to capture nematodes.

Microbial parasitism in soil and other natural habitats is exhibited by species of the bacterial genus *Bdellovibrio* and by some viruses. The characteristics that distinguish a parasite from a predator have been defined as: (1) the parasite is usually smaller than the host; (2) the necessity for direct physical or metabolic interaction and a high specificity of interaction between the parasite and the host; and (3) a restriction of the habitat of the parasite to the host organism (Atlas and Bartha 1987). Viruses are obligate intracellular parasites of bacterial, fungal, algal, plant, and animal hosts. Of these viruses, the ecology of bacterial viruses (bacteriophages) in soil has been the most studied (e.g., Stotzky 1986). Bacteriophages have been linked to the decline of bacterial populations in soil (Casida and Liu 1974) and in aquatic environments (Roper and Marshall 1974).

The importance of bacteriophages in transferring genetic material among bacteria in soil and other habitats is beginning to be elucidated. Bacteriophage P1 has been shown to transduce strains of *Escherichia coli* added to sterile and nonsterile soil (Germida and Khachatourians 1988; Zeph et al. 1988), and transduction of *Pseudomonas aeruginosa* has been demonstrated in lake water (Morrison et al. 1978; Saye et al. 1987) and soil (L.R. Zeph and G. Stotzky, unpublished observations) microcosms (Stotzky 1989; Stotzky et al. 1990). Thus, in addition to their possible importance in controlling the size of bacterial populations in soil and other natural habitats, bacteriophages also appear to have a role in natural selection processes in the environment.

Bdellovibrio species are considered parasites, as they can penetrate the cell wall of a host bacterium, multiply in the periplasmic space at the expense of the bacterium, and lyse the cell to release progeny (e.g., Shilo 1984; Starr and Seidler 1971; Stolp and Starr 1963). Only Gram-negative bacteria appear to be hosts for this parasite. *Bdellovibrio* species are unique in their highly effective system of motility that results in rapid movement and appears to assist in the penetration of prey bacteria. Species of *Bdellovibrio* have been isolated in high numbers from soil and sewage, and they are also found in natural aquatic environments. In many instances, however, the major predatory activity in soil and other natural habitats is associated with protozoa or bacteria other than *Bdellovibrio*, suggesting that the environmental factors in soil and other habitats influence the predator groups that are operative at any given time.

More is currently known about the activity of these microbial predators and parasites in the laboratory than in the environment. Many of the microorganisms described above probably do not operate effectively as obligate predators in soil and other natural habitats, and thus, their precise effects on prey populations are insufficiently understood. However, knowledge of whether specific GEMs serve as prey cells after their release to the environment would be useful in predicting the perpetuation and efficacy of the recombinant DNA in the GEMs.

Interactions between microbes in soil are numerous, and the interactions are probably in constant flux. Positive interactions enable species to occupy microhabitats that were previously unsuitable and uninhabitable when these species were present alone. However, when growing together, they are able to utilize existing resources more efficiently, as the result of their combined metabolic capabilities. Negative interactions limit population densities and, thereby, prevent overpopulation, which could modify the microhabitat or the species (Atlas 1986).

6.3 MONITORING GEMs IN SOIL

The major concern about the introduction of GEMs into soil and other natural environments should be their potential adverse ecological impacts, especially on the homeostasis of these environments. However, the development of sensitive detection techniques with which to monitor the survival and growth of, and genetic transfer by, GEMs in these environments is necessary before meaningful ecological data can be obtained. Soil is a complex environment, and the indigenous microbiota is not only complex in its range of species but also vast (10^6 to 10^9 bacteria/g soil). Hence, it is difficult to monitor the survival and establishment and the growth of and gene transfer by GEMs introduced into soil. The use of selective media (e.g., MacConkey agar (MAC) for Gram-negative bacteria), the incorporation of cycloheximide to inhibit fungi, and the expression of some natural phenotypic characteristics (e.g., red colonies of lactose-positive *E. coli* on MAC) are helpful in reducing the spectrum of bacteria enumerated from soil. Other phenotypic characteristics, expressed either by the host or by the novel genetic material or its vector, can be useful in some instances. For example, the presence of chromosomally borne genes for resistance to nalidixic acid and other antibiotics can be exploited in the isolation of the host, and plasmid-borne genes for the catabolism of toluene, chlorobenzoates, 2,4-dichlorophenoxyacetate (2,4-D), and other esoteric compounds or genes that confer resistance to antibiotics or heavy metals can be exploited for the isolation of host-plasmid systems containing either of these groups of genes. However, the isolation of specific hosts and vectors based on resistance, either chromosomal or vector-borne, to many commonly used antibiotics (e.g., streptomycin, chloramphenicol, tetracycline) or to some heavy metals, such as mercury, has limitations, as bacteria in many soils now exhibit resistance to these antimicrobial agents. In some soils, the "background" resistance to these antimicrobial agents can be as high as 50% of the bacteria isolated. The use of auxotrophic markers is of little value in the direct isolation of specific bacteria from soil, but these markers, in conjunction with those used for the initial isolation, can be valuable in confirming the identity of the GEMs or DNA of interest.

Serological methods, based usually on the expression of specific surface antigens, also require the isolation of the bacteria of interest on media that are relatively selective (i.e., that eliminate or reduce the growth of most unrelated soil bacteria). The bacteria are purified and then challenged with a spectrum of polyclonal or monoclonal antibodies raised against the bacteria of interest. The reactions between the bacteria and the (usually polyclonal) antisera are then scored as being either: (1) nonreactive; (2) cross-reactive with many sera without distinctive features and common to all isolates; (3) cross-reactive with distinctive features between isolates and a few antisera; and (4) specifically reactive with a single antiserum. Although this method can be highly specific in some instances, the surface antigens of some bacteria change during growth, and the antigens may be different in soil than in the pure cultures that were used to raise the antigens. Moreover, serological data do not always agree with identifications based on DNA probes and other characteristics of the bacteria (e.g., Schofield et al. 1987).

Perhaps the most critical methods for confirming the identity of GEMs and the novel DNA of interest are the use of DNA probes and gel electrophoresis. In the former, a portion of DNA that is highly specific for a unique DNA sequence in the chromosome of the host or for a sequence in the vector, preferably in the novel DNA, is labeled with either a radioactive nuclide (e.g., ^{32}P) or a chromogenic system (e.g., biotin-streptavidin-alkaline phosphatase-dyes). The probe is hybridized with DNA from cells of bacteria that have been isolated from soil, usually on a selective medium, and immobilized and lysed on a nitrocellulose or nylon membrane. Those bacteria whose DNA gives a hybridization signal ("light-up") contain the DNA of interest (e.g., Stotzky et al. 1990). The application of the polymerase chain reaction (PCR) technique should facilitate the preparation of large quantities of highly specific probes, as well as of target DNA (e.g., Saiki et al. 1988; Steffan and Atlas 1988). DNA probes have been used in conjunction with a most-probable-number technique to study the survival of *Rhizobium leguminosarum* and *Pseudomonas putida* in soil (Fredrickson et al. 1988).

In gel electrophoresis, the bacteria that are presumed to contain the DNA of interest are lysed, and their isolated DNA is electrophoretically separated on an agarose gel. If the gene of interest is carried on a vector, partial confirmation can be made by determining the size of the vector by comparing its electrophoretic mobility with that of known molecular mass standards. To confirm that the gene of interest is present in the vector or in the chromosome, the DNA is cleaved by different restriction endonucleases, subjected to electrophoresis as above, the resultant "fingerprint" compared with molecular mass standards, and the fragment that carries the DNA of interest is identified. Final confirmation may require DNA-DNA hybridization with a labeled probe. For methodologies, see, for example,

Ausubel et al. (1987), Jain et al. (1988), Maniatis et al. (1982), and Stotzky et al. (1990).

All of the above methods require isolation of the bacteria of interest from soil. Such isolation is not always possible, and some introduced GEMs may become so debilitated in soil that they will not grow on selective isolation media. Consequently, methods that isolate total bacterial DNA from soil and then hybridize this DNA with specific DNA probes may be valuable in monitoring the survival and establishment and the growth of and gene transfer by GEMs in soil (e.g., Holben et al. 1988; Ogram et al. 1987; Sommerville et al. 1989; Steffan et al. 1988). Unfortunately, these direct DNA methods have a detection sensitivity of only about 4×10^4 cells/g soil, whereas some of the bacterial isolation methods have a sensitivity approaching 2×10^1 cells/g soil (e.g., Devanas and Stotzky 1988; Devanas et al. 1986; Stotzky et al. 1990). The choice of methods will depend on the information and the sensitivity required. Moreover, methods that are more sensitive, specific, and rapid, as well as less expensive, will undoubtedly be developed in the near future.

6.4 GENE TRANSFER IN SOIL

The majority of studies on the transfer of genetic information in soil, especially with GEMs, has been conducted with bacteria, where DNA can be transferred in situ by conjugation (cell-to-cell contact), transduction (via a bacteriophage), and transformation (uptake of "naked" DNA by an intact cell). Although these phenomena have been demonstrated in a wide spectrum of Gram-positive and Gram-negative bacteria in the laboratory (i.e., in pure culture), there is sparse information on their occurrence in soil and other natural habitats (Stotzky 1989; Stotzky and Babich 1986; Trevors et al. 1987).

Most studies on gene transfer in soil have been conducted in the laboratory, because of the potential risks associated with the release of an untested GEM to the environment. A variety of terrestrial microcosms that purportedly simulate field conditions have been used. These microcosms include: extremely simple systems that inoculate a GEM into sterile soil added to a sterile nutrient broth in test tubes (Walter et al. 1987); sterile or nonsterile soil in a test tube, flask, or other container (e.g., Stotzky et al. 1990); multiple containers of nonsterile soil enclosed within a larger container (e.g., Stotzky et al. 1990); more complex systems that involve undisturbed soil cores of varying size that are brought into the laboratory with minimum disturbance of the structure and biotic composition of the soil (e.g., Bentjen et al. 1989; Fredrickson et al. 1989; Hicks et al. 1990; Van Voris 1988); either disturbed or undisturbed soils that are cropped and maintained within chambers that enable the control of temperature, relative humidity, light/dark cycles, and other environmental variables (e.g., Armstrong et al. 1987; Gile et al. 1982; Knudsen et al. 1988). Examples of microcosms with different degrees of complexity and the rationales for their use have been discussed (e.g., Atlas and Bartha 1981; Gillett 1988; Greenberg et al. 1988; Hicks et al. 1990; Johnson and Curl 1972; Pritchard 1988; Pritchard and Bourquin 1984; Stotzky et al. 1990). Guidelines for the use of soil core microcosms, with descriptions of various core designs, sampling procedures, and statistical analyses, have been published (Van Voris 1988).

The first studies on the transfer of genetic information in soil were conducted only in sterile soil (Weinberg and Stotzky 1972), as methods for conducting such studies in nonsterile soil had not been developed. Studies conducted in sterile soil have little relevance to what occurs in nonsterile soil. Hence, extrapolations from results obtained in sterile soil to what presumably occurs in nonsterile soil should be viewed with skepticism, as should studies conducted in soil extracts, wherein microhabitats have been disrupted and physicochemical characteristics altered. Nevertheless, studies in sterile soil can be informative, when coupled with parallel studies in nonsterile soil, as techniques for subsequent use in nonsterile soil can be developed. Moreover, the importance of the indigenous microbiota and the effects of surfaces and some other physicochemical characteristics on survival, establishment, growth, and gene transfer can be estimated in sterile soil.

Many of the early studies on gene transfer in soil were conducted with *E. coli*, which is not an autochthonous member of the microbiota of soil. There were numerous plausible reasons why *E. coli* was used as the model bacterium in these studies: e.g., (1) the genetics of *E. coli* were better defined than those of other bacteria; (2) numerous strains with a spectrum of chromosomal alterations or containing different plasmids were readily available; (3) there had been extensive genetic engineering of *E. coli* strains for use in a variety of industrial applications; (4) because of the successful experience with genetic engineering for industrial applications, it appeared reasonable to assume that strains of *E. coli* engineered to perform specific functions would be used initially for releases to the environment; (5) *E. coli* is "sexually promiscuous" and has been shown to transfer plasmid-borne genetic information to over 40 genera of Gram-negative bacteria and even to some Gram-positive bacteria; and (6) although primarily an inhabitant of the gastrointestinal tract of many, but not all, animals, *E. coli* is increasingly found in fresh and estuarine waters and in soils in urban and agricultural areas, probably as the result, in large part, of the presence of human beings (e.g., Stotzky 1989). The tendency now is to conduct studies on survival, growth, establishment, and gene transfer with bacterial species that are autochthonous in soil, especially as the genetics of these species become more defined.

A major impetus for studies on the transfer of genetic information in natural habitats has been the concern about the release of GEMs to the environment, especially to soil, for agricultural (e.g., enhanced nitrogen fixation, biological pest control, reduction in frost damage) and other (e.g., degradation of recalcitrant organic pollutants) purposes. The concern is that these released GEMs will not just perform the functions for which they have been engineered but that they may also perturb the homeostasis of soil and cause unanticipated and undesirable changes in the activity, ecology, and population dynamics of the soil microbiota.

6.5 EFFECTS OF PHYSICOCHEMICAL FACTORS OF SOIL ON GENE TRANSFER

The fate of genetic material in soil and other natural habitats is ultimately dependent on the survival, establishment, and growth of the microbial hosts that house the genetic material in these habitats. The survival, establishment, and growth of the hosts, which in this chapter are restricted to bacteria, are, in turn, dependent on their genetic constitution and on the biological and physicochemical characteristics of the habitats (see Table 6–1). Detectable transfer of genetic information, regardless of the mode of transfer, usually requires sufficiently high populations of donors (whether bacteria, phages, or transforming DNA) and recipients. Moreover, the ability to predict the fate of introduced GEMs and the potential transfer of their novel genetic information to indigenous microorganisms in different types of soil and other habitats would be enhanced if the knowledge of the effects of different physicochemical characteristics could be related to the actual characteristics of the recipient habitats. Unfortunately, insufficient studies have been conducted on the effects of physicochemical and biological factors on the activity, ecology, and the population dynamics of microbes and on their ability to transfer genetic information in soil and other natural habitats (e.g., Curtiss 1976; Freter 1984; Stotzky 1974, 1986, and 1989; Stotzky and Babich 1986; Stotzky and Krasovsky 1981).

6.5.1 Energy Sources

The presence of available carbon and energy sources may enhance the transfer of genetic information, as such sources appear to increase the population densities of the donors and recipients and the subsequent growth of the recombinant bacterium. This effect is considerably less pronounced in nonsterile than in sterile soil, as the indigenous microbiota quickly utilizes and mineralizes the added substrates. If the donor or recombinant organism contains a gene or genes that enable it to metabolize a carbonaceous compound that cannot be metabolized by other members of the indigenous biota (e.g., a recalcitrant xenobiotic), the organism may have a temporary selective advantage as the result of its ability to utilize the compound (e.g., Chatterjee et al. 1981). However, this advantage is probably lost after the specific compound is mineralized or transformed to an unavailable form. Moreover, the concentration of such recalcitrant compounds in soil is usually low and less than the concentration of natural organic materials that are available to essentially all the indigenous microbiota. However, the addition to soil of a compound that only an introduced GEM can utilize should enhance the survival of that GEM in soil, and no further addition of the compound after it has been degraded should reduce the level of the GEM. This may be an efficient method with which to restrict and control the population densities of some GEMs in soil.

6.5.2 Temperature

Temperatures near the optimum growth temperature appear to be necessary for the efficient in vivo transfer of genetic information in laboratory strains of bacteria, although this is not always the case (e.g., Stotzky and Babich 1986). For example, the rate of maximal conjugal transfer of the R plasmid R1drd-19 in strains of *E. coli* decreased progressively as the temperature was decreased from 37 to 17°C, and no transfer was detected at 15°C (Singleton and Anson 1981). The rate of transfer of the plasmid pRD1, which was derived from a clinical isolate of *Pseudomonas* and maintained in the laboratory, from an *E. coli* K12 donor to an *E.coli* K12 recipient decreased progressively as the temperature was reduced from 37 to 15°C; when *Erwinia herbicola* was the recipient, transfer occurred even at 12.5°C (Kelly and Reanney 1984). The plasmids pWK1 and pWK2, which confer resistance to antibiotics and mercury, were transferred optimally in vitro from a species of *Citrobacter* and of *Enterobacter* isolated from soil to *E. coli* at 28°C, and frequencies were greatly reduced at 15 and 37°C. In contrast, the transfer of a plasmid conferring resistance to kanamycin from a strain of *Proteus vulgaris*, isolated from the human urinary tract, to *E. coli* was about five orders of magnitude higher at 25 than at 37°C (Terawaki et al. 1967). The in vitro transfer frequency of a plasmid conferring resistance to streptomycin and tetracycline from an *E. coli* isolate from sewage to an *E. coli* isolate from creek water was highest at 25°C and lowest at 35°C, with frequencies at 15, 20, and 30°C being intermediate. The frequency of transfer in raw sewage in situ was also higher at 22.5 than at 29.5°C (Altherr and Kasweck 1982).

Plasmids present in enterobacteria isolated from human beings, fecally polluted rivers, and sewage treatment plants were differentiated on the basis of their thermosensitivity: "thermotolerant" plasmids were transferred equally well from 22 to 37°C, whereas "thermosensitive" plasmids were transferred at high frequencies at 22 or 28°C, but at low frequencies at 37°C. Only 3.1% of the 775 conjugative antibiotic-resistance plasmids evaluated were thermosensitive (Smith et al. 1978).

The maintenance of some plasmids by their host cells in vitro is also dependent on temperature, and many plasmids are lost above the optimum growth temperature (e.g., Stotzky and Babich 1986). For example, the Ti plasmid of *Agrobacterium tumefaciens* (Watson et al. 1975) and the nodulation plasmid, pW22, of *Rhizobium trifolii* (Zurkowski and Lorkiewicz 1979) were lost when the host cells were grown at 37°C. In contrast, the survival of *E. coli* J5(RP4) and JC5466(pRD1) were reduced to a greater extent in soil maintained at 20°C than at 4°C (Schilf and Klingmüller 1983).

There have been few controlled studies on the effects of temperature on the transfer of genetic information in soil. It appears, however, that conjugal transfer of both plasmid- and chromosomal-borne genes occurs in both sterile and nonsterile soil at temperatures lower than those necessary for optimal transfer in vitro; e.g., transfer in *E. coli* in soil occurred at temperatures between 15 and 27°C, considerably below the requisite temperatures in vitro (e.g., Krasovsky and Stotzky 1987; Trevors 1987b; Weinberg and Stotzky 1972), and in *Bacillus subtilis* at 15°C as well as at 27°C (van Elsas et al. 1987). Transduction of *E. coli* occurred in soil at 25 to 27°C (Germida and Khachatourians 1988; Zeph et al. 1988). Transformation of *B. subtilis* in sterile soil occurred at 37°C (Graham and Istock 1978), in sterile and nonsterile soil at 25°C (G.H. Lee and G. Stotzky, unpublished observations), in a simulated sterile marine system containing sea sand at 23°C (Aardema et al. 1983; Lorenz et al. 1988), and in suspensions of montmorillonite at 33°C (G. Stotzky and A. Golard, unpublished observations) and at 25°C (Khanna and Stotzky 1990). Unfortunately, no comparative studies have been conducted on the effects of temperature on transduction and transformation in soil.

Inasmuch as temperature in soil cannot be conveniently controlled in situ, and as soil temperatures, at least below the top few cm, do not fluctuate more than a few degrees C in any annual season, this physicochemical characteristic may not be very important in influencing gene transfer between microbes in soil. However, the annual range in fluctuation in soil temperature may be an important consideration when constructing a GEM for introduction into soil.

6.5.3 pH

The hydrogen ion concentration is an important physicochemical characteristic of soil that is amenable to control in situ, but its effect on gene transfer in soil has been insufficiently studied. Conjugal transfer of plasmids in *E. coli* was restricted in vitro to pH 6 to 8.5 (Curtiss 1976). In soil, both intra- and interspecific plasmid transfer were not detected until the bulk pH (pH_b) was adjusted to 6.8 with $CaCO_3$ (M.A. Devanas and G. Stotzky, unpublished observation), and conjugal transfer of chromosomal genes was also detected only at pH values near neutrality (Krasovsky and Stotzky 1987; Weinberg and Stotzky 1972) (pH values above neutrality were not evaluated in these studies). Transduction of *E. coli* by phage P1 was higher in a soil with a pH of 7.9 than in a soil with a pH of 6.8 (Germida and Khachatourians 1988). However, as these soils also differed in texture, organic matter content, and other physicochemical characteristics, differences in the transduction frequencies cannot be attributed solely to these differences in pH. The effect of pH on transformation in soil has apparently not been studied.

The effects of the pH of soil on gene transfer can be both direct (e.g., on survival and growth of the parentals) and indirect (e.g., on growth of the competitive and amensalistic indigenous microbiota; on pH-pI [isoelectric point] relations involved in the adsorption of transducing phages, transforming DNA, and DNase on soil particles). For example, the adsorption of bacteriophages and other viruses on clay is pH-dependent (Lipson and Stotzky 1987; Stotzky et al. 1981 and 1986), and the adsorption of DNA on sea sand increased as the pH was increased from 5 to 9 (Lorenz and Wackernagel 1987). In contrast, the adsorption of transforming DNA from *B. subtilis* on montmorillonite decreased as the pH was increased (Khanna and Stotzky 1990). The pH determines the sign of the net surface charge of amphoteric materials (e.g., enzymes, bacteria, bacteriophages) and the negative charge of ionizable materials (e.g., DNA). Hence, the adsorption of these materials on soil particles, especially on most types of clay minerals, which have a pH-independent negative charge, will be influenced by the soil pH, which, in turn, will influence their persistence, activity, and genetic transfer (Stotzky 1986). Consequently, controlled studies on the effects of pH on genetic transfer in soil should be conducted.

6.5.4 Water Content

The few data that are available on the effects of water content and tension on the transfer of genetic information in soil indicate that transfer frequencies are higher when the soil water tension is near or at the optimum for microbial growth (i.e., water tension equals −33 kPa). For example, plasmid transfer and survival of the donors, *E. coli* J5(RP4) and JC5466(pRD1), and the exconjugants, Enterobacteriaceae strain 1(pRD1) and *Pseudomonas fluorescens* (pRD1), were higher in soil maintained at 16% water than when the soil was allowed to dry to 4% (Schilf and Klingmüller 1983). Similarly, plasmid transfer in sterile soil was greater between strains of *B. subtilis* at 20–22% water (which is equivalent to −33 kPa) than at 8% water (van Elsas et al. 1987); between strains of *E. coli* in sterile soil at 80 than at 20% of the water-holding capacity (Trevors 1987a); and between *E. coli* and other enterobacteria and *P. aeruginosa* in nonsterile soils at a water tension of −33 kPa (24 to 26% water) than at 16% water (M.A. Devanas and G. Stotzky, unpublished observations). No studies appear to have been conducted on the effects of water content or tension on transduction and transformation in soil. Inasmuch as microbial growth in soil is optimum at −33 kPa, genetic transfer, regardless of the mechanism of transfer, is probably maximal at this tension, which can be maintained by irrigation in soil in situ.

6.5.5 Oxygen and E_h

At soil water tensions above −33 kPa, the content of oxygen and the E_h will be reduced, as the pore space becomes saturated (e.g., Stotzky 1974 and 1986). The effects of oxygen content (pO_2) and E_h on the transfer of genetic information in soil has apparently not been studied, and the results from the few studies conducted in vitro are contradictory: e.g., the conjugal transfer of chromosomal genes between *E. coli* was similar under aerobic and anaerobic conditions (Stallions and Curtiss 1972); the expression of antibiotic resistance by 45 different plasmids in *E. coli* was the same under aerobic and anaerobic conditions, although the formation of sex pili was reduced under anaerobic conditions (Burman 1977); and the frequency of conjugal transfer of R plasmids by two donor strains of *E. coli* isolated from human feces was reduced 10- to 1000-fold under anaerobic conditions (Moodie and Woods 1973).

6.5.6 Ionic Composition

The ionic composition of the soil solution affects the activity of water (a_w) and surface interactions among and between soil constituents, especially between clay minerals and bacteria, bacteriophages, DNA, and proteins (Stotzky 1986). Consequently, the ionic composition probably affects gene transfer, and this aspect requires study in soil. The in vitro transfer of plasmid R1drd-19 in *E. coli* was apparently stimulated by concentrations of NaCl similar to those present in estuaries (Singleton 1983), and the survival of antibiotic-sensitive fecal coliforms was reduced in seawater, whereas that of R-plasmid-containing strains of *E. coli* was unaffected (Smith et al. 1974).

6.5.7 Electromagnetic Radiation

Although light probably affects only microbes residing on the surface of soils, this physicochemical factor can be important in arid and semi-arid soils, where photosynthesis in algal crusts, both procaryotic and eucaryotic, is probably the major source of carbon and energy (e.g., Skujins 1984). Hence, the transfer of genes that confer resistance to ultraviolet radiation may enhance the survival of microbial surface dwellers (Marsh and Smith 1969). The importance of electromagnetic radiation on gene transfer in soil and other natural habitats has apparently not been studied.

6.5.8 Surfaces

The effects of surfaces, especially those of clay minerals, on transfer of genetic information in soil have been studied to a greater extent, albeit also insufficiently, than those of other physicochemical characteristics of soil. Montmorillonite appears to enhance conjugal transfer of both plasmid- and chromosomal-borne genes in both sterile and nonsterile soil, whereas kaolinite appears to have essentially no effect (Devanas and Stotzky 1988; Krasovsky and Stotzky 1987; van Elsas et al. 1987). In contrast, colloidal montmorillonite reduced the in vitro transfer of plasmid R1drd-19 in *E. coli* by several orders of magnitude (Singleton 1983). This apparent paradox between in vitro and in vivo effects of clay minerals may be the result of free clay particles in vitro blocking sites on the bacterial surface necessary for gene transfer, whereas few free clay particles exist in soil, as they are maintained in a relatively stable form in cutans and domains (Stotzky 1986).

Particles, especially clay minerals, can exert both direct and indirect effects on microbial activities, including transfer of genetic information, in soil. However, the mechanisms by which some clays enhance conjugal gene transfer in soil are not clear. Montmorillonite has been shown to stimulate growth of bacteria, in part by maintaining a suitable pH for sustained growth in microhabitats, and to reduce the growth of fungi, in part by complexing siderophores necessary for iron transport (Stotzky 1986). Van Elsas et al. (1987) suggested that the enhanced transfer of a plasmid in *B. subtilis* in soil amended with montmorillonite was not the result of an effect of the clay on pH but of "a modification of the physicochemical soil environment, possibly modifying cellular physiology or promoting cell-to-cell contact" or because the "clay apparently protected the recipient population." Unfortunately, no details were presented on the mechanisms of such modifications and protection. Moreover, it is also not clear which physicochemical properties of clays (e.g., cation-exchange capacity, specific surface area, surface charge density, nature of the charge-compensating cations) are responsible for their effects on conjugation and survival of parentals and exconjugants in soil.

The frequency of transduction in soil was not affected by montmorillonite, although the survival of the transducing phage P1 was increased (Zeph et al. 1988). Transduction by phage P1 was greater in a sandy (8% clay) than in a silty clay loam (21% clay) soil (Gemida and Khachatourians 1988), but as these soils also differed in pH, organic matter content, and other characteristics, and the types of clay present were not described, the differences in transduction frequencies cannot be attributed solely to differences in clay content in these two soils. The effects of clays and other surfaces on transformation in soil have been insufficiently studied. However, both sea sand (Aardema et al 1983; Lorenz et al. 1988) and montmorillonite (Khanna and Stotzky 1990; G. Stotzky and A. Golard, unpublished observations) reduced the in vitro frequency of transformation in *B. subtilis*.

It should be emphasized that the value of studies conducted on the effects of one physicochemical or biological factor of soil at a time may be limited, as a change in one factor can result in changes in numerous other biological and physicochemical factors, and several factors can interact to affect gene transfer. For example, the in vitro conjugal transfer of plasmid R1drd-19 in *E. coli* was inhibited more by deviations in pH from the optimum of 6.9 when the temperature was simultaneously decreased from 37 to 17°C (Singleton and Anson 1983). Studies on the effects of interactions between multiple physicochemical and biological characteristics are needed not only with respect to the transfer of genetic information but on all aspects of microbial ecology in soil and other natural habitats. Nevertheless, even the few studies that have been conducted on the effects of these characteristics individually on gene transfer in vitro and in soil and other natural habitats indicate that in vitro studies of conjugation, transduction, and transformation, which are usually conducted under standardized and optimal growth conditions, are not always adequate predictors of gene transfer in natural habitats, where these environmental characteristics fluctuate continually and in concert.

6.6 CONCLUSIONS

Until the verification of the existence of genetic recombination in bacteria by Avery, MacLeod, and McCarty in 1944, diversification and adaptation in bacteria was reasonably explained by mutation and Darwinian selection. It is, therefore, readily understandable why the discovery of genetic recombination in bacteria was exciting and appealing to evolutionary scientists and microbial ecologists: bacteria had another method for genotypic adaptation in natural habitats, in addition to chance and undirected mutation, followed by selection of those few mutants that were better adapted than their parent cell to their surroundings. However, although frequencies of mutation are low (10^{-6} to 10^{-9}) and most mutants are probably less fit than the parent cell, there is no convincing evidence that gene transfer and subsequent fitness of the recipient of the new gene(s) have a higher frequency in soil and other natural habitats. In one of the few studies that has demonstrated the transfer of genetic material in nonsterile soil and fresh water in situ, the frequencies were also only 1×10^{-9} and 3.3×10^{-8}, respectively, with the wide-host-range plasmid pRD1 (Schilf and Klingmüller 1983). Moreover, there is no experimental evidence that genetic transfer occurs routinely and successfully in soil and other natural habitats, despite the contention of some bacterial geneticists, based almost entirely on observations made in pure culture, that gene flow between bacteria, especially between Gram-negative bacteria, is continual and rapid. However, there is considerable empirical evidence (e.g., the increase in antibiotic- and heavy-metal-resistant bacteria in many environments) that gene transfer does sometimes occur outside of the in vitro, pure culture environment.

Although the transfer of genetic information among bacteria has been studied extensively in pure culture, there is insufficient information on the frequency of transfer, whether by conjugation, transduction, or transformation, and on the survival and activity of the recombinant bacteria in soil and other natural environments that contain high numbers of other microorganisms not involved in the transfer. The relatively few studies that have been conducted in soil—and it must be emphasized that even these studies have been conducted primarily in the laboratory in microcosms of varying degrees of complexity or under greenhouse conditions, as few field releases have been authorized—suggest that it can occur in soil. However, insufficient information is available on how transfer and survival are affected by the physicochemical and biological characteristics of soil; on the numbers of donors and recipients necessary for detectable gene transfer; on the numbers of recombinant bacteria that can be accurately detected; on the probability that low levels of recombinant bacteria, especially below the level of detection, can multiply sufficiently to become a significant portion of the soil microbial population; and on numerous related questions.

The major lack of knowledge, however, is in the area of the potential effects that recombinant bacteria could have on the structure and function of soil and other natural habitats; e.g., what kinds of genes need to be transferred and how many recombinant bacteria need to be functioning per unit volume of soil to result in detectable changes in the activity, ecology, and population dynamics of microorganisms in soil? Even if a GEM introduced into soil survives and transfers its novel genetic information to indigenous microbes, there should be little cause for concern unless the novel genetic information, either in the introduced GEM or in an autochthonous recipient to which it has been transferred, produces some unexpected and untoward impacts. Unfortunately, with existing knowledge and methodology, it is difficult to predict the occurrence, extent, and severity of such impacts.

Studies on the effects of adding various strains of *E. coli*, *Enterobacter cloacae*, *P. putida*, and *P. aeruginosa*, with and without plasmids carrying antibiotic-resistance genes, to soil have not shown any consistent and lasting effects on the gross metabolic activity (as measured by CO_2 evolution), the transformation of fixed nitrogen, the activity of soil enzymes (phosphatases, arylsulfatases, dehydrogenases), or the species diversity of the soil microbiota (Doyle et al. 1988; Jones et al. 1988). However, when 2,4-D was added to a soil that had been inoculated with a strain of *P. putida* containing a plasmid that coded for the partial degradation of 2,4-D, the rate of CO_2 evolution was retarded and the number of fungal propagules detected was reduced (Doyle et al. 1989). In other preliminary studies, the introduction into soil of a strain of *Streptomyces lividans* that contained a plasmid carrying a lignin peroxidase gene from the chromosome of *Streptomyces viridosporus* enhanced the rate of mineralization of soil carbon (as measured by CO_2 evolution) during the 30-day incubation period, especially when the soil was amended with lignocellulose (Wang and Crawford 1988). More such studies are obviously needed, especially with GEMs that have been engineered to perform specific enzymatic functions in soil. Studies are also needed on the potential ecological effects of the accumulation in soil of the products of genes that have been introduced into new hosts not in order to perform a specific enzymatic function but to produce a toxic product (e.g., the delta endotoxins of strains of *Bacillus thuringiensis*, which are synthesized by soil bacteria and transgenic plants into which the genes coding for the toxins have been introduced for the control of insects). Studies are also needed on the epidemiology in soil and other natural habitats of biological-control agents that have been engineered to cause diseases in unwanted plants, animals, and microorganisms.

Another factor to consider when introducing recombinant DNA into natural habitats is the possible occurrence of unanticipated effects that cannot be predicted from the information encoded on the DNA (i.e., pleiotropic effects). For example, increased production of capsular slime has been reported after the transformation of plasmid DNA into *Azotobacter vinelandii* (Glick et al. 1986) and *Azotobacter beijerinckii* (Owen and Ward 1985). The insertion of a 0.9-Mda cDNA fragment that codes for a yolk protein in *Drosophila grimshawi* into the *amp* region of pBR322 resulted in extreme mucoidy in the host strain, *E. coli* HB101(C357), when it was reisolated from soil, whereas isolates of HB101(pBR322) were nonmucoid (Devanas and Stotzky 1986 and 1988). The transcription of the *Drosophila* gene through the *amp* promoter may have resulted in the overproduction of a protein that was nonfunctional in *E. coli* HB101 but that saturated the activity of a regulatory protease, coded for by the *lon* gene, and produced a mucoid phenotype (Stueber and Bujard, 1982). More lactose-positive mutants were also observed in HB101(C357) than in HB101(pBR322) reisolated from soil (Devanas and Stotzky 1986 and 1988). The expression of some cloned genes has been shown to produce other morphological and biochemical changes, such as filament formation and increased cell fragility (Carrier et al. 1983); elevated mutation frequencies in the presence of an R plasmid (Frigo 1985); unpredicted phenotypic changes in phytopathogenic bacteria that contained plasmid pRD1 carrying genes for nitrogen fixation and resistance to three antibiotics—these changes included increased virulence, altered utilization of amino and organic acids, and enhanced resistance not only to the three antibiotics, but also to chloramphenicol for which no resistance was present in the parent strain (Kozyrovskaya et al. 1984); and greatly reduced utilization of citrate by *Erwinia herbicola* (pBR322), attributable to the tetracycline-resistance gene (J. Armstrong, personal communication).

The occurrence of such pleiotropic effects indicates that investigators studying gene transfer in soil and other natural habitats should be alert for such unrelated, unanticipated, and unpredictable effects and that the possibility of their occurrence should be considered when GEMs constructed for a specific purpose are to be introduced into these habitats. Consequently, initial environmental releases of GEMs should be done cautiously, only after extensive testing, on a case-by-case basis, and only to soils and other natural habitats wherein containment, decontamination, and mitigation can be successfully executed, in the event that the predictions were incorrect or incomplete.

It is obvious, based on the few studies that have been conducted on the transfer of genetic information in situ, that more research is necessary to determine the frequency, location, and effects of physicochemical and biological factors on gene transfer in soil and other natural habitats. The attainment of this knowledge will not only be of immense academic interest to microbial and macrobial ecologists, as well as to evolutionary theorists, but also to the areas of risk assessment and regulation of the release of GEMs to soil and other natural habitats.

REFERENCES

Aardema, B.W., Lorenz, M.G., and Krumbein, W.E. (1983) *Appl. Environ. Microbiol.* 46, 417–420.

Alexander, M. (1971) *Microbial Ecology*, Wiley, New York.

Alexander, M. (1977) *Introduction to Soil Microbiology*, 2nd ed., Wiley, New York.

Alexander, M. (1981) *Ann. Rev. Microbiol.* 35, 113–133.

Altherr, M.R., and Kasweck, K.L. (1982) *Appl. Environ. Microbiol.* 44, 838–843.

Armstrong, J.L., Knudsen, G.R., and Seidler, R.J. (1987) *Curr. Microbiol.* 15, 229–232.

Atlas, R.M. (1986) in *Bacteria in Nature*, vol. 2, *Methods and Special Applications in Bacterial Ecology* (Poindexter, J.S., and Leadbetter, E.R., eds.), pp. 339–370, Plenum Press, New York.

Atlas, R.M., and Bartha, R. (1987) *Microbial Ecology: Fundamentals and Applications*, 2nd ed., Benjamin Cummings, Menlo Park, CA.

Ausubel, F.M., Bent, R., Kingston, R.E., et al. (1987) *Current Perspectives in Molecular Biology*, Greene Publishing and Wiley-Interscience, New York.

Barron, G.L. (1977) *The Nematode-Destroying Fungi*, Canadian Biological Publications, Guelph, Canada.

Bentjen, S.A., Fredrickson, J.K., Van Voris, P., and Li, S.W. (1989) *Appl. Environ. Microbiol.* 55, 198–202.

Brock, T.D. (1971) *Bacteriol. Rev.* 35, 38–58.

Burman, L.G. (1977) *J. Bacteriol.* 131, 69–75.

Burnham, J.C., Heath, E.C., and Fraleigh, P.C. (1988) in *Abstr. Ann. Mtng. Amer. Soc. Microbiol.*, p. 298, American Society for Microbiology, Washington, D.C.

Carrier, M.J., Nugent, M.E., Tacon, W.C.A., and Primrose, S.B. (1983) *Trends Biotechnol.* 1, 109–113.

Casida, L.E., Jr. (1988) *Microbial Ecol.* 15, 1–8.

Casida, L.E., Jr. (1989) *Appl. Environ. Microbiol.* 55, 1857–1859.

Casida, L.E., Jr., and Liu, K.-C. (1974) *Appl. Microbiol.* 28, 951–959.

Chatterjee, D.K., Kellogg, S.T., Furukawa, K., Kilbane, J.J., and Chakrabarty, A.M. (1981) in *Recombinant DNA* (Walton, A.G., ed.), pp. 199–212, Elsevier, Amsterdam.

Clark, F.E. (1965) in *Ecology of Soil-Borne Plant Pathogens* (Baker, K.F., and Snyder, W.C., eds.), pp. 339–348, University of California Press, Berkeley.

Clements, F.E., and Shelford, V.E. (1939) *Bioecology*, Wiley, New York.

Curtiss, R., III (1976) *Ann. Rev. Microbiol.* 30, 507–533.

Danso, S.K.A., Keya, S.O., and Alexander, M. (1975) *Can. J. Microbiol.* 21, 884–895.

Devanas, M.A., Rafaeli-Eshkol, D., and Stotzky, G. (1986) *Curr. Microbiol.* 13, 269–277.

Devanas, M.A., and Stotzky, G. (1986) *Curr. Microbiol.* 13, 279–283.

Devanas, M.A., and Stotzky, G. (1988) in *Developments in Industrial Microbiology*, vol. 29, *J. Indust. Microbiol.*, suppl. no. 3, (Pierce, G.E., ed.), pp. 287–296. Elsevier Science Publishers, Amsterdam.

Doyle, J., Jones, R., Broder, M., and Stotzky, G. (1988) *Abstr. Ann. Mtng. Amer. Soc. Microbiol.*, p. 283, American Society for Microbiology, Washington, D.C.

Doyle, J., Short, K., and Stotzky, G. (1989) *Abstr. Ann. Mtng. Amer. Soc. Microbiol.*, p. 152, American Society for Microbiology, Washington, D.C.

Fredrickson, A.G., and Stephanopoulis, G. (1981) *Science* 213, 972–979.

Fredrickson, J.K., Bentjen, S.A., Bolton, H., Jr., Li, S.W., and Van Voris, P. (1989) *Can. J. Microbiol.* 35, 867–873.

Fredrickson, J.K., Bezdicek, D.F., Brockman, F.J., and Li, S.W. (1988) *Appl. Environ. Microbiol.* 54, 446–453.

Freter, R. (1984) in *Current Perspectives in Microbial Ecology* (Klug, M.J., and Reddy, C.A., eds.), pp. 105–114, American Society for Microbiology, Washington, D.C.

Frigo, S.M. (1985) *Rev. Microbiol.* (Sao Paulo) 16, 255–259.

Germida, J.J., and Khachatourians, G.G. (1988) *Can. J. Microbiol.* 34, 190–193.

Gile, J.D., Collins, J.C., and Gillett, J.W. (1982) *J. Agric. Fd. Chem.* 30, 295–301.

Gillett, J.W. (1988) in *Ecotoxicology—Problems and Approaches* (Levin, S.A., Harwell, M., Kelly, J., and Kimball, K.D., eds.), pp. 367–410, Springer-Verlag, New York.

Glick, B.R., Brooks, H.E., and Pasternack, J.J. (1986) *Can. J. Microbiol.* 32, 145–148.

Graham, J.B., and Istock, C.A. (1978) *Mol. Gen. Genet.* 166, 287–290.

Grant, W.D., and Long, P.E. (1981) *Environmental Microbiology*, Wiley, New York.

Greenberg, E.P., Poole, N.J., Pritchard, H.A.P., Tiedje, J.M., and Corpet, D.E. (1988) in *Release of Genetically-Engineered Micro-Organisms* (Sussman, M., Collins, C.H., Skinner, F.A., and Stewart-Tull, D.E., eds.), pp. 265–274, Academic Press, London.,

Griffin, D.M. (1969) *Ann. Rev. Phytopath.* 7, 289–310.

Habte, M., and Alexander, M. (1978) *Soil Biol. Biochem.* 10, 1–6.

Hicks, R.J., Stotzky, G., and Van Voris, P. (1990) *Adv. Appl. Microbiol.* 35, 197–253.

Higgins, I.J., and Burns, R.G. (1975) *The Chemistry and Microbiology of Pollution*, Academic Press, London.

Holben, W.E., Jansson, J.K., Chelm, B.K., and Tiedje, J.M. (1988) *Appl. Environ. Microbiol.* 54, 703–711.

Jain, R.K., Burlage, R.S., and Sayler, G.S. (1988) *Crit. Rev. Biotech.* 8, 33–84.

Jensen, H.L., and Swaby, R.J. (1941) *Nature* (London) 141, 147–148.

Johnson, L.F., and Curl, E.A. (1972) *Methods for Research on the Ecology of Soil-Borne Plant Pathogens*, Burgess Publishing Co., Minneapolis.

Jones, R., Broder, M., Doyle, J., and Stotzky, G. (1988) *Abstr. Ann. Mtng. Soil Sci. Soc. Amer.*, p. 219, Soil Science Society of America, Madison, WI.

Kelly, W.J., and Reanney, D.C. (1984) *Soil Biol. Biochem.* 16, 1–8.
Khanna, M., and Stotzky, G. (1990) *Abstr. Ann. Mtng. Amer. Soc. Microbiol.*, p. 320, American Society for Microbiology, Washington, D.C.
Knudsen, G.R., Walter, M.V., Porteous, L.A., et al. (1988) *Appl. Environ. Microbiol.* 54, 343–347.
Kozyrovskaya, N.A., Gvozdyak, R.I., Muras, V.A., and Kordyum, V.A. (1984) *Arch. Microbiol.* 137, 338–343.
Krasovsky, V.N., and Stotzky, G. (1987) *Soil Biol. Biochem.* 19, 631–638.
Liang, L.N., Sinclair, J.L., Mallory, L.M., and Alexander, M. (1982) *Appl. Environ. Microbiol.* 44, 708–714.
Lipson, S.M., and Stotzky, G. (1987) in *Human Viruses in Sediments, Sludges, and Soils* (Rao, V.C., and Melnick, J.L., eds.), pp. 198–229, CRC Press, Boca Raton, FL.
Lorenz, M.G., Aardema, B.W., and Wackernagel, W. (1988) *J. Gen. Microbiol.* 134, 107–112.
Lorenz, M.G., and Wackernagel, W. (1987) *Appl. Environ. Microbiol.* 53, 2948–2952.
Maniatis, T., Fritsch, E.F., and Sambrook, J. (1982) *Molecular Cloning. A Laboratory Manual*, Cold Spring Harbor Laboratory, Cold Spring Harbor, NY.
Marsh, E.B., and Smith, D.H. (1969) *J. Bacteriol.* 100, 128–139.
Moodie, H.S., and Woods, D.R. (1973) *J. Gen. Microbiol.* 76, 437–440.
Morrison, W.D., Miller, R.V., and Sayler, G.S. (1978) *Appl. Environ. Microbiol.* 36, 724–730.
Odum, E.P. (1959) *Fundamentals of Ecology*, W.B. Saunders Co., Philadelphia.
Ogram, A., Sayler, G.S., and Barkay, T. (1987) *J. Microbiol. Meth.* 7, 57–66.
Owen, D.J., and Ward, A.C. (1985) *Plasmid* 14, 162–166.
Pritchard, H.A.P. (1988) in *The Release of Genetically-Engineered Micro-Organisms* (Sussman, M., Collins, C.H., Skinner, F.A., and Stewart-Tull, D.E., eds.), pp. 265–274, Academic Press, London.
Pritchard, H.A.P., and Bourquin, A.W. (1984) *Adv. Microbial Ecol.* 7, 133–215.
Richards, B.N. (1987) *Microbiology of Terrestrial Ecosystems*, Longman Scientific & Technical, Essex, England, and Wiley, New York.
Roper, M.M., and Marshall, K.C. (1974) *Microbial Ecol.* 4, 279–290.
Saiki, R.K., Gelfand, D.H., Stoffel, S., et al. (1988) *Science* 239, 487–491.
Saye, D.J., Ogunseitan, O., Sayler, G.S., and Miller, R.V. (1987) *Appl. Environ. Microbiol.* 53, 987–995.
Schilf, W., and Klingmüller, W. (1983) *Recomb. DNA Techn. Bull.* 6, 101–102.
Schofield, P.R., Gibson, A.H., Dudman, W.F., and Watson, J.M. (1987) *Appl. Environ. Microbiol.* 53, 2942–2947.
Shilo, M. (1984) in *Current Perspectives in Microbial Ecology* (Klug, M.J., and Reddy, C.A., eds.), pp. 334–339, American Society for Microbiology, Washington, D.C.
Singleton, P. (1983) *Appl. Environ. Microbiol.* 46, 756–757.
Singleton, P., and Anson, A.E. (1981) *Appl. Environ. Microbiol.* 42, 789–791.
Singleton, P., and Anson, A.E. (1983) *Appl. Environ. Microbiol.* 46, 291–292.
Skujins, J. (1984) *Adv. Microbial Ecol.* 7, 49–91.
Smith, P.R., Farrell, E., and Dunican, K. (1974) *Appl. Microbiol.* 27, 983–984.
Smith, H.W., Parsell, Z., and Green, P. (1978) *J. Gen. Microbiol.* 190, 37–47.
Sommerville, C.C., Knight, I.T., Staube, W.L., and Colwell, R.R. (1989) *Appl. Environ. Microbiol.* 55, 548–554.
Stallions, D.R., and Curtiss, R., III (1972) *J. Bacteriol.* 111, 294–295.

Starr, M.P., and Seidler, R.J. (1971) *Ann. Rev. Microbiol.* 25, 649–675.
Steffan, R.J., and Atlas, R.M. (1988) *Appl. Environ. Microbiol.* 54, 2185–2191.
Steffan, R.J., Goksoyr, J., Bej, A.K., and Atlas, R.M. (1988) *Appl. Environ. Microbiol.* 54, 2908–2915.
Stolp, H., and Starr, M.P. (1963) *Antoine van Leeuwenhoek J. Microbiol. Serol.* 29, 217–248.
Stotzky, G. (1974) in *Microbial Ecology* (Laskin, A.I., and Lechevalier, H., eds.), pp. 57–135, CRC Press, Boca Raton, FL.
Stotzky, G. (1986) in *Interactions of Soil Minerals with Natural Organics and Microbes* (Huang, P.M., and Schnitzer, M., eds.), pp. 305–428, Soil Science Society of America, Madison, WI.
Stotzky, G. (1989) in *Gene Transfer in the Environment* (Miller, R.V., and Levy, S.B., eds.), pp. 165–222, McGraw-Hill, New York.
Stotzky, G., and Babich, H. (1986) *Adv. Appl. Microbiol.* 31, 93–138.
Stotzky, G., Devanas, M.A., and Zeph, L.R. (1990) *Adv. Appl. Microbiol.* 35, 57–169.
Stotzky, G., and Krasovsky, V.N. (1981) in *Molecular Biology, Pathogenicity, and Ecology of Bacterial Plasmids* (Levy, S.B., Clowes, R.C., and Koenig, E.L., eds.), pp. 31–42, Plenum Press, New York.
Stotzky, G., Schiffenbauer, M., Lipson, S.M., and Yu, B.H. (1981) in *Viruses and Wastewater Treatment* (Goddard, M., and Butler, M., eds), pp. 199–204, Pergamon Press, Oxford.
Stueber, D., and Bujard, H. (1982) *EMBO* 1, 1399–1404.
Terawaki, Y., Takayasu, H., and Akiba, T. (1967) *J. Bacteriol.* 94, 687–690.
Trevors, J.T. (1987a) *Water Air Soil Poll.* 34, 409–414.
Trevors, J.T. (1987b) *Bull. Environ. Contam. Toxicol.* 39, 74–77.
Trevors, J.T., Barkay, T., and Bourquin, A.W. (1987) *Can. J. Microbiol.* 33, 191–198.
van Elsas, J.D., Govaert, J.M., and van Veen, J.A. (1987) *Soil Biol. Biochem.* 19, 639–647.
Van Gemerden, H. (1974) *Microbial Ecol.* 1, 104–119.
Van Voris, P. (1988) in *Annual Book of ASTM Standards*, vol. 11.04, pp. 743–755, American Society for Testing and Materials, Philadelphia.
Veldkamp, H., van Gemerden, H., Harder, W., and Laanbroek, H.J. (1984) in *Current Perspectives in Microbial Ecology* (Klug, A.J., and Reddy, C.A., eds.), pp. 279–280, American Society for Microbiology, Washington, D.C.
Walter, M.V., Barbour, K., McDowell, M., and Seidler, R.J. (1987) *Curr. Microbiol.* 15, 193–197.
Wang, Z., and Crawford, D.L. (1988) *Abstr. First Intl. Conf. on the Release of Genetically-engineered Micro-organisms*, p. 13, REGEM Ltd., Cardiff, U.K.
Watson, B., Currier, T.C., Gordon, M.P., Chilton, M.D., and Nester, E.W. (1975) *J. Bacteriol.* 123, 255–264.
Weinberg, S.R., and Stotzky, G. (1972) *Soil Biol. Biochem.* 4, 171–180.
Williams, F.M. (1980) in *Contemporary Microbial Ecology* (Ellwood, D.C., Hedger, J.N., Latham, M.J., Lynch, J.M., and Slater, J.H., eds.), pp. 349–375, Academic Press, London.
Wireman, J.W., and Dworkin, M. (1975) *Science* 189, 516–522.
Zeph, L.R., and Casida, L.E., Jr. (1986) *Appl. Environ. Microbiol.* 52, 819–823.
Zeph, L.R., Onaga, M.A., and Stotzky, G. (1988) *Appl. Environ. Microbiol.* 54, 1731–1737.
Zurkowski, W., and Lorkiewicz, Z. (1979) *Arch. Microbiol.* 123, 195–201.

CHAPTER

7

Genetic Exchange and Genetic Stability in Bacterial Populations

Conrad A. Istock

At an informal, open meeting held at the U.S. National Science Foundation (NSF) in the spring of 1989, a population geneticist remarked that population biology has much to contribute to future assessments of the risks of biotechnology. An NSF official immediately responded sharply with a rhetorical question: "Why would anyone even think there might be any risks?" It was an awkward moment to digress into a substantive debate of such a big issue. All the scientists in the room inclined to such a debate apparently let the urge to do so pass, in favor of getting on with the discussion of what to do about the current low level of support for population research.

Perhaps someone should have returned at least a cursory challenge to the attitude behind that rhetorical question, for it is a common and eerie attitude shared by many people, including some research biologists. It reveals a lack of understanding of the full scope of the scientific investigations needed to increase our understanding of the natural world as we weigh the

---

I am grateful to K. Duncan, J. Bell, R. Pfister, M. Pantastico, S. Raouf, and N. Ferguson for the many ways in which they participated in the *Bacillus* investigations of my laboratory, and to J. Bell and K. Duncan for their comments and corrections on the manuscript of this chapter. These studies were supported by grants from the National Science Foundation, BSR 819076, the U.S. National Institutes of Health R01 GM36471, and BRSG grant 828975.

risks and benefits of biotechnology. The attitude of that NSF official strikes me as eerie because it completely ignores or consciously denies a recurrent progression in the application of many recent technologies, namely: euphoric promotion and application; protracted denial of early signs of trouble; emergence of undeniable evidence of serious deleterious effects; and subsequent regulation or even abandonment of the technology. That such a pattern has accompanied most of the great technological innovations in the 20th century is still not generally enough appreciated. The lesson to be learned is that most if not all advanced technologies can and should be assessed exhaustively prior to their application. Thereby we may be able to avoid in the future the unexpected personal, environmental, and social traumas of the past when new technologies were quickly marketed and used in what amounted to large-scale, unplanned experiments performed on society and the world at large. Some well-documented examples include: lead emissions to the atmosphere and smog from gasoline fuel; release of refrigerants, radioactive elements, toxic organic wastes, and asbestos; use of tobacco, fertilizers, pesticides, food additives, electric power, antibiotics, and dangerous drugs of many kinds; heavy metal contamination and poisoning; and coal miner's lung disease. Climatic effects of CO_2 emissions will probably provide another case of far-reaching proportions. Even the semi-conductor revolution now seems destined to engender its mixture of mixed blessings. Is biotechnology next, or can we do better this time?

I am not recommending that we all become latter-day Luddites eschewing all the benefits of technology, but merely that we apply our scientific acumen to reducing the chances of unexpected deleterious outcomes before engaging in the widespread application of future technologies. In the case of biotechnology, particularly as it applies to the genetic properties and processes of bacteria, there are both theoretical and empirical reasons to recommend caution and careful assessment of potential risks. I will try to provide a coherent justification for such a cautious approach in this chapter. I also hope that some of the intellectual excitement associated with studies of the sexual proclivities of bacteria and the organization of biological diversity among bacteria, as well as with studies of microbial ecology, will be apparent.

Visions of the potential benefits and hazards of recombinant DNA technology during the 1970s usually led to controversy. Biologists themselves organized their own "debates" (Jackson and Stich 1979; Yoxen 1983). Controversy and debate help to organize conceptual frameworks and sharpen hypotheses, but they cannot decide where a reasonable approximation of truth lies in scientific matters; this is the province of scientific investigation. We owe much to the participants in the controversies of the 1970s, but we have also moved away from them to some extent. The largely inchoate fears that arose in the 1970s and which led to the guidelines of the U.S. National Institutes of Health that now regulate recombinant DNA research are still with us to a considerable extent, but other concerns based on more readily definable phenomena have arisen since. Many of these later concerns are analogous to those of other environmental problems; they involve questions about potential alterations of processes in the natural world once the products or processes of biotechnology are widely applied. Unlike earlier environmental problems, however, the ones that biotechnology might engender involve replicating and physiologically active entities. Thus the products of biotechnology themselves can be self-propagating; this fact leads to some interesting new questions, listed below, about the natural world—questions which lie squarely in the domain of population biology. I will discuss the extent to which we have some answers to these questions at the close of this chapter.

1. What avenues and barriers for genetic transmission exist in nature?
2. How likely are exotic genes to be transferred within and between species in nature?
3. What are the associated rates of both intraspecific and interspecific gene transfer?
4. Will the existing differentiations among species become blurred if intermediate, higher homology, genetic sequences are introduced into nature or will natural selection root out such sequences? (One might refer to this possibility as the problem of the genetic and evolutionary mixmaster.)
5. How readily can genes with new physiological functions be expressed when they reach various recipient genomes in nature?
6. How stable will exotic gene expression be when such genes become new additions to the previously evolved genomes of wild species and encounter the welter of selective forces imposed by the surrounding ecological community?

The casual argument that concern about such questions is unnecessary because nature has already tried all possible genetic combinations can be dismissed. Even if all genetic combinations have been tried during evolution (and surely they have not) they could not have all been tried in all environments and ecological associations and across all the phyletic lineages of the last three billion years of life's history. Quite simply, not all genes were present in all places at all times. Similarly, all possible genomes to be obtained by universal genetic recombination among extant organisms could not have yet occurred in all present-day environments. Probably less than 1% of such recombinant genomes have actually come to exist, though no one really knows. Though species boundaries are quite frequently indistinct within closely related groups of species, there remain many vast phyletic differentiations, which presently create complete barriers to genetic exchange within the earth's biological diversity (Duncan et al. 1989). However, as we venture ever deeper into applications of recombinant DNA technology and create a variety of procedures for the reconstruction of organisms, it will be possible to bridge natural barriers to genetic exchange (Heinemann and Sprague 1989). As products of biotechnology enter the natural world, be they novel DNA sequences, entire genes, suites of genes, or complete organisms, these entities will interact with existing natural pathways and barriers for genetic exchange. Many of the early genetically engineered entries are certain to be bacteria and bacterial genes.

Bacteria are frequently thought to be more promiscuous in gene transfer than eucaryotes, especially when plasmids are involved. A plasmid is a closed circular DNA molecule that serves as an accessory chromosome in a bacterial cell. Bacteria are also asexual in their reproduction. This juxtaposition of easy genetic exchange and the potential for prolonged asexuality, combined with the central place of bacteria in much of biotechnology, makes them particularly important for questions of genetic exchange, and for studies of the stability of expression of genes transferred within and between species. My exploration of these topics in this chapter will rely heavily on studies in my laboratory of genetic exchange among bacteria in the genus *Bacillus*, but will also provide a partial review of the literature concerning bacterial genetic exchange. My main purpose is to cover enough examples to support some general conclusions, and to point toward some avenues of promising research for the future.

7.1 SEXUALITY IN BACTERIA

Three forms of bacterial sexuality are currently known: transformation, transduction, and conjugation. *Transformation* is the direct transfer of DNA from one cell to another through the surrounding medium and its subsequent integration into, and expression by, the recipient organism. Either chromosomal DNA or plasmid DNA can be transferred by the process of transformation. In fact, it is unknown whether transfer of plasmids occurs by the transformation pathway in nature, though such transfer is possible through laboratory manipulation. *Transduction* is the genetic exchange of host bacterial DNA mediated by a bacteriophage (a bacterial virus). The primary accomplishment of the interaction of a phage and bacterium is to propagate the phage, which can only replicate inside a bacterium. However, during the process of phage transmission, chromosomal or plasmid DNA of the host bacterium may also be transferred (i.e., transduced) as well. *Conjugation* is primarily plasmid exchange guided by genes on a conjugative plasmid that cause two bacterial cells to make contact between their cell surfaces and exchange plasmid DNA. Thread-like structures called pili may or may not be required to link the mating cells and may also serve as a conduit for the DNA. The plasmid is an accessory circular chromosome supplemental to the main and much larger circular chromosome of the bacterium containing the plasmid. Unlike a eucaryotic chromosome, the plasmid may replicate and relocate to another cell independently of the main bacterial chromosome. Conjugative plasmids may carry genes with important bacterial functions such as nitrogen fixation, antibiotic resistance, or pathogenesis to plants, animals, or other bacteria. Other nonconjugative plasmids or host chromosomal genes may be transferred during bacterial conjugation.

The definitions of and distinctions among the three modes of bacterial genetic exchange seem clear, but we should be cautious here. The construction of these definitions is based largely on laboratory manipulations, not on any direct assessment of the way they function spontaneously in nature. Other modes of exchange may exist, as well.

7.1.1 Transformation

Bacterial transformation was discovered six decades ago in *Pneumococcus* (Griffith 1928; Avery et al. 1944) and nearly three decades ago in *B. subtilis* (Spizizen 1958; Anagnostopolos and Spizizen 1961; Anagnostopolos and Crawford 1961; Young and Spizizen 1961; Ephrati-Elizur and Fox 1964). *Natural transformation*, by which I mean a spontaneous process is probably quite widespread among bacteria (Stewart and Carlson 1986). It occurs in at least some species of the following genera: *Pneumococcus, Bacillus, Neisseria, Streptococcus, Pseudomonas, Haemophilus, Micrococcus*, and probably others (Catlin 1960; Gwinn and Thorne 1964; Smith et al. 1981; Reanney et al. 1983; Stuy 1985). Our knowledge is still at the stage where both the taxonomic range of the process and the consistency of the details of the process across this wide taxonomic range are far from complete. Presently, it is impossible to say how much natural transformation differs among different species and genera, and how much it differs from artificial transformation in liquid culture where the transforming DNA is extracted chemically from the donor bacterial strain. The paucity of knowledge about natural transformation forces us to reason largely from the more completely known details of transformation in liquid culture, though such extrapolations to natural transformation may be faulty. The process of transformation is not the same in Gram-positive and Gram-negative bacteria. One major difference is the form of the DNA transferred; it is taken up as single-stranded in the Gram-positives and as double-stranded in the Gram-negatives (Ingraham et al. 1983). Reviews of liquid culture transformation can be found in Smith et al. (1981) and Dubnau (1982).

Natural transformation requires the release of replicated, chromosomal DNA by one cell followed by its uptake by another cell. It is best known for the Gram-positive bacterium *Bacillus subtilis*. At both its origin and terminus, the circular chromosome of *B. subtilis* is attached to the cell membrane, and possibly to the cell wall as well. We might think of the process as one commencing with the release of a newly replicated copy of the chromosome from a single cell as it emerges from a resting *B. subtilis* spore; the free chromosomal DNA can perhaps be thought of as a bacterial gamete. Chromosomal replication during outgrowth from the spore proceeds bidirectionally for 180° from origin to terminus. Simultaneously, the chromosome is released into the surrounding medium. Using synchronous cultures, such a process was revealed by the progressive appearance of markers in map order on the transforming DNA recoverable from the medium when the process of chromosomal replication was sequentially stopped after longer and longer times. After about 90 minutes, entire copies of the chromosome are free in the medium (Borenstein and Ephrati-Elizur 1969; Henner and Hoch 1982). Other studies have also indicated that *B. subtilis* actively releases transforming DNA (Ephrati-Elizur 1968; Singha and Iyer 1971; Streips and Young 1974; Orrego et al. 1978; Stewart et al. 1983). Transforming DNA can also be released when a cell dies and breaks open. Though no one has demonstrated it, it is theoretically possible that the chromosome of a host bacterium would be released when phage cause lysis of the host, and this free chromosomal DNA might subsequently transform a recipient cell. Redfield (1988) has modelled the process of genetic exchange involving transformation by DNA from lysed cells.

During transformation, free DNA attaches to receptor sites, of which there are 30–70 on the surface of the *B. subtilis* cell. After attachment, the donor DNA is nicked, made single-stranded, and transported into the cell. Within the cell the incoming DNA is protected from endonucleases by an "eclipse complex" apparently composed of the incoming DNA and a DNA-binding protein, then moved to the recipient cell's chromosome where it may be substituted for a part of the chromosome (Duncan et al. 1978). During transformation, an SOS-like DNA repair system is invoked (also called an SOB repair system; Bresler et al. 1968; Love and Yasbin 1984). The repair enzymes presumably resolve the heterodimer formed of donor and recipient DNA. Thus a recombinant chromosome is fashioned whenever the heterodimer is resolved in favor of the donor strand. Rates of transformation can be measured for single marker loci, or for multiple, linked or unlinked, marker loci. As the transformed cell goes on through many rounds of binary fission it produces a colony of cells loosely analogous to a multicellular organism. Thus, with the active release of a "gamete" and a corpus of cells in the colony, transformation is perhaps the most eucaryote-like of the bacterial forms of genetic transmission.

A central result from our early laboratory experiments with *B. subtilis* populations in soil culture was that there occurs a spontaneous, rapid, and pervasive exchange of genetic information (DNA) between the cells of two reciprocally marked strains in such a population. These experiments started with an inoculum of spores induced to germinate synchronously by heating. We found that such exchange can quickly create a large field of recombinant genetic variation that in turn responds to strong, directional, natural selection (Graham and Istock 1978, 1979, and 1981). A single recombinant genotype and phenotype rose to high frequency in most of these experiments.

However, the transformation-like process we found in experiments with strains of *B. subtilis* in soil is not completely identical to transformation in liquid culture, where DNA obtained by chemical extraction is taken up and incorporated by cells carefully brought to competence in the late log phase of population growth. In the soil system, the transfer of DNA and the encoded marker variants could not be blocked by treatment with DNase or an excess of heterologous (calf thymus) DNA, as is the usual experience with transformations conducted in broth cultures (but see Orrego et al. 1978). Our finding that DNase will not block the transfer has been verified by Lorenz et al. (1988); they also found facilitation of gene transfer within *B. subtilis* when the colonies were attached to sand grains. Given this failure to block the process with DNase or heterologous DNA, we distinguish the transformation process in soil culture from the analogous one in broth by referring to it as "natural transformation" (Stewart et al. 1983; Stewart and Carlson 1986). The conceptual distinction here is between the process engaged in by cells free in the soil environment as opposed to that in the contrived liquid culture procedure. The liquid culture procedure amounts to a "forced mating" of the bacteria. In some experiments we use intermediate situations where genetically marked, chemically extracted, donor DNA is presented to recipient cells in soil. Cells of a single recipient strain are transformed with fairly high frequencies when they encounter such "artificially" provided donor DNA (Graham and Istock 1978 and 1979). This experimental model simulates a situation in which transforming DNA might enter a natural environment from some biotechnological manufacturing process which produces "waste DNA."

The natural transformation of *B. subtilis* in soil differs from the standard laboratory process in other ways: cells competent to undergo transformation are present throughout the life of a population in soil, creating a continuous stream of observable transformants. We found the frequencies of transformation are about 10- to 100-fold lower than those obtained in laboratory transformation with saturating DNA (Graham and Istock 1978 and 1979; but see Lorenz et al. 1988). Other ways in which natural transformation in soil populations may differ include: integration of longer stretches of DNA (Graham and Istock 1978), association with spore germination (Graham and Istock 1978; Orrego et al. 1978), involvement of cell to cell contact (Graham and Istock 1978; Orrego et al. 1978), and possibly functional and biochemical differences between actively released ("excreted") and chemically extracted DNA (Streips and Young 1974).

7.1.2 Transduction

Occasionally the replication and packaging of virulent bacteriophage particles goes awry (Ingraham et al. 1983). Instead of the cleaving of evenly sized pieces of double-stranded DNA from a replicated concatemer along which the phage genome is repeated many times, double-stranded DNA of the host chromosome is cleaved instead and packaged in a phage head. The host DNA may then be transferred to another bacterium where it will either be integrated into the recipient chromosome (complete transductant) or left free in the recipient cytoplasm (abortive transductant). This process is called generalized transduction. Abortive transductants may remain stably diploid, but they always form tiny colonies because they cannot adequately replicate the transducing DNA and spread it to their progeny. Abortive transductants are about 10 times more abundant than complete transductants (Ingraham et al. 1983). Temperate bacteriophages, which integrate at specific sites into the chromosome of the host bacterium, similarly accomplish the transfer of host genes, but only for genes flanking the site of the phage insertion. This process is called specialized transduction. Generalized transduction was discovered by Zinder and Lederberg (1952) using *Salmonella typhimurium* and phage P22.

Several studies have demonstrated that both plasmid and chromosomal genes can be transferred by spontaneous transduction (Morrison et al. 1979; Saye et al. 1987). Using strains of *Pseudomonas aeruginosa* and several different bacteriophages, these studies demonstrated transduction in sterile and nonsterile lake water. Some of the experiments took place in small chambers suspended in a freshwater lake. The presence of the natural microbial community lowered the rates of transduction, but did not eliminate the process of genetic exchange altogether.

Genetic exchange by transduction is also known for *Bacillus* (Hemphill and Whiteley 1975; Throne 1962 and 1968; Romig and Brodetsky 1961; Lovett 1972; Thorne and Kowalski 1976; Ruhfel et al. 1984). Transduction differs from transformation in *Bacillus* in that DNA fragments that are as much as 10 to 20 times longer may be transferred, as was found when phage PBS1 was used to construct the now-extensive chromosomal map of *B. subtilis* (Henner and Hoch 1982; Piggot and Hoch 1985). However, transduction may be hindered by restriction and modification systems (Trautner et al. 1974; Wilson and Young 1976; Saito et al. 1979; Jentsch 1983).

7.1.3 Conjugation

Conjugation is best known from studies with the F plasmid, which replicates in Gram-negative *Escherichia coli, Salmonella typhimurium*, and other enteric bacteria. It is generally considered an example of a plasmid with a narrow host range. The F plasmid is a double-stranded circular DNA molecule. It is a conjugative plasmid because it can cause hosts and potential hosts to make the cell-to-cell contact necessary for transmission of the plasmid to new host cells. Following cell contact between conjugal pairs of F^+ (male) and F^- (female) cells, the self-transmission of the plasmid is guided by 13 genes forming a single long operon that occupies about one-third of the plasmid genome. The transfer process requires about 100 minutes at 37°C in *E. coli* (Ingraham et al. 1983). Such plasmid transmission is very efficient in liquid suspensions of *E. coli* F^+ and F^- cells in the laboratory, with virtually all the F^- cells acquiring the plasmid, and thus being converted to F^+, or "male," cells. During conjugation some DNA from the F^+ donor cell chromosome may also be transferred with the plasmid, but at a much lower frequency, on the order of 10^{-7}. Conjugation also occurs in some Gram-positive bacteria such as *Streptococcus faecalis*, but the process may be quite different from that in Gram-negative bacteria (Ingraham et al. 1983).

Conjugative plasmids with broad host ranges, such as R751, also exist. These plasmids cause conjugation between many species of bacteria, and even between Gram-negative and Gram-positive bacteria (Theil and Wolk 1987; Piffaretti et al. 1988; Brisson-Noel et al. 1988).

Conjugative transfer of single and multiple antibiotic resistances has been reported by Talbot et al. (1980) using strains of *Klebsiella* isolated from human sputum and bovine mastitis infections. In experiments with two simulated, nonclinical habitats (radish roots and sawdust suspensions), genetic exchange was measured at rates of 10^{-5} to 10^{-8} transconjugants per donor cell. Exchange between strains of *K. pneumoniae* and between strains of *K. pneumoniae* and *K. oxytoca* were found. These simulated habitats reflect the fact that *Klebsiella* species occur naturally and frequently on market vegetables, and in sawdust from sawmills. In turn, the sawdust may be used as bedding for domestic animals. *Klebsiella* populations may rise by 1,000-fold within the bedding under diseased cattle. Talbot et al. (1980) suspect that there are paths of genetic exchange from human clinical sources and sites of care for diseased animals to nonclinical populations on vegetables and elsewhere, both being situations where antibiotics are heavily used.

7.1.4 Some Bacteria Have Multiple Modes of Sexuality

Various species of the Gram-negative genus *Pseudomonas* exchange chromosomal genes by all three of the sexual processes of bacteria—conjugation, transformation, and transduction (Khan and Sen 1967; Chakrabarty et al. 1968; Mylroie et al. 1978; Hermann et al. 1979; Stewart et al. 1983; Carlson et al. 1983). Plasmids may also be transmitted via conjugation and transformation. Genetic exchange by conjugation in *Pseudomonas* has been found under various artificial conditions (Dunn and Gunsalus 1973; Yen and Gunsalus 1981). Transformation has been demonstrated several times in broth culture or with "filter matings" (Khan and Sen 1967; Mylroie et al. 1978). Plasmid transformation has also been shown for the naphthalene-degradative plasmid *nah*7 plasmid (Johnston and Gunsalus 1977).

Genetically engineered *Pseudomonas* species are beginning to be developed for various agricultural applications. One example involves the use of ice-minus strains of *P. syringae* and *P. fluorescens* to reduce frost damage on crops. The cloned delta-endotoxin gene from *Bacillus thuringiensis* has been inserted into the chromosome of *P. fluorescens* using plasmid-mediated transformation in the hope of creating a living pesticide. *Pseudomonas* species now figure significantly in most efforts to create bacteria to clean up pollution and degrade toxic waste. The degradative functions are typically encoded on plasmids (Wheelis 1975; Heinaru et al. 1978). These plasmids can be manipulated by both in vitro and in vivo forms of genetic engineering. In addition, the structures and functions of the plasmids themselves as well as their capacities for horizontal transfer within and between species raise basic questions about their origin, evolution, and population biology (Pemberton 1983). Such genetic engineering efforts are being designed to fight soil and water contamination from oil, halogenated hydrocarbons, polyaromatic hydrocarbons, pesticides, herbicides, and heavy metals (U.S. Congress, Office of Technology Assessment Report 1988).

7.2 THE POSSIBILITY OF GENETIC EXCHANGE BETWEEN SPECIES

Perhaps environmental applications of single, genetically engineered strains would raise fewer problems if the strains released were totally isolated from the pathways of sexuality that may operate in natural populations of bacteria. In the case of *Bacillus subtilis*, the evidence from microcosm studies suggests that naturally occurring cells of this species could be transformed if DNA from a genetically engineered strain were released into the soil either actively or by cell lysis. Thus the genes of an engineered strain might flow into a natural population. If the species of *Bacillus* were completely isolated, the release of such genes would go no farther than the one species. However, we now have evidence that the species boundaries of at least some *Bacillus* species are not impenetrable barriers to gene flow. They can be crossed by what appears to be natural transformation.

In a variety of *laboratory* studies, genetic exchange via transformation and transduction *in liquid culture* has been found to breach the *Bacillus* species boundaries defined by bacterial taxonomists (transformation—Marmur et al. 1963; Dubnau et al. 1965; Wilson and Young 1972; Harford and Mergeay 1973; Seki et al. 1975; Harris-Warwick and Lederberg 1978a and b; transduction—Lovett 1972; Ruhfel et al. 1984). Aronson and Beckman (1987) have reported both chromosomal and plasmid transfers between *B. thuringiensis* and *B. cereus*. I emphasize liquid culture above because the normal habitat of *B. subtilis* and many other *Bacillus* species is soil, though many species also occur in fresh water. Reanney and Teh (1976) have suggested that transducing phages with very broad host ranges may exist in soil and cause gene exchange among all resident *Bacillus* species. I doubt that this will prove to be true, but the necessary direct experimentation has not been done.

We have studied gene transfer between *Bacillus* species using small, sterile, soil cultures. Sterile soil cultures are not completely realistic, but they offer a reasonable bridge between liquid culture or petri plate experiments and nature. We have found, for example, that gene transfer can be a spontaneous and efficient process within and between *Bacillus* species under the rather natural setting created in soil when many spores germinate at once. We have also done experiments with unsterilized soils, and will probably do so in the future. However, since the interpretation of results from such experiments is difficult, much more experience is needed before experiments in nonsterile soil will be useful. Experiments in sterile soils provide a tractable assessment of the gene transfer processes that *can* happen in soil. Pemberton (1983) makes important and sweeping conclusions about the relative effectiveness of each of the modes of sexuality in Gram-negative soil bacteria. Yet, not one analysis of these processes in soil is part of the empirical basis for his conclusions. Soil microcosms are now recognized as an important tool in risk assessment associated with field applications of biotechnology (Pritchard and Bourquin 1984).

7.3 INTERSPECIFIC GENE TRANSFER AND ITS CONSEQUENCES

In some recent studies we focused on genetic exchange between *Bacillus* species via transformation. The standard laboratory strains and wild strains chosen for these experiments do not carry plasmids. These experiments were conducted in the laboratory and employed both broth culture transformation and spontaneous transformation in soil culture. We found four surprising results, which are described below. These results are presented more completely in Duncan et al. (1989).

7.3.1 Studies with Laboratory Strains

We found that two *Bacillus* species known to share 7–15% DNA-DNA homology can readily exchange chromosomal DNA, with the result that multiple phenotypic characteristics are also exchanged, and other characters are altered in expression. The species used were *Bacillus subtilis* and *B. licheniformis*. The earliest of these experiments used two standard laboratory strains: a derivative of *B. subtilis* 168 and *B. licheniformis* M28. To achieve interspecific broth transformations it was necessary to develop new procedures because the existing transformation procedures used for each of these species separately are very different. However, in soil culture experiments, we found that these standard strains were able to carry out genetic exchange spontaneously. A large number of control experiments were done to make certain that we could accurately recognize species hybrids. All possibility of confusion due to mutational changes or mixed colonies of the two strains was ruled out. We found that *intraspecific* exchange was about 10- to 1000-fold more frequent than *interspecific* exchange in liquid culture transformation experiments. We chose to carry out our first experiments with standard laboratory strains in order to anchor these studies in the solid foundation of the research with these strains done by many other authors.

It was Seki et al. (1975) who first demonstrated the low DNA-DNA homology between the two species. They reported values ranging from 7 to 15% homology in comparisons of *B. subtilis* strain 168 with 16 strains of *B. licheniformis*. Three additional comparisons of *B. subtilis* strain W23 with *B. licheniformis* ranged from 9–11% DNA-DNA homology. The values obtained when the same *B. licheniformis* strains were tested with *B. licheniformis* strains IFO12107 and S207 ranged from 64–116% homology (31 comparisons, the percentage can exceed 100% due to the method of standardization). Priest (1981) provides an additional value of 14% homology between a *B. subtilis* Marburg strain and *B. licheniformis* strain 9945a. The standard convention for bacterial species is a difference of about 70% DNA-DNA sequence homology.

7.3.2 Studies with Wild Strains

The remaining three results of our recent studies come from tests for spontaneous genetic exchange between wild isolates of *B. subtilis* (T3A14) and *B. licheniformis* (Te48) in soil culture. The strains were isolated from soil samples taken at the Tumamoc Hill preserve of the Desert Laboratory on the outskirts of Tucson, Arizona. Figure 7–1 shows the dynamics of exchange between these strains in two soil culture experiments. The wild strains are clearly capable of transformation. The decrease in the number of interspecies hybrids with time suggests that they may be competitively inferior to the parental strains.

Thirteen hybrids were used for the more extensive analyses yielding the results shown in Table 7–1; they were the only hybrids from the initial screening on selective plates to be fully characterized for all the 46 characters shown in the table, and should represent an unbiased sample of the products of hybridization between the parental strains. Three of the hybrids had the colony type and general marker background of *B. subtilis*, the other 10 resembled *B. licheniformis*. We found that about 9% of the traits scored were altered by interspecific DNA exchange.

7.3.3 Phenotypic Instability in Species Hybrids

About half of the traits changed by spontaneous gene exchange between *B. subtilis* and *B. licheniformis* were phenotypically unstable, meaning that on repeated serial culture on nonselective agar plates, isolated colonies switched from one parental species character state to that of the other species (see

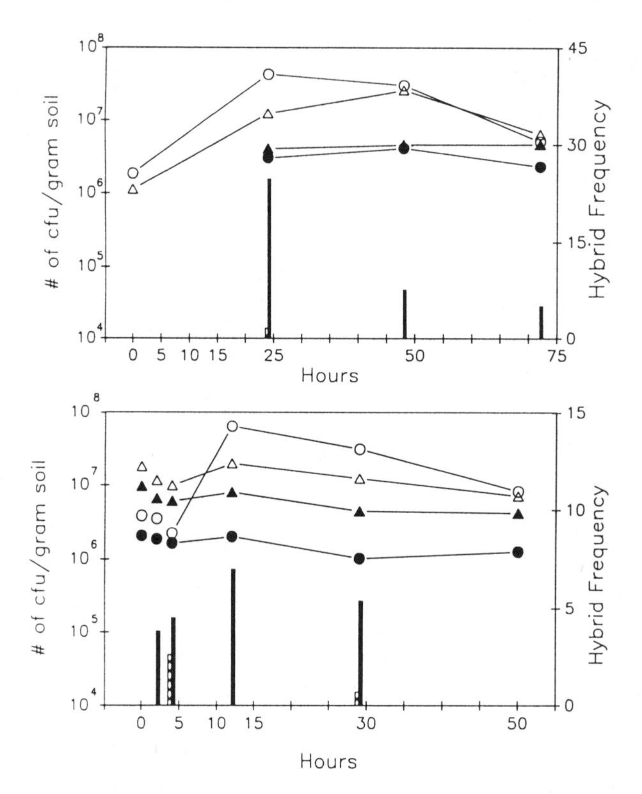

FIGURE 7–1 Dynamics of spontaneous hybridization in two different soil microcosm experiments with wild *Bacillus* strains. The upper curves and left ordinate scale in both graphs plot the time course of the densities of total colony forming units (cfu, open symbols) and spores (solid symbols) as the average of two replicate microcosms inoculated with wild parental strains. Circles denote the densities of colonies resembling *B. subtilis* T3A14; triangles denote the densities of colonies resembling *B. licheniformis* Te48. The frequency bars at the bottom and the right ordinate indicate the frequencies of hybrids at each sample time. The frequencies of hybrids were calculated as the percentage of a given colony morphology, T3A14-like (hatched bars) or Te48-like (solid bars), whose markers tested as hybrids rather than parentals. Note the occurrence of a very small percent of T3A14-like hybrids at day 24 in the upper graph. In general, hybrids with *B. licheniformis* as the recipient parent predominated in these experiments.

Table 7–1). The molecular structure of these hybrid genomes and the molecular events causing phenotypic instability are currently unknown.

When phenotypic instability ensues among these *B. subtilis* × *B. licheniformis* hybrids, it resolves itself during ensuing cell replications with a 3.3-to-1 bias in favor of the trait originally carried by the recipient species parent, i.e., the parental strain that supplied its characteristic colony morphology and presumably most of its genome. A quantitative summary of

TABLE 7–1 Character Profiles of *B. subtilis* T3A14, *B. licheniformis* Te48, and 13 Interspecies Hybrids[1]

Character	*B. subtilis* *T3A14*	*Interspecies Hybrids Strain No.* 1	2	3	4	5	6	7	8	9	10	11	12	13	*B. licheniformis* *Te48*
Antibiotic resistance:															
Erythromycin	−[2]	w−	−	+−	+	+	+	+	+	+	+	+	+	+	+
Chloramphenicol	−	−	−	−	+	+	+	+	+	+	+	+	+	+	+
Growth on carbon sources on agar plates:															
Sodium propionate	−	−	−+	−	+	−+	−+	−w	+	+	w+	+	*	+	+
Tagatose	−	−w	w+	+w	+	+	+	+	+	+	+	+	*	+	+
Glu-min	+	+	+	+	+	+	+	+	+	+	+	+	−	+	+
Rapid API tests:															
ONPG	+	+	+	+	−+	+	−+	+	+	+	+	+	+	+	+
ADH	−	−	−	+−	−	+	+	+	+	+	+	+	+	+	+
Citrate	+	+−	+	−	−	+	−+	+	+	−	−	+	−	−+	+
Urease	−	−	−	−	−	+	−+	+d	+d	+−	+	d	+d	+	+
Gelatin	+	+	+	+	−+	+	w+	+	+	+	+	+	+	+	+
Galactose	−(+)	−	d	d−	+	+d	+	+	+	+	+	+	+	+	+
Sorbose	−	−	−	−	+	+d	+	+	−+	−+	−	−+	−	−	−(+)
Rhamnose	−	−	−	+−	+	+	+	+	+	+	+	+	+	+	+
Sorbitol	+	+	+	+	−	−	−	−	−	−	−	−	+	−	−
Lactose	−	−	−	−	+d	−	+−	−	+−	d	d−	−+	+	+d	−(d)
Melibiose	−	−	+	−d	−	−	−	−	−	−	−	−	+	−	−
Raffinose	+(d)	−	+	d−	−	−	−d	−d	d	+d	−d	−d	+	+	d(−)
β-gentiobiose	−	−	+	−	+	+	+	+	+	+	+	+	+	+	+
Turanose	−(d)	−	+	d	+	+	+	+	+	+	+	+	+	+	+
Tagatose	−	−	−	+−	+	+	+	+	+	+	+	+	+	+	+

Sensitivity to bacteriophage:															
THO	s	s	s	s	s	s	r	r	s	s	s	r	r	s	s
SP02	s	s	s	s	r	r	r	r	r	r	r	r	r	r	r
SP10	r	r	r	r	r	r	r	r	r	r	r	r	s	r	r
Colony morphology	S	S	S	S	U	U	U	U	LU	L	U	U	U	UL	L

[1] Initial and subsequent character profiles for the parental strains and 13 hybrids are shown. The initial character state for each hybrid is shown first to the left in its main column. If the character changed after reculturing and retesting, the subsequent character state is shown to the right of the original observation. The vertical line separates hybrids according to colony morphology and the presumed recipient parental genome. From Duncan et al. 1989).

[2] Symbols: S, *subtilis* T3A14 colony morphology; U, an unusual colony morphology similar to the minority morphology of Te48; L, typical *licheniformis* Te48 colony morphology; + = growth, w = weak growth, − = no growth, d = delayed growth appearing after 6 days (normal growth is in 1–2 days); s = sensitive; r = resistant; Glu-min = growth on glucose minimal medium; ONPG is orthonitrophenol-galactopyranoside; ADH is arginine dihydrolase. The * indicates that these two scorings are redundant with the auxotrophy recorded as—for glu-min for hybrid 12. THO is a bacteriophage isolated from soil samples at the same site where the parental strains were isolated. SP02 and SP10 are standard laboratory bacteriophages. A (d), (−), or (+) means that under repeated testing of the parental strain this character occasionally fluctuates.

the types of character changes, and their stability observed in this experiment is shown in Table 7–2. It is possible that all cells in the same hybrid clone undergo the same genomic changes at the molecular level, i.e., "correction" toward the recipient parent phenotype is the result of a purely molecular process. It is also possible that natural selection may be involved. Selection due to differential division rates might allow corrected cells to crowd out less-corrected ones.

About 2% of the altered characteristics of these species hybrids were new characteristics—ones not shown by either parent. These de novo characteristics include a number of metabolic characteristics, as well as resistance or sensitivity to several bacteriophages (see Table 7–1).

It is impossible to conclude from our genetic data that the heterospecific DNA coding for the foreign genes expressed in a species hybrid was physically integrated into the recipient chromosome, though this is our assumption, and electrophoresis of total DNA of hybrids did not reveal any extrachromosomal DNA. Harris-Warrick and Lederberg (1978b) suggested that the integration of the DNA of *B. subtilis* into the *B. globigii* chromosome under artificial transformation was made possible by flanking regions of

TABLE 7–2 Classification and Frequencies of the Types of Character State Conditions of the 13 Hybrids[1]

Character State	*Number Observed*	*Frequency*
1. Unstable deviations	26	0.0446
Subsequently corrected to most-probable recipient	(20)[2]	
Subsequently corrected to wrong character for most-probable recipient	(6)	
2. Stable deviations	26	0.0446
Wrong character state for most-probable recipient	(11)	
Character state is "unusual", i.e., different from either parent	(15)	
3. Character never changed, and was correct for most-probable recipient parent; parental strains differed for these characters	141	0.242
4. Character never changed; parental strains were the same for these characters	390	0.669
Total	583	1.000

[1] From Duncan et al. 1989. Colony morphology is not included.
[2] Parentheses indicate subtotals within categories 1 and 2.

DNA homology. The discovery by Niaudet et al. (1985) that heterologous stretches of plasmid DNA flanked by homologous *B. subtilis* sequences can be integrated into the *B. subtilis* chromosome by artificial transformation gives further reason to believe that integration is also occurring in our interspecies hybrids, and that this type of recombination may not be a particularly difficult feat even when the recombining DNAs have little sequence homology.

The correction process we have observed perhaps offers one way in which species distinctness could be reinforced in nature, even though genetic exchange occurs rather steadily. The correction process would be particularly effective if its underlying mechanism results in the complete elimination of heterospecific DNA sequences. From an evolutionary perspective, it thus becomes important to understand the physical changes that occur in the hybrid genome when correction occurs. Any process short of a precise excision of the foreign DNA would cause remnants of heterospecific sequences to accumulate in species gene pools. If the heterospecific DNA is integrated into the chromosome initially, an imperfect excision process during correction could both leave heterospecific sequences and delete homospecific ones. If the process of correction is due to chromosomal rearrangement or repression of foreign gene expression, rather than excision, we would expect large amounts of DNA to be exchanged horizontally between species in the course of time. The level of phenetic distinctness observed generally among *Bacillus* species suggests that cumulative mixing of genomes has not occurred. With more closely allied forms such as *B. subtilis* and *B. licheniformis*, where interspecific genetic exchange is clearly possible, it becomes important to ascertain whether species intermediates arise in natural populations.

It is striking that the phenotypic instability of our interspecific hybrids appears similar to that following intraspecific protoplast fusion, i.e., artificial creation of diploid *B. subtilis* through fusion of two cells (Schaeffer et al. 1976; Hotchkiss and Gabor 1980 and 1985; Guillen et al. 1982, 1983, and 1985; Sanchez-Rivas et al. 1988), where the molecular basis is thought to be noncomplementing diploidy. At this point we do not think that protoplast fusion per se occurs in soil or in broth transformations between species, because it is unlikely that the bacteria would lose their cell walls, or survive such a loss in a soil environment. However, the possibility of natural protoplast fusion cannot be rejected on the basis of present information; it would be a new form of bacterial sexuality.

Merodiploidy, in which heterologous DNA is inserted to form a duplication in the recipient chromosome, is known to produce genetic instability. The directed insertion of heterologous DNA into the *B. subtilis* chromosome has been achieved (Duncan et al. 1978; Young 1983) using a segment of DNA homologous with the *B. subtilis* chromosome. Similarly, the conserved region of ribosomal RNA genes near the origin of replication of both the *B. subtilis* and *B. licheniformis* chromosomes (Marmur et al.

1963; Seki et al. 1975; Perlak and Thorne 1981; Osawa and Tokui 1978) provide a duplicated region of naturally high sequence homology (Gottlieb and Rudner 1985). However, there is evidence that transformation into this region *intraspecifically*, within *B. subtilis*, may sometimes be unstable for insertion of rRNA operons (Widom et al. 1988). Some of the antibiotic resistance markers that we have used (*rfm*, *spc*, and *fus*) to examine both genetic exchange and the subsequent instability of species hybrids also reside in this region of the chromosomal map. Merodiploidy and hybrid instability have been found in hybrids from intergeneric crosses of enteric bacteria (Luria and Burrous 1957; Baron et al. 1968; and Lehner and Hill 1985). Partial diploidy must presumably be involved in at least some of the hybrid instability we have observed. How else could the hybrids switch between parental traits?

Anderson and Roth (1977) review several cases where hybrid instability and correction to the recipient parent phenotype have been observed among interspecific hybrids. They offer a model to account for the formation of merodiploidy and its subsequent instability resolved in favor of the recipient parent genome. They imagine a situation in which a duplicated region with two copies of some marker on the donor DNA has two small regions of homology with a single small region in the recipient's unduplicated DNA with its different marker. The region of homology is on only one side of the marker in both donor and recipient, and thus the duplication gives the donor two such regions of homology. The assumption of homology on only one side of each marker is made because we are considering crosses among species of generally low sequence homology. On transformation, the duplicated donor DNA provides two regions of homology bounding one copy of the donor marker, and it recombines into the recipient chromosome by splitting its single region of homology and forming a merodiploid containing tandem copies–one each of the donor and recipient marker genes–and now two regions of homology. Subsequently, recombination within the merodiploid hybrid strain produces both triplications with two copies of the donor marker and one of the recipient marker and revertants to the original recipient parent state with a single copy of its original gene. No regulation of gene expression is involved in the model.

The Anderson and Roth (1977) model could account for some of our observations of character state change including eventual recovery of parental phenotypes, but it cannot fully account for the *variability* among the forms of hybrid instability we have observed. There are many cases in our data where the hybrid was initially minus for some character but the plus state–either from the original recipient or the donor–appeared subsequently. A merodiploid that simultaneously carried markers for both plus and minus would never test minus initially if both marker genes were expressed. Some additional form of "regulation" is necessary first to silence expression of the plus marker and later to remove such control. The model also does not account for cases in which the character state of the hybrid differs from both parents, as was observed with hybrids of the wild strains. Since transformation in soil populations can involve fairly extensive stretches of the chromosome (Graham and Istock 1978), recombination based on two widely spaced regions of homology could lead to the formation of a merodiploid hybrid and to subsequent instability and the recovery of a parent phenotype.

The capacity for introgression, or lasting gene flow, between these two *Bacillus* species is particularly interesting because it involves organisms of such low DNA sequence homology. However, at the field site where these strains were isolated we also find numerous isolates that type unequivocally as either *B. subtilis* or *B. licheniformis*. Therefore, the process of introgression between the species certainly does not run to completion at this field site. Our experimental data, combined with the observed distinctness of wild isolates of *B. subtilis* and *B. licheniformis*, suggest that a conflict may exist between the capacity to recombine across species boundaries and the selective forces in nature that reinforce the distinctness of these species. We have not yet tried to screen for intermediates. Previously, investigators faced with the prospect of introgression among well-marked forms of bacteria, via transformation or transduction, developed the genospecies concept (Ravin 1963; Wilson and Young 1972): the genospecies was seen as an evolving gene pool larger than any one constituent species and delimited by the full system of sexual genetic exchange among bacterial species (Reilly 1976).

7.3.4 More Extreme Cases of Hybrid Formation

Recently, Heinemann and Sprague (1989) have reported an extraordinary example of an engineered, spontaneous, plasmid exchange between *Escherichia coli* and the yeast *Saccharomyces cervisiae*. This transfer of a plasmid (YEp13, a derivative of pBR322) from a procaryote to a eucaryote was made possible by including an appropriately engineered conjugative plasmid in the *E. coli* cells. The trick is to have the mobilization proteins that are encoded in the transfer (*tra*) operon of the conjugative plasmid in *E. coli* be of the same compatibility group as the origin of transfer (*ori*T) of YEp13. These mobilization proteins are thought to literally drag plasmid DNA from one cell to another. YEp13 also contains a replication sequence that allows it to replicate in yeast cells. Both a broad-host-range plasmid (pDPT51, a derivative of R751) and a derivative of the narrow-host-range F plasmid of *E. coli* were engineered to serve as helper plasmids to transmit YEp13 to yeast cells. The transmission of YEp13 was detected both by a genetic marker that changed the yeast cells from ones requiring leucine (leu$^-$) to cells that did not require leucine (leu$^+$), and by restriction fragment analyses, which revealed that plasmid DNA identical to YEp13 had appeared in the leu$^+$ yeast cells. Presumably the conjugative plasmid also was transferred to the yeast cells, but lacking a replication sequence recognized by yeast it was not replicated, although Heinemann and Sprague suggest that some plasmid DNA might occasionally be recombined into the chromosome of the new host. Considerable evidence was also presented which indicates that the process of transfer is indeed the same as bacterium-to-bacterium conjugation.

Heinemann and Sprague speculate that transfers between bacteria and plants and animals have occurred naturally during evolution, and that this phenomenon may have played a significant role in evolution and account for some difficulties in resolving phylogenetic trees. They conclude: "Thus it appears that the conjugation functions encoded by the conjugative plasmids are capable of forming contacts competent for DNA transfer with organisms separated by large evolutionary distances." This intriguing suggestion, if such transfers have been reasonably common, might imply a one-way process with genetic information flowing only from procaryotes to eucaryotes and thereby editing eucaryotic evolution from time to time. If Heinemann and Sprague are right in this conjecture, bacteria are sexually promiscuous on a scale much larger than was previously surmised. Further molecular steps would be necessary, however, for the process to have a lasting effect. For example, the procaryotic DNA would have to recombine into eucaryotic chromosomes and it would have to be expressed. At present, we do not know just what accommodations at the molecular level would be required, and are possible, for the eucaryotic genetic machine to make use of procaryotic DNA sequences.

Another example of a procaryote-to-eucaryote transfer of DNA involves the transfer of a portion of the Ti (tumor-inducing) plasmid of *Agrobacterium tumifaciens* to several plant species (Zambryski et al. 1989; Buchanan-Wollaston et al. 1987). This is an evolved form of natural genetic engineering. The Ti sequences do become stably integrated into the plant chromosomes and cause phenotypic effects on plant root development known as crown-gall disease. Some of the steps in DNA metabolism involved in the transfer of Ti sequences resemble conjugation—the DNA enters the plant cells as a linear single-stranded molecule formed by nicking and separating the original, circular, double-stranded plasmid molecule.

What makes the Heinemann and Sprague (1989) demonstration so dramatic, though, is the fact that it is a system that, after it is engineered by recombinant DNA techniques into the right molecular configuration, is able to conduct DNA exchange spontaneously between intact cells. The same property exists in the case where modified F plasmids have become broad-host-range plasmids capable of transmission to *Pseudomonas aeruginosa* (Guiney 1982). Feitelson and Lederberg (1980) reported that crude lysates of plasmids from *Staphylococcus aureus* could transform *B. subtilis* to drug resistance, and they concluded: "We have shown that purified DNA is not necessary for plasmid transformation between widely divergent bacterial species and thus offer no basis to doubt that heterospecific gene transfers can occur in nature." Though, in both these latter cases, at least, the transfers are only among procaryotes, unlike that of Heinemann and Sprague (1988) or the Ti plasmid example. I must add that the certainty expressed by Feitelson and Lederberg in their extrapolation to nature is not a certainty which I share even a decade later; there is a great deal more to be studied before we will really know the extent and consequences of bacterial sexuality in nature.

Another striking example of interspecies hybrid formation is the report by Richaume et al. (1989) of spontaneous plasmid transfer between *E. coli* and *Rhizobium fredii* in sterile soil culture. They found that specific ecological conditions characterized by clay content, organic matter concentration, soil moisture content, pH, and temperature influence the rate of plasmid exchange.

A final example involves chromosomal gene exchange from streptococci to *E. coli* (Brisson-Noel et al. 1988). This exchange again breaches the time-honored Gram-positive to Gram-negative distinction among bacteria. Brisson-Noel and coworkers also reached an important evolutionary conclusion germane to our topic: "... the identity of DNA sequences [technically the *erm*BC gene] originating in streptococci and found in *E. coli* provides evidence for an in vivo horizontal transfer of genetic information between these phylogenetically remote pathogenic bacteria." The direction of the transfer was deduced from the fact that the *erm*BC gene has the same proportion of DNA bases in both *Streptococcus* and *E. coli* (G + C content = 33 mol%); this is the same value as the *Streptococcus* genome, while the background genome of *E. coli* is quite different in base composition (G + C content = 50 mol%). Hence the gene probably originated in *Streptococcus*.

7.4 CONCLUSIONS

The evolutionary framework which increasingly surfaces in the foregoing discussion underscores the importance of a population-biology perspective to any program of risk assessment in biotechnology. Population biology provides a corpus of theory and empirical experience necessary, in combination with knowledge and methods from molecular biology, to answer the questions posed in the introduction to this chapter. Where are we now in obtaining answers to these questions, at least as they pertain to the bacterial forms of life? The questions I posed are:

1. What avenues and barriers for genetic transmission exist in nature?
2. How likely are exotic genes to be transferred within and between species in nature?
3. What are the associated rates of both intraspecific and interspecific gene transfer?

4. Will the existing differentiations among species become blurred if intermediate, higher homology, genetic sequences are introduced into nature or will natural selection root out such sequences?
5. How readily can genes with new physiological functions be expressed when they reach various recipient genomes in nature?
6. How stable will exotic gene expression be when such genes become new additions to the previously evolved genomes of wild species and encounter the welter of selective forces arising within the surrounding ecological community?

It is certain that avenues for gene transfer of several kinds exist among some bacteria. We have evidence, from experiments done in simulated microcosms and in chambers placed in lakewater, that all three of the modes of bacterial sexuality probably can occur spontaneously in nature: as demonstrated for transformation within and between *Bacillus* species, transduction in *Pseudomonas*, and conjugation within and between species of *Klebsiella*. These processes can lead to chromosomal genetic recombination in both the population and molecular genetic senses; i.e., genetic variation will be reshuffled and DNA sequences will be interchanged. There are many observations suggesting that it is possible for chromosomal gene transfers to occur between species of bacteria with moderately low DNA-DNA sequence homology, but no field studies of these phenomena exist yet. Many more experiments are required before we will understand how widespread such sexual exchanges of genes are among the whole of the bacterial world, or between bacteria and evolutionarily distant forms of life, and under what conditions and at what rates these various exchanges might be expected to occur. Fairly widespread exchange of plasmids seems likely, based on the ease with which they transfer in so many laboratory situations, but we are far from having any grasp of what the transfer opportunities, rates of transfer, or extent of gene expression in a range of host strains and species might be like in nature. Again, there are no direct field studies of the processes of plasmid exchange. The consequences of introducing hybrid DNA sequences into mixed species assemblages are largely unexplored. The patterns of gene expression and phenotypic and genetic stability and instability that follow genetic exchange are becoming better known from a variety of laboratory studies. But only the smallest beginnings have been made toward evaluating ecological competition and natural selection as forces opposing or facilitating processes of gene transfer. We need to do many kinds of studies directed at better understanding the molecular basis of bacterial sexuality and the molecular processes that create stability and instability following horizontal gene transfers.

At present we are in a most interesting situation with regard to our knowledge of the extent of recombination in natural populations of bacteria. A variety of extensive studies done with *E. coli* and *Shigella* species (Selander and Levin 1980; Levin 1981; Caugant et al. 1981 and 1983; Whittam et al. 1983 and 1984; Ochman et al. 1983; Achtman et al. 1983), with *Neisseria meningitidis* (Caugant et al. 1986a and b, 1987a and b), and with *Legionella* species (Selander et al. 1985) have concluded that little recombination occurs within or between populations of these organisms. The genetic population structure of these pathogenic or commensal organisms is said to be "clonal." All the evidence presented in these studies is from protein electrophoretic analyses of allozymic variation. Such evidence is indirect, and the screening of a small number of allozymes actually sampled only a small portion of the genome of each isolate, though it amounts to a great deal of hard work (see discussion of this point in Caugant et al. 1987). Thus, such studies might not detect modest levels of genetic recombination, but there is another possibility. Recombination may be quite frequent on a cell-to-cell basis under appropriate conditions for mating, but such recombination is masked almost completely because most of the time reproduction by binary fission—clonal reproduction—puts an overwhelmingly clonal stamp on the populations. Such functional clonality makes the populations of these bacteria assume a "clonal structure," but it does not rule out the co-occurrence of a capacity for, and practice of, sexual exchange. However, it prevents one from measuring genetic recombination. A selective advantage could cause sexuality to be retained in such populations if it makes it possible for clones lagging behind in binary reproduction to "steal the genes of current success" from the currently flourishing clones found in the same habitat. This assumes that all clones regularly experience periods of lagging growth. It is interesting that the laboratory protocols to make bacteria competent for transformation typically involve bringing the cells into a lagging growth phase. Clearly, direct studies of spontaneous sexuality are needed using these same organisms for which a clonal population structure is claimed. The question is a central one because, equally clearly, we are faced here with a key issue for understanding what opportunities actually exist for the spread of new genes in wild populations of pathogenic or obligately commensal bacteria, and for understanding whether a clonal population structure might impede or facilitate the spread of new genetic types or new genetic sequences. It may well turn out that pathogenic and commensal bacteria are predominantly clonal. At the same time, we need to find out if free-living bacteria also have clonal population structures.

The source of the eeriness which strikes me when I hear someone assume that biotechnology has no dangers should now be clear. In our understanding of the natural lives and the genetic properties and processes of bacteria, we are at a beginning. Far beyond this point lies the depth of scientific understanding of these matters sufficient to allow anyone to speak with confidence about the safety or hazards of particular interventions using the techniques of recombinant DNA technology, or even those using imaginative applications of more classical procedures for genetically modifying organisms. This same depth of understanding will also guide us toward many of the most tractable and efficacious applications of recombinant DNA technology.

For the time being, there is much work to be done in the laboratory and the field studying natural populations of bacteria, and the consequences of their proclivities toward sexuality.

REFERENCES

Achtman, M.A., Mercer, A., Kusecek, B., et al. (1983) *Infect. Immun.* 39, 315–335.

Anagnostopolos, C., and Crawford, S.P. (1961) *PNAS* 47, 378–391.

Anagnostopolos, C., and Spizizen, J. (1961) *J. Bacteriol.* 81, 741–746.

Anderson, R.P., and Roth, J.R. (1977) *Ann. Rev. Microbiol.* 31, 473–505.

Aronson, A.I., and Beckman, W. (1987) *Appl. Env. Microbiol.* 53, 1525–1530.

Avery, O.T., McLeod, C.M., and McCarty, M. (1944) *J. Exp. Med.* 79, 137–158.

Baron, L.S., Gemski, P., Johnson, E.M., and Wohlhieter, J.A. (1968) *Bacteriol. Rev.* 32, 362–669.

Borenstein, S., and Ephrati-Elizur, E. (1969) *J. Mol. Biol.* 45, 137–152.

Bresler, S.E., Kareneva, R.A., and Kushev, V.V. (1968) *Mol. Gen. Genet.* 102, 257–268.

Brisson-Noel, A., Arthur, M., and Courvalin, P. (1988) *J. Bact.* 170, 1739–1745.

Buchanan-Wollaston, V., Passiatore, J.E., and Cannon, F. (1987) *Nature* 328, 172–175.

Carlson, C.A., Pierson, L.S., Rosen, J.J., and Ingraham, J.L. (1983) *J. Bact.* 153, 93–99.

Catlin, B.W. (1960) *J. Bact.* 79, 579–590.

Caugant, D.A., Bovre, K., Gustad, R., et al. (1986a) *J. Gen. Microbiol.* 132, 641–652.

Caugant, D.A., Froholm, L.O., Bovre, K., et al. (1986b) *PNAS* 83, 4927–4931.

Caugant, D.A., Levin, B.R., Lidin-Janson, G., et al. (1983) in *Progress in Allergy*, vol. 33 (Hanson, L.A., Kallos, P., and Westphal, O., eds.), pp. 203–227, S. Karger AG, Basel.

Caugant, D.A., Levin, B.R., and Selander, R.K. (1981) *Genetics* 98, 467–490.

Caugant, D.A., Mocca, L.F., Frasch, C.E., et al. (1987a) *J. Bact.* 169, 2781–2792.

Caugant, D.A., Zollinger, W.D., Mocca, L.F. et al. (1987b) *Infect. Immun.* 55, 1503–1512.

Chakrabarty, A.M., Gunsalus, C.F., and Gunsalus, I.C. (1968) *PNAS* 60, 168–175.

Dubnau, D., Smith, I., Morell, P., and Marmur, J. (1965) *PNAS* 54, 491–498.

Dubnau, D.A. (1982) *The Molecular Biology of the Bacilli*, vol. 1, *Bacillus subtilis*, Academic Press, New York.

Duncan, C.H., Wilson, G.A., and Young, F.E. (1978) *PNAS* 75, 3664–3668.

Duncan, K.E., Istock, C.A., Graham, J.B., and Ferguson, N. (1989) *Evolution* 43, 1585–1609.

Dunn, N.W., and Gunsalus, I.C. (1973) *J. Bact.* 114, 974–979.

Ephrati-Elizur, E. (1968) *Genet. Res. Camb.* 11, 83–96.

Ephrati-Elizur, E., and Fox, M.S. (1964) *Nature* 192, 433–434.

Feitelson, J.S., and Lederberg, J. (1969) *J. Bacteriol.* 116, 545–547.

Gottleib, P., and Rudner, R. (1985) *Intl. J. Syst. Bact.* 35, 244–252.

Graham, J., and Istock, C. (1981) *Evolution* 35, 954–963.

Graham, J.B., and Istock, C.A. (1978) *Mol. Gen. Gent.* 166, 287–290.

Graham, J.B., and Istock, C.A. (1979) *Science* 204, 637–639.
Griffith, F. (1928) *J. Hyg.* 27, 113–159.
Guillen, N., Amar, M., and Hirschbein, L. (1985) *EMBO* 4, 1333–1338.
Guillen, N., Gabor, M.H., Hotchkiss, R.D., and Hirschbein, L. (1982) *Mol. Gen. Gent.* 185, 69–74.
Guillen, N., Sanchez-Rivas, C., and Hirschbein, L. (1983) *Mol. Gen. Genet.* 191, 81–85.
Guiney, D.C. (1982) *J. Mol. Biol.* 162, 699–703.
Gwinn, D.D., and Thorne, C.B. (1964) *J. Bacteriol.* 87, 519–526.
Harford, N., and Mergeay, M. (1973) *Molec. Gen. Genet.* 120, 151–155.
Harris-Warrick, R.M., and Lederberg, J. (1978a) *J. Bact.* 133, 1237–1245.
Harris-Warrick, R.M., and Lederberg, J. (1978b) *J. Bact.* 133, 1246–1253.
Heinaru, A.L., Duggleby, C.J., and Broda, P. (1978) *Mol. Gen. Genet.* 160, 347–351.
Heinemann, J.A., and Sprague, G.F., Jr. (1989) *Nature* 340, 205–206.
Hemphill, H.E., and Whiteley, H.R. (1975) *Bacteriol. Rev.* 39, 257–315.
Henner, D.J., and Hoch, J.A. (1982) in *The Molecular Biology of the Bacilli*, vol. 1, *Bacillus subtilis* (Dubnau, D.A., ed.), pp. 1–33, Academic Press, New York.
Hermann, M., Garg, G.K. and Gunsalus, I.C. (1979) *J. Bact.* 137, 28–34.
Hotchkiss, R.D., and Gabor, M.H. (1980) *PNAS* 77, 3553–3557.
Hotchkiss, R.D., and Gabor, M.H. (1985) in *The Molecular Biology of the Bacilli*, vol. 2, (Dubnau, D., ed.), pp. 109–149, Academic Press, New York.
Ingraham, J.L., Maaloe, O., and Neidhardt, F.C. (1983) *Growth of the Bacterial Cell*, Sinauer Associates, Inc., Sunderland, MA.
Jackson, D.A., and Stich, S.P., eds. (1979) *The Recombinant DNA Debate*, Prentice-Hall, Inc., Englewood Cliffs, NJ.
Jentsch, S. (1983) *J. Bact.* 156, 800–808.
Johnston, J.B., and Gunsalus, I.C. (1977) *Biochem. Biophys. Res. Com.* 75, 13–18.
Khan, N.C., and Sen, S.P. (1967) *J. Gen. Microbiol.* 49, 210–209.
Lehner, A.F., and Hill, C.W. (1985) *Genetics* 110, 365–380.
Levin, B.R. (1981) *Genetics* 99, 1–23.
Lorenz, M.G., Aardema, B.A., and Wackernagel, W. (1988) *J. Gen. Microbiol.* 134, 107–112.
Love, P.E., and Yasbin, R.E. (1984) *J. Bact.* 160, 910–920.
Lovett, P.S. (1972) *J. Virol.* 47, 743–752.
Luria, S.E., and Burrous, J.W. (1957) *J. Bact.* 74, 461–476.
Marmur, J., Seaman, E., and Levine, J. (1963) *J. Bact.* 85, 461–467.
Morrison, W.D., Miller, R.V., and Sayler, G.S. (1979) *Appl. Envir. Microbiol.* 36, 724–730.
Mylroie, J.R., Fiello, D.A., and Chakrabarty, A.M. (1978) *Biochem. Biophys. Res. Commun.* 82, 281–288.
Niaudet, B., Janniere, L., and Ehrlich, S.D. (1985) *J. Bact.* 163, 111–120.
Ochman, H., Whittam, T.S., Caugant, D.A., and Selander, R.K. (1983) *J. Gen. Microbiol.* 129, 2715–2726.
Orrego, C., Arnaud, M., and Halvorson, H.O. (1978) *J. Bact.* 134, 973–981.
Osawa, S., and Tokui, A. (1978) *Mol. Gen. Genet.* 164, 113–129.
Pemberton, J.M. (1983) in *International Review of Cytology*, vol. 84, (Bourne, G.H., and Danielli, J.F., eds), pp. 155–183, Academic Press, New York.
Perlak, F.J., and Thorne, C.B. (1981) in *Sporulation and Germination* (Levinson, H.S., Sorenshein, A.L., and Tipper, D.J., eds.), pp. 78–82, American Society of Microbiology, Washington, D.C.

Piffaretti, J., Arina, A., and Frey, J. (1988) *Molec. Gen. Genet.* 212, 215–218.

Piggot, P.J., and Hoch, J.A. (1985) *Microbiol. Rev.* 49, 158–179.

Priest, F.G. (1981) in *The Aerobic Endospore-forming Bacteria: Classification and Identification* (Berkeley, R.C.W., and Goodfellow, M., eds.), pp. 33–58, Academic Press, New York.

Pritchard, P.H., and Bourquin, A.W. (1984) in *Advances in Microbial Ecology*, vol. 7 (Marshall, K.C., ed.), pp. 133–215, Plenum Press, New York.

Ravin, A.W. (1963) *Am. Nat.* 97, 307–318.

Reanney, D.C., Gowland, P.C., and Slater, J.H. (1983) in *Microbes in Their Natural Environments*, 34th *Symposium of the Soc. for Gen. Microbiol.* (Slater, J.H., Whittenbury, R., and Wimpenny, J.W.T., eds.), pp. 379–421, Cambridge Univ. Press, Cambridge, U.K.

Reanney, D.C., and Teh, C.K. (1976) *Soil Biol. Biochem.* 8, 305–311.

Redfield, R.J. (1988) *Genetics* 119, 213–221.

Reilly, B.E. (1976) in *Microbiology—1976* (Schlessinger, D., ed.), pp. 228–237, American Society Microbiology, Washington, D.C.

Richaume, A., Angle, J.S., and Sadowsky, M. (1989) *Appl. Envir. Microbiol.* 55, 1730–1734.

Romig, W.R., and Brodetsky, A.M. (1961) *J. Bact.* 82, 135–141.

Ruhfel, R.E., Robillard, N.J. and Thorne, C.B. (1984) *J. Bact.* 157, 708–711.

Saito, H., Shibata, T., and Ando, T. (1979) *Mol. Gen. Genet.* 170, 117–122.

Sanchez-Rivas, C., Karmazyn-Campelli, C., and Levi-Meyrueis, C. (1988) *Mol. Gen. Genet.* 214, 321–324.

Saye, D.J., Ogunseitan, O., Sayler, G.S., and Miller, R.V. (1987) *Appl. Env. Microbiol.* 53, 987–995.

Schaeffer, P, Cami, B., and Hotchkiss, R.D. (1976) *PNAS* 73, 2151–2155.

Seki, T., Oshima, T., and Oshima, Y. (1975) *Intern. J. Syst. Bact.* 25, 258–270.

Selander, R.K., and Levin, B.R. (1980) *Science* 210, 545–547.

Singha, R.P., and Iyer, V.N. (1971) *Biochim. Biophys. Acta.* 232, 61–71.

Smith, H.O., Danner, D.B., and Deich, R.A. (1981) *Ann. Rev. Biochem.* 50, 41–68.

Spizizen, J. (1958) *PNAS* 44, 1072–1078.

Stewart, G.J., and Carlson, C.A. (1986) *Ann. Rev. Microbiol.* 40, 211–235.

Stewart, G.J., Carlson, C.A., and Ingraham, J.L. (1983) *J. Bact.* 156, 30–35.

Streips, U.N., and Young, F.E. (1974) *Mol. Gen. Genet.* 133, 47–55.

Stuy, J.H. (1985) *J. Bact.* 162, 1–4.

Talbot, H.W., Yamamoto, D.K., Smith, M.W., and Seidler, R.J. (1980) *Appl. Environ. Microbiol.* 39, 97–104.

Thiel, T., and Wolk, P.C. (1987) *Methods Enzymol.* 153, 232–243,

Thorne, C.B. (1962) *J. Bact.* 83, 106–111.

Thorne, C.B. (1968) *J. Virol.* 2, 657–662.

Thorne, C.B., and Kowalski, J.B. (1976) in *Microbiology—1976* (Schlessinger, D., ed.), pp. 303–314, American Society for Microbiology, Washington, D.C.

Thorne, C.B., and Stull, H.B. (1966) *J. Bact.* 91, 1012–1020.

Trautner, T.A., Pawlek, B., Bron, S., and Anagnostopolos, C. (1974) *Mol. Gen. Genet.* 131, 181–191.

U.S. Congress, Office of Technology Assessment (1988) *New Developments in Biotechnology—Field-Testing Engineered Organisms: Genetic and Ecological Issues*, OTA-BA-350, Washington, D.C.

Wheelis, M.L. (1975) *Ann. Rev. Microbiol.* 29, 505–524.

Whittam, T.S., Ochman, H., and Selander, R.K. (1983) *PNAS* 80, 1751–1755.
Whittam, T.S., Ochman, H., and Selander, R.K. (1984) *Mol. Biol. Evol.* 1, 67–83.
Widom, R.L., Jarvis, E.D., LaFauci, G., and Rudner, R. (1988) *J. Bact.* 170, 605–610.
Wilson, G.A., and Young, F.E. (1972) *J. Bact.* 111, 705–716.
Wilson, G.A., and Young, F.E. (1976) in *Microbiology—1976* (Schlessinger, D., ed.), pp. 350–357, American Society for Microbiology, Washington, D.C.
Yen, K.-M., and Gunsalus, J.C. (1981) *PNAS* 79, 874–878.
Young, F.E., and Spizizen, J. (1961) *J. Bact.* 86, 392–400.
Young, M. (1983) *J. Gen. Micro.* 129, 1497–1512.
Yoxen, E. (1983) *The Gene Business*, Oxford University Press, New York.
Zambryski, P., Tempe, J., and Schell, J. (1989) *Cell* 56, 193–201.
Zinder, N.D., and Lederberg, J. (1952) *J. Bact.* 64, 679–689.

PART III

Mathematical Models in Biotechnology Risk Assessment

CHAPTER

8

Models for the Population Dynamics of Transposable Elements in Bacteria

Richard Condit

Transposable elements are segments of DNA capable of moving to new sites in the genome. They are widespread among both eucaryotes and procaryotes, probably occurring in all genomes. Besides being replicated during each generation as part of normal DNA synthesis, transposable elements (or transposons) also replicate during transposition, leaving a copy at the original site while inserting at a new site. This ability to transfer between genomes has led to concern about releasing genetically engineered organisms because of the possibility that introduced genes might be mobilized into transposons, onto plasmids, and spread into new organisms (Levy and Marshall 1988). So understanding the basic population dynamics of transposable elements—the processes that allow them to invade new bacterial strains and the factors that control their copy number in genomes—are important to biotechnology risk assessment. We should base decisions about the importance of gene transfer in genetically engineered microorganisms on a firm understanding of basic population processes.

Two alternative theories account for how transposons spread into new bacterial strains. According to one, transposons are parasitic DNA and can invade new populations despite the fact that they are detrimental to the genome in which they reside (Doolittle and Sapienze 1980; Orgel and Crick 1980; Campbell 1981). The major alternative is that transposons improve the fitness of their host and so might be termed "mutualistic DNA." In bacteria, the latter hypothesis is particularly attractive because many transposons carry with them (during transposition) traits that are useful to the host bacterium, such as genes for antibiotic resistance or heavy metal resistance (Cohen 1976; Tanaka et al. 1983). Not all bacterial transposons carry beneficial genes, but there are other ways that they might benefit their host, such as by causing beneficial mutations (Reynolds et al. 1981; Chao et al. 1983).

One goal of population studies of transposable elements is to develop an experimental and theoretical system for testing the above hypotheses, and mathematical models are a major component of this system. The purpose of this paper is to review the use of mathematical models in studies of the population processes that account for the abundance, distribution, and movement of transposons in and between bacterial populations. There have been two modeling studies explicitly aimed at the study of bacterial transposons–one by Sawyer and Hartl (1986), and the other by myself, Frank Stewart, and Bruce Levin (Condit et al. 1988). I will not consider here the substantial literature on models for eucaryotic transposons (such as Hickey 1982; Charlesworth and Charlesworth 1983; Ginzburg et al. 1984).

Sawyer and Hartl were interested in the copy number of transposable elements in bacteria. Since transposons can replicate within a genome, copy number should build up through evolutionary time, until detrimental effects of the elements balance the increase. The Sawyer and Hartl model seeks to predict equilibrium copy number distributions based on various assumptions about transposition and its effect on fitness. Condit et al. (1988) were interested in a different issue–the invasion of a transposon into a new bacterial population due to its ability to transpose into new genomes.

Before continuing with a discussion of the models, it is necessary to provide a brief review of transposon biology and the terminology associated with it. In addition, discussion of the Condit et al. model requires some background in plasmid biology.

8.1 BACKGROUND AND TERMINOLOGY

In bacteria, transposable elements that carry no genes other than the two necessary for transposition are called "insertion sequences" and are given names that begin with "IS"–such as IS*4* and IS*10*. Elements that carry genes besides the basic two are called "Tn elements"–such as Tn*3*, Tn*5*, and Tn*10* (Campbell et al. 1977).

Most theoretical considerations of transposon biology assume that transposition is a *replicative* process, where transfer to a new site leads to an increase in copy number of the transposon. Indeed, where direct evidence is available, transposition is replicative (Klaer et al. 1980; Read et al. 1980).

It has been suggested, however, that some transposons are *conservative*, meaning that they move to a new site without leaving a copy behind (Berg 1977). From a population perspective, this is a very important distinction and one that must be considered in constructing a model. To avoid confusion, I define replicative and conservative strictly in terms of copy number—replicative transposition means a cell with one copy gets two, while conservative transposition leaves it with one—not in terms of molecular mechanisms. Transposons can also be *excised* from a genome, reducing by one the number of copies in the cell (Berg 1977; Bottstein and Kleckner 1977; Egner and Berg 1981; Iida et al. 1983).

In theory, rates of transposition (or excision) can be defined as the probability that an individual cell carrying a transposon will undergo transposition (or excision) per unit of time. Alternatively, this can be viewed as the proportion of cells in which the event occurs in the same time interval. In practice, accurate measurement of these rates can prove elusive.

Transposition is often *regulated* by the transposon itself. Multiple copies of an element within a cell tend to inhibit the transposition process, so that the rate of transposition may decline (Kleckner 1981).

Transposons are generally capable of moving between two sites within the same chromosome, or between two different pieces of DNA in the *same* cell, such as a plasmid and a chromosome, or two plasmids. Most transposons cannot move into *new* cells without the help of some other mechanism for DNA exchange. (A few transposons are able to move between cells, but they are exceptional and will not be considered here; see Gawron-Burke and Clewell 1982.) In eucaryotes, fusion of gametes from different individuals ("sex") provides such a mechanism, and this happens reliably once each generation. Gene transfer is less reliable and less well understood in procaryotes, occurring by one of three mechanisms—transduction, transformation, or conjugation. We chose to model the last one as the vehicle for transposon movement in our work (Condit et al. 1988).

Conjugation occurs when conjugative plasmids move between bacterial cells. Conjugative plasmids are small circles of DNA which inhabit bacteria, replicating and partitioning to daughter cells during cell division, and also capable of infecting new cells via conjugation (or mating). A cell with plasmids physically contacts one without, and the plasmid replicates and transfers between cells—two plasmid-bearing cells result. Plasmids can also be lost from individual cells through a process called *segregation*. Of particular significance for the present subject, plasmids often carry transposable elements.

8.2 MODELS

8.2.1 A Basic Model

As a simple illustration, which serves to introduce the structure of models for bacterial transposons, I first describe a very basic model: consider a set of interacting, time-dependent, dynamical systems that represent a mixture of populations of bacterial cells, some without transposons and others with one or more elements. Let N_i be the population density of cells carrying i copies of a transposable element, and let Ψ_i be the growth rate of this i^{th} population. Transposition in the N_{i-1} population and excision in the N_{i+1} population create N_i cells, while either transposition or excision in the N_i population destroys N_i cells (the "N" symbols designate cell types as well as densities of those cell populations). Then the rate of change of each population (for $i>1$) can be written:

$$\dot{N}_i = \Psi_i N_i + \delta N_{i-1} - (\epsilon + \delta)N_i + \epsilon N_{i+1} \quad (1)$$

Here and in subsequent equations, a dot over a letter represents differentiation with respect to time. Transposition rate is represented by δ, and excision rate by ϵ.

Ψ_i is an important term because it defines any fitness effect of the transposable element. For example, if $\Psi_i<\Psi_0$ for all $i>0$, then the transposon is deleterious. In more familiar population genetics terminology, the selective coefficient s_i of the N_i population would be defined such that $\Psi_i = \Psi_0(1 - s_i)$.

The reason equation 1 cannot be applied when $i = 0$ and $i = 1$ is because cell type N_0 has no transposons and cannot be converted to type N_1 via transposition. Additional assumptions are needed to deal with the infection process; these will be treated below.

Both models considered here—that of Sawyer and Hartl and that of Condit et al.—are developed from this simple framework. But since they asked different questions, their analyses were quite different. Condit et al. examined this system when transposon-bearing populations are rare, and Sawyer and Hartl examined equilibria. In addition, as will become evident, there are other important differences between the models, particularly with regard to how the uninfected cells N_0 get their first transposon.

8.2.2 Sawyer and Hartl—A Model for Copy Number Distribution

Sawyer and Hartl (1986) were most interested in the dynamics of the copy number of a transposon in a population already carrying the element, and what regulates copy number. Copy number should be a result of the interplay between transposition, which drives copy number up, and excision and the fitness effect of the transposon, which tend to force it down. To simplify matters, however, Sawyer and Hartl ignored excision, since its rate is usually much lower than the transposition rate (Egner and Berg 1981; Foster et al. 1981b; Kleckner 1981).

8.2.2.1 Design of the Model Although their model is fundamentally very similar to the basic one described above, it differs in a few particulars. The growth rate of each population N_i is expressed as $\Psi - \Delta_i$, where Ψ is constant and $\Delta_0 = 0$; hence, $\Psi - \Delta_i = \Psi_i$ of equation 1. The death rate caused by i transposons is given by Δ_i. Excision is ignored altogether. The problem of the infection of new cells is overcome in the simplest way possible: a single rate constant μ is defined as the infection rate.

This model leads to a series of differential equations:

$$\begin{aligned} \mathring{N}_0 &= \Psi_0 N_0 - \mu N_0 \\ \mathring{N}_1 &= \Psi_1 \mathrm{N}_1 + \mu N_0 - \delta_1 N_1 - \Delta_1 N_1 \\ \mathring{N}_i &= \Psi_i N_i + \delta_{i-1} N_{i-1} - \delta_i N_i - \Delta_i N_i. \end{aligned} \tag{2}$$

Note that absolute population size is not regulated in this model—it either grows or shrinks indefinitely. The only concern is with relative population sizes. Sawyer and Hartl also considered a model where population size is kept constant, in which the death of transposon-bearing cells is always accompanied by the creation of an equal number of transposon-free cells. This leads to only slightly different equilibria from the unregulated model given in equation 2.

The distribution of copy number of transposons is given by the relative populations of N_i. The equilibrium distribution will depend on transposition and death rates—δ_i and Δ_i—and how each changes with copy number i. If transposition is unregulated, δ increases linearly with i, and if each transposon copy causes the same decrement in fitness, then Δ increased linearly with i. On the other hand, if transposition were regulated, then δ might increase less than linearly, or even decrease with i.

Sawyer and Hartl's analysis consisted of finding equilibrium distributions of copy number for given examples of functions for δ_i and Δ_i. They found the equilibria by analyzing the system as a Markov chain, that is, each cell can be viewed as travelling a probabilistic pathway through different copy numbers. For example, a cell with one copy is converted to either one with zero, one, or two with defined probabilities, and then converted again, etc. The derivations of equilibria of Markov chains are beyond the scope of this chapter (see Sawyer and Hartl 1986).

An alternative procedure for finding equilibria would be to use the differential equations (see equation 2). N_{i+1}/N_i is constant at an equilibrium distribution of copy number, which is true when $\mathring{N}_i N_{i+1} = \mathring{N}_{i+1} N_i$. In general, this does not lead to explicit solutions for $\mathrm{N}_{i+1}/\mathrm{N}_i$, but the technique can be used for particular functions of Δ_i and δ_i. In the regulated model, where population sizes are constant at equilibrium, the distribution of copy number can be found by setting all derivatives to zero. This method can lead to explicit solutions for N_{i+1}/N_i.

8.2.2.2 Results. For the purpose of illustrating how the model works and some of its conclusions, I have calculated equilibrium copy number distributions for a couple of fitness and transposition functions used by Sawyer et al. (1987), using the equations of the model with no population regulation as given in their paper:

1. Without any fitness detriment caused by the transposon ($\Delta = 0$), and without excision, copy number can only increase. Without any regulation of transposition rate, increase continues until all cells have an infinite number of copies, or until a copy number at which transposition ceases. This is a "null hypothesis" that spawned the modeling analysis. Clearly, in nature, something limits the copy number of genes that are capable of replicating themselves within the genome. There are other circumstances that lead to unlimited build-up of copy number, for example, if $\mu > \Delta_1 + \delta_1$ and $\Delta_i < \delta_i$.
2. The simplest model considered by Sawyer and Hartl is where δ and Δ are constant and independent of copy number. This represents strong regulation of transposition rate, since δ_i does not increase at all. The equilibrium distribution of copy number is geometric, with the ratio N_{i+1}/N_i a constant for all $i \geq 1$ (Figure 8–1). This is a rather nonintuitive result: since high copy number results in no greater mortality, and transposition creates more and more copies, one might expect copy number to build up indefinitely. The reason for a geometric decline in the abundance of cell types with higher copy number can be explained as follows: The density of any cell type N_i declines due to two forces—transposition to make N_{i+1} and death—but is created by only one event—transposition in the N_{i-1} population. Thus, for the relative sizes of N_i to be at equilibrium, N_{i-1} must be greater than N_i.

   The stationary distribution generated for any particular pair of functions δ_i and Δ_i and for the parameter μ depends only on their relative, not their absolute, values. This should be intuitive. All that matters is whether or not transposition rate is fast enough to overcome a fitness cost it engenders—the relative strength of the two forces. Higher parameter values will increase the speed with which the equilibrium is reached, though.
3. A simple alternative is when transposition and mortality rates are proportional to i: then $\Delta_i = i\Delta$ and $\delta_i = i\delta$ (Δ and δ are constants), and there is no regulation of transposition rate. In this situation, the ratio N_{i+1}/N_i is not constant at equilibrium, as in the model above, but declines to an asymptote as i increases. As illustrated in Figure 8–1, it is possible to generate a bimodal distribution of copy number with these functions, with modes at $i = 0$ and some $i > 0$.
4. Sawyer et al. (1987) considered four other functions for δ_i and Δ_i: harmonic, root, inverse root, and quadratic. Since each model required one function for each of the two variable parameters, a total of 36 models could have been analyzed (six functions for each parameter). Only nine were actually tested though, those that were most reasonable biologically.

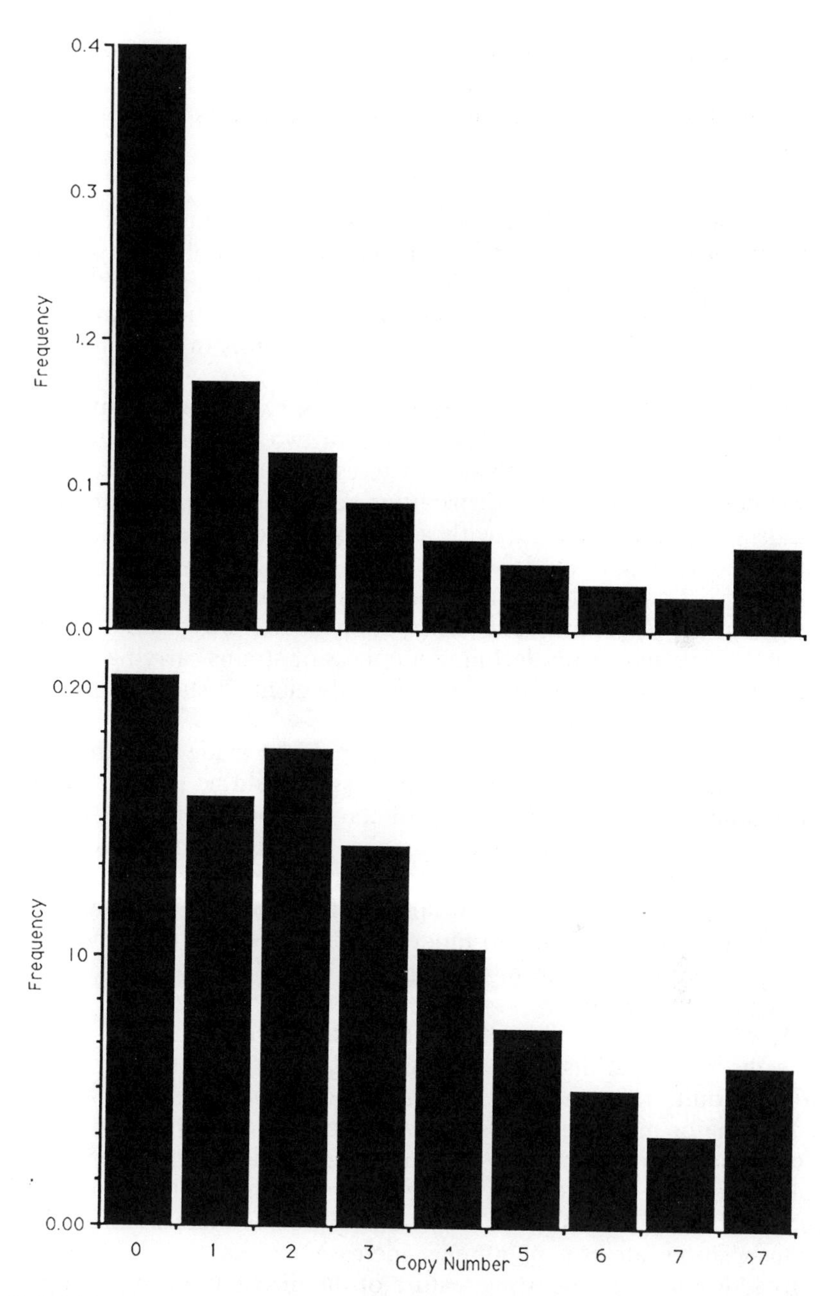

FIGURE 8–1 Theoretical distributions of transposon copy number at equilibrium, shown as the proportion of strains with various copy numbers. Top: constant model (transposition rate δ and mortality due to transposition Δ do not vary with copy number); $\delta = 10^{-6}$ and $\Delta = 10^{-4}$; infection rate $= \mu = 3 \times 10^{-5}$. Bottom: linear model; $\delta = 10^{-4}i$ and $\Delta = 6 \times 10^{-5}i$ (increasing linearly with copy number i); infection rate $\mu = 7 \times 10^{-5}$. These parameters were chosen in order to create a bimodal distribution. (The equation used to calculate the equilibrium distribution is given in Sawyer et al. 1987, p. 56; note that the variable symbols are different.)

Equilibrium distributions resulting from these models will not be given here.

8.2.2.3 Applications and Conclusions. One of the advantages of this modeling approach was that it was aimed at generating data of a sort that could be collected from natural populations. In fact, Sawyer and Hartl devised the model with the explicit intent of comparing theoretical copy number distributions with data collected on seven IS elements in 71 strains of *Escherichi coli* (Sawyer et al. 1987).

To fit models against the data, it was necessary first to estimate parameter values; the models allowed estimation of two parameters—μ/Δ and δ/Δ. A goodness-of-fit-test was then used to establish whether there was significant deviation between the distribution generated by any one model and the data for one IS element; with a total of nine models and seven IS elements, 63 tests were made.

Figure 8–2 shows the distribution of copy number for two insertion elements in the 71 strains. The distribution of IS*5* is typical. The largest copy number is zero, with declining numbers of strains carrying more and more elements. IS*3* is unusual, being the only element showing a bimodal distribution.

Unfortunately, no strong conclusions could be made from the model-fitting exercise. Nearly all of the models tested could be used to construct distributions that fit data from any of the insertion elements. The only exception was the bimodal IS*3*, for which only two models provided a reasonable fit. Since most models fit the data, it appears that the overall framework provides a useful description of the population dynamics of transposons. However, since all models fit, the approach could not distinguish the quantitative form of relationships between copy number and fitness or transposition rate.

However, one might suggest useful qualitative conclusions just by comparing the shapes of distributions from different elements. IS*4*, IS*5*, and IS*30* were quite rare—most strains had none of these elements, and very few had one or more. In contrast, IS*1* was common, with few strains uninfected and some strains with as many as 27 copies. It thus appears that regulation of copy number is much weaker in IS*1* than in IS*4*, IS*5*, or IS*30*. Either detrimental fitness effects are weaker for IS*1*, or there is less regulation of transposition rate as copy number increases.

In addition, one interesting feature of the distribution of insertion sequences in *E. coli* is that they never fit Poisson distributions, as *Drosophila* transposons do. Sawyer and Hartl (1986) argue that this is indicative of the lack of recombination between strains of *E. coli*.

8.2.2.4 Limitations of the Model. Two fundamental assumptions of this model need to be underscored in order to assess the generality of the conclusions:

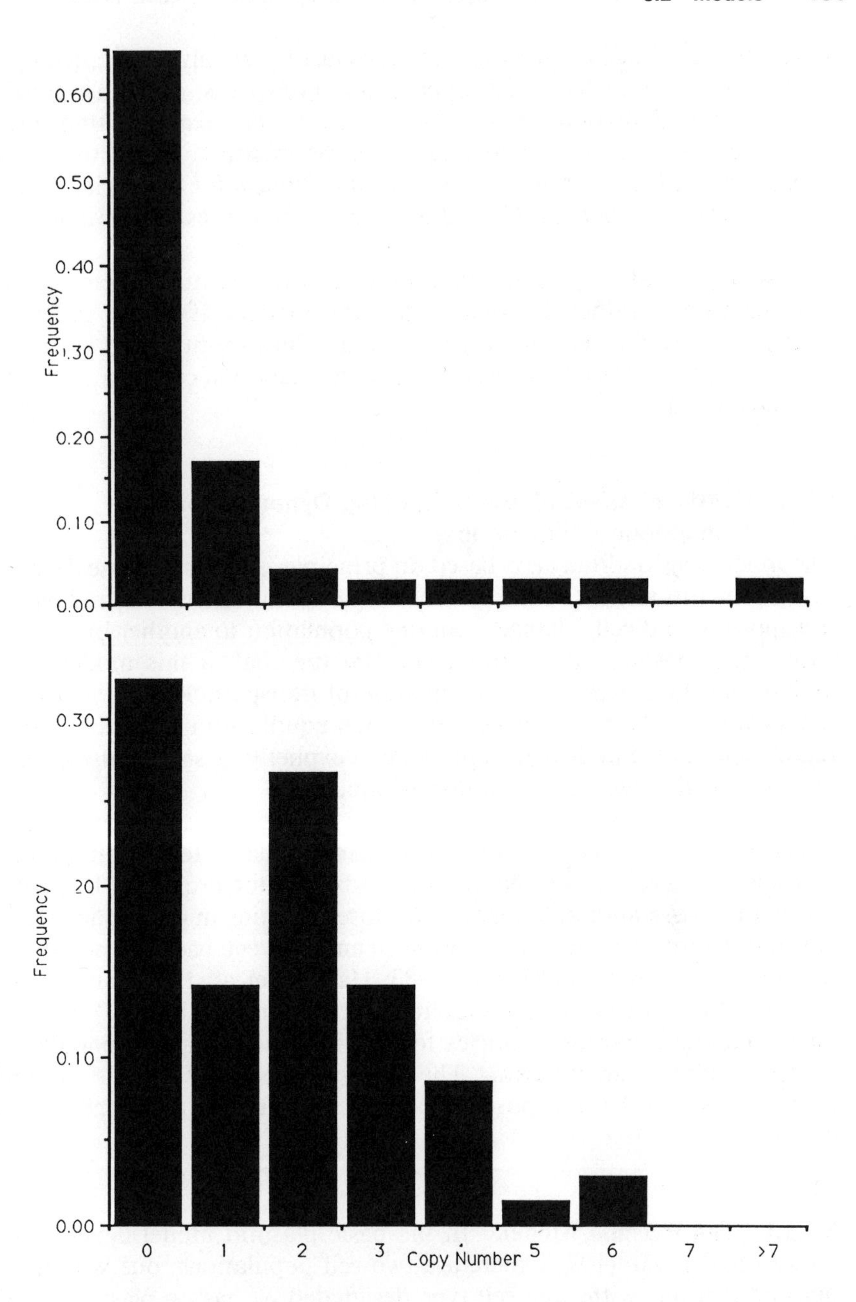

FIGURE 8–2 Actual copy number distributions for two insertion elements; frequency is the proportion of 71 strains with each copy number, as reported in Sawyer et al. (1987). Top: IS*5*. Bottom: IS*3*.

1. In any model-fitting exercise, one's conclusions are always limited by the scope of the models tested. Even if one model fits, one must consider that untested alternatives may have fit better. For example, transposons were assumed to be detrimental in terms of fitness in all the models examined by Sawyer et al. (1987). Could models for beneficial transposons (which Sawyer and Hartl did create but never tested) have provided better fits?
2. It was necessary to assume that all 71 strains examined were identical in parameter values in order to test the models. If strains differed in important ways, the statistical tests of goodness-of-fit would be invalid. The technique has no power to evaluate regulation of copy number *within* a single strain.

8.2.3 Condit et al.—A Model Describing Dynamics of Transposon Populations

The model by Condit et al. is based on principles similar to those described in the previous section. There is a mixture of populations with and without transposons, and cells change from one population to another by infection with a transposon, transposition, etc. But the goal of this model was to understand the dynamics of a population of transposable elements (within a population of bacterial cells), rather than equilibrium densities. For this reason, we wanted to develop a model that explicitly described the infection process and the dynamics of transposition.

Most transposons are unable to move between cells, so they can only infect new cells if there is some mechanism available for genome mixing. In bacteria, there are two reasonable candidates for providing this sort of "sexuality"—plasmids and phage. We chose plasmids in our model because plasmids commonly carry transposons in and between bacterial populations (Cohen 1976; Datta and Hughes 1983; Hawkey et al. 1985), and because models for their dynamics are available (Stewart and Levin 1977). Although phage *may* be important vehicles for transposons in natural populations, evidence that they are is lacking. The following model for plasmid dynamics is so intrinsic to the transposon model that it must be described in some detail.

8.2.3.1 The Plasmid Model. In the basic plasmid model developed by Stewart and Levin (1977), there are two cell populations, one without the plasmid and one with. The cell type designated N_0 has no plasmids, while type N_p does. Plasmid copy number is ignored—either a cell is infected, or it is not. Infection is assumed to happen at a rate proportional to the product of the two population densities, since plasmid transfer requires contact of N_0 and N_p cells. The rate constant of transfer is γ, defined as the probability per unit time that a single plasmid-free recipient will be infected if a single plasmid-bearing donor is present. Also, assume cells lose plasmids at a constant rate τ. Then the dynamics of the system are described by the following equations:

$$\begin{aligned}\mathring{N}_0 &= \Psi_0 N_0 - \gamma N_0 N_p + \tau N_p \\ \mathring{N}_p &= \Psi_p N_p + \gamma N_0 N_p - \tau N_p. \end{aligned} \tag{3}$$

The rate of growth of each population Ψ can be used to define a fitness cost of carrying the plasmid; define this as s, where $\Psi_p = \Psi_0(1 - s)$. If $s<0$, the plasmid confers a benefit. In addition, population regulation can be built in by making Ψ a function of a limiting resource concentration. Using chemostat models with saturation kinetics (Monod 1949) is a convenient way to model population regulation; readers should consult other works for details (Novick 1955; Dykhuizen and Hartl 1983).

A model for the dynamics of two interacting plasmids can easily be created by extending the basic model (Condit and Levin 1990): population N_1 carries plasmid 1, N_2 carries plasmid 2, and N_{12} carries both. Inclusion of the latter population is justified by reality—cells do routinely carry more than one plasmid type. A new conjugation parameter (γ_p) is needed to describe plasmid transfer into cells already carrying a plasmid, and a new segregation parameter (τ_p) for loss of one plasmid type from cells with two. Then:

$$\begin{aligned}\mathring{N}_0 &= \Psi_0 N_0 \quad - \gamma N_0(N_1 + N_2 + N_{12}) + \tau(N_1 + N_2) \\ \mathring{N}_1 &= \Psi_1 N_1 \quad + \gamma N_0 N_1 + \gamma N_0 N_{12}/2 - \tau N_1 \quad - \gamma_p N_1(N_2 + N_{12}/2) + \tau_p N_{12}/2 \\ \mathring{N}_2 &= \Psi_2 N_2 \quad + \gamma N_0 N_2 + \gamma N_0 N_{12}/2 - \tau N_2 \quad - \gamma_p N_2(N_1 + N_{12}/2) + \tau_p N_{12}/2 \\ \mathring{N}_{12} &= \Psi_{12} N_{12} \quad + \gamma_p N_{12}(N_1 + N_2)/2 + 2\gamma_p N_1 N_2 - \tau_p N_{12} \end{aligned} \tag{4}$$

These equations are written in three parts. The first section to the right of the equal sign describes cell division; the second section describes dynamics between cells with plasmids and plasmid-free cells (identical to equation 3 above); the third section describes dynamics between cells with one plasmid and cells with two.

When cell type N_{12} acts as a donor, one of the two plasmids is transferred but not both; each is equally likely to be transferred (hence the terms $\gamma_p N_1 N_{12}$ and $\gamma_p N_2 N_{12}$ are divided by two). Segregation is assumed not to favor either plasmid, so that cells N_1 and N_2 are created equally often from N_{12}. In all our plasmid models, plasmid exchanges between identical cell types are not written into the equations because they do not cause cell transitions; however, one can demonstrate their occurrence experimentally and assume that they always occur.

As above, plasmid copy number is ignored. Cell type N_{12} has some of type 1 and some of type 2; segregation of one type means that all copies of that plasmid are lost. Neglecting copy number at this stage represents a loss of reality, since a cell with four copies of type 1 and one copy of type 2 would segregate quite differently from one with the opposite arrangement.

However, tallying copy number would add burdensome detail unnecessary for a description of the fundamental features.

Under many circumstances, it is reasonable to ignore plasmid-free cells, N_0. In experimental populations with plasmids present and persisting, cells without plasmids are usually not detectable, and probably have density below 10^{-3} of the plasmid-bearing population. This simplification will be used in the model of transposable elements. Later I will discuss the impact of plasmid-free cells, since in our previous work we did incorporate them in the model.

8.2.3.2 The Transposon Model. One of the fundamental differences between our model and the earlier work by Sawyer and Hartl is that we do not consider multiple copies of transposons on chromosomes or plasmids—a piece of DNA either has transposons or it does not. This simplification is acceptable because our goal was to describe the dynamics of infection, so movement of transposons within cells was not relevant. Sawyer et al. were interested in equilibrium copy numbers, so they had to consider intra-cell transposition and the build-up of copy number.

The transposon model of Condit et al. (1988) is essentially a model of two plasmids, as described above, where one of the plasmids carries the transposon and the other does not. The only complication is that transposition onto the chromosome must be included. Let plasmids without a transposon be number 1 while those with the element are number 2. Cells with a transposon on the chromosome are designated M, while those without are still N. Ignoring plasmid-free cells, the model requires six cell types: N_1, N_2, N_{12}, M_1, M_2, and M_{12}. All but cell type N_1 carry the transposon.

Let transposition occur at a constant rate δ per unit time in the population of cells carrying the element, and excision at rate ϵ. Assuming that transposition is replicative, then the cell transitions brought about by transposition are:

$N_2 \rightarrow M_2$ (transposon on plasmid moves to chromosome);
$N_{12} \rightarrow M_{12}$ (transposon on plasmid moves to chromosome);
$N_{12} \rightarrow N_2$ (transposon on plasmid 2 moves to other plasmid, turning it into plasmid 2);
$M_1 \rightarrow M_2$ (transposon on chromosome moves to plasmid); and
$M_{12} \rightarrow M_2$ (transposon on chromosome or on plasmid moves to other plasmid).

Other transposition events lead to a build-up in copy number of the element and are ignored.

Excision causes the following cell transitions:

$N_2 \rightarrow N_1$ (loss of transposon from plasmid);

$N_{12} \rightarrow N_1$ (loss of transposon from plasmid);
$M_1 \rightarrow N_1$ (loss of transposon from chromosome);
$M_{12} \rightarrow N_{12}$ (loss of transposon from chromosome);
$M_{12} \rightarrow M_1$ (loss of transposon from plasmid);
$M_2 \rightarrow N_2$ (loss of transposon from chromosome); and
$M_2 \rightarrow M_1$ (loss of transposon from plasmid).

In this model, it has been necessary to make assumptions about plasmid copy number: the above transitions are based on the assumption of only one copy of each plasmid per cell. Cell types N_{12} and M_{12} have one of both. It would be preferable to maintain generality and not consider copy number, but it does not seem feasible without adding considerable complexity.

If transposition is conservative, the cell transitions caused by transposition must be changed. The only relevant transfer occurs between plasmid and chromosome, causing interchange of cell types: $N_{12} \leftrightarrow M_1$ and $N_2 \leftrightarrow M_1$. In addition, transposition in M_2 acts just like an excision event, with either plasmid or chromosome losing its copy of the element.

To simplify writing the equations, define a population $P = (N_{12}/2 + M_{12}/2 + N_2 + M_2)$ as the density of all cells that are capable of donating the transposon-bearing plasmid (plasmid 2), and likewise, $Q = (N_{12}/2 + M_{12}/2 + N_1 + M_1)$ for cells donating plasmid 1. γ_1 and γ_2 are rate constants for transfer of plasmids 1 and 2, respectively. Otherwise, the assumptions are the same as those behind equation 4, and the resulting set of differential equations is:

$$
\begin{aligned}
\mathring{N}_1 &= \Psi N_1 - \gamma_2 N_1 P + \tau N_{12}/2 + \epsilon(N_2 + N_{12} + M_1) \\
\mathring{N}_{12} &= \Psi \mathrm{N}_{12} + \gamma_1 N_2 Q + \gamma_2 N_1 P - \tau N_{12} + \epsilon M_{12} - 2\delta N_{12} - \epsilon N_{12} \\
\mathring{N}_2 &= \Psi N_2 - \gamma_1 N_2 Q + \tau N_{12}/2 + \delta N_{12} + \epsilon M_2 - \epsilon N_2 - \delta N_2 \\
\mathring{M}_1 &= \Psi M_1 - \gamma_2 M_1 P + \tau M_{12}/2 + \epsilon M_2 + \epsilon M_{12} - \delta M_1 - \epsilon M_1 \\
\mathring{M}_{12} &= \Psi M_{12} + \gamma_2 M_1 P + \gamma_1 M_2 Q - \tau M_{12} + \delta N_{12} - 2\delta M_{12} - 2\epsilon M_{12} \\
\mathring{M}_2 &= \Psi M_2 - \gamma_1 M_2 Q + \tau M_{12}/2 + \delta N_2 + \delta M_1 - 2\delta M_{12} - 2\epsilon M_2
\end{aligned} \tag{5}
$$

Equations for an eight-cell model that includes plasmid-free cells can be found in Condit et al. (1988).

We have found two useful ways to analyze the equations. One is to use computer simulations based on Euler approximations, varying all parameters systematically to assess the impact of each on solutions to the equations. The other is to consider the situation where a transposon has just been introduced into a new population and is rare compared to the transposon-free cells, N_1. In this circumstance, the density of cell type N_1 can be treated as constant, and products of any two of the rare cell densities can be ignored. The set of equations is then linear and first order, and can be solved analytically, providing one uses a chemostat scenario where total cell density is approximately at equilibrium. Both approaches lead to the results discussed in the next section.

8.2.3.3 Results of the Transposon Model. We have been primarily interested in the fate of transposable elements that do not carry any useful function and so are either selectively neutral or deleterious. I return to the alternative scenario, beneficial transposons, below. All these results refer to chemostat models.

A replicative transposable element that is selectively neutral will invade a population of bacterial cells, providing there are conjugative plasmids present (Figure 8–3). In the absence of transposition ($\delta = 0$), otherwise identical conditions do not lead to invasion (see Figure 8–3), and if plasmid transfer is set to zero, no invasion occurs. The same result is obtained whether the plasmids are maintained by transfer or by benefitting their host, as long as some plasmid transfer occurs. When transposons are rare, their rate of invasion is constant per capita, so that the density of transposon-bearing cells increases exponentially (appearing linear on a semi-log plot, see Figure 8–3). We symbolize this invasion rate by I_0, where the "0" refers to the zero fitness effect of the element.

The rate at which a transposon invades turns out to be rather simple to predict, at least given the seeming complexity of the model. To formulate this prediction, it is useful to define an intuitively simple and important concept—"plasmid turnover" (symbolized by ρ), which is the total probability that any given cell in a population will exchange a plasmid per unit time. In a population without any transposon and a single plasmid type (cell N_1), the number of conjugation events per unit time (symbolized as C) is $\gamma_p(N_1)^2$, since all conjugation events happen between cells of the same type (these events were not included in equations 3, 4, and 5). Plasmid turnover is C/N_1 or $\gamma_p N_1$. This is the probability per unit time that a piece of plasmid DNA will find itself associated with a novel piece of chromosomal DNA, and likewise for chromosomes.

If transposition rate is much less than plasmid turnover, then the invasion rate I_0 will exactly equal the transposition rate. Conversely, if plasmid turnover is much less than the transposition rate, then I_0 will equal the plasmid turnover rate. That is

$$\text{if } \delta \ll \rho, \text{ then } I_0 = \delta,$$
$$\text{if } \rho \ll \delta, \text{ then } I_0 = \rho. \qquad (6)$$

Since ρ and δ were defined per unit time, invasion rate here takes the same units (time^{-1}). This is a simple and intuitively appealing result. Invasion of the transposable element requires that it infects new cells, which requires two steps—transposition and plasmid transfer. Like a chemical reaction, the rate of invasion is controlled by the slower of the two steps on which it depends.

The fact that two steps are necessary for invasion may not be immediately evident. At first consideration, it might appear that once a transposon is on a plasmid, then only plasmid transfer is necessary for invasion to occur, as if a nontransposing element could invade by hitch-hiking on a

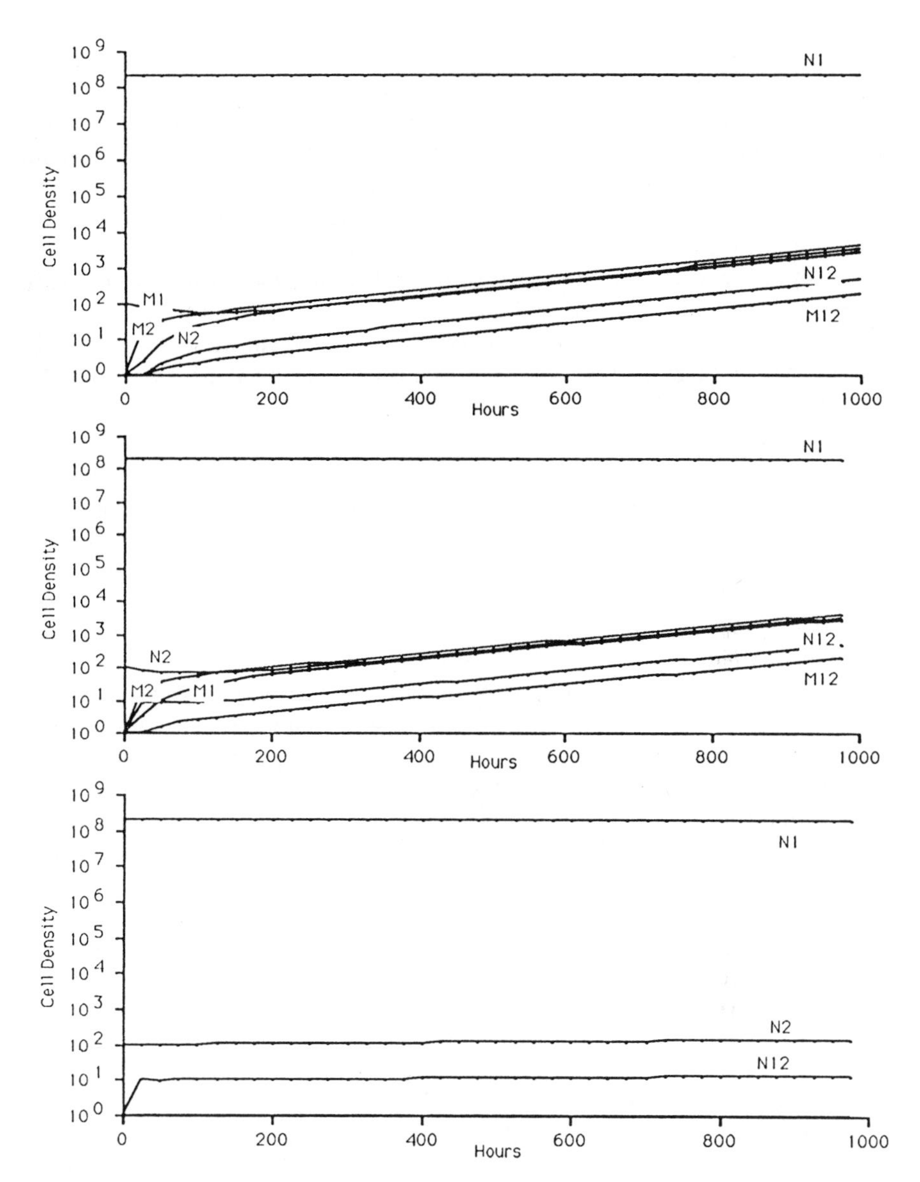

FIGURE 8–3 Simulations of changes in population density, based on equation 5 in the text. All populations maintained according to chemostat dynamics so that the total population reached an equilibrium of 1.9×10^8 cell ml^{-1}. Transposon-free cells (N_1) are at equilibrium prior to introduction of transposons at density of 10^2 cell ml^{-1}. In all three runs, $\gamma_1 = \gamma_2 = 10^{-10}$ ml cell^{-1} hr^{-1}; $\tau = 0.45$ hr^{-1}; $\epsilon = 0$ hr^{-1}; and there are no differences in fitness among the six cell types. Top: Transposon is introduced on the chromosome in cell type M_1, and transposition rate $\delta = 10^{-2}$. This value for δ is unrealistically high just to illustrate the invasion process; if δ were given a more realistic value (10^{-5} or lower), the increase would be visible only after about 10^6 hours (more than 100 years). Middle: Identical parameters, but the transposon is introduced on the plasmid in cell type N_2. Bottom: As above, but the transposition rate is set to zero.

plasmid. The reason this is wrong is that the plasmid population is assumed to be at equilibrium prior to the appearance of the transposon. Therefore, every plasmid transfer event, which adds one plasmid copy, must be matched by a segregation event, which eliminates one copy; if this were not true, then the number of plasmids would not be at equilibrium. If transposition never occurred, the transposon-bearing plasmid could not increase in abundance, since it must segregate as often as it transfers. But by transposing to the chromosome when arriving in a new cell, the element is protected against plasmid segregation. Thus transposition is necessary for the element to gain in frequency.

This conclusion, that invasion rate is controlled by the slower of two critical steps, provides a basic illustration of the functioning of the model—how a transposable element will behave in a model system of plasmids. Several other conclusions can be drawn by extending the analysis, and are justified further in Condit et al. (1988):

1. A conservative transposon that is selectively neutral or deleterious can never invade.
2. Even if a cell carrying a transposon suffers a fitness cost, it may still invade, providing the cost is not too high. Define the selective disadvantage as s, where $\Psi_t = \Psi_1(1 - s)$, Ψ_1 is the growth rate of the cell population without transposons, and Ψ_t is the growth rate for all cell types with the transposon. Then a simple conclusion can be drawn regarding the threshold level for s: if a transposon with no fitness cost has invasion rate I_0, then the same element will still invade with any fitness burden $s<I_0$. In fact,

$$I_s = I_0 - s. \tag{7}$$

(Actually, it is necessary to match units in order to achieve equality. Invasion rate has been defined per unit time interval, and the s defined here has units of "per generation," so a correction for generation time is necessary; see Condit et al. 1988.)
3. If plasmid-free cells are added to the population, then all of the above conclusions hold, providing the term for plasmid turnover, ρ, is adjusted. When plasmid-free cells, N_0, are available, $\rho = [\gamma N_0 N_1 + \gamma_p (N_1)^2]/(N_0 + N_1)$. As above, this is the total number of conjugation events divided by the total number of cells.

In addition, I have repeated many of the analyses of transposon invasion using an alternative to the chemostat approach described by Condit et al. (1988). In the alternative, a small population of cells is provided with a large amount of resource at time zero, grows exponentially until the resource is depleted, and a small fraction of the cells are then transferred to fresh resource. This is known as serial transfer and is a common experimental tool for microbiologists (Atwood et al. 1951). Unlike a chemostat, bacterial populations under serial transfer are not at equilibrium, which is not conducive to mathematical analysis of the differential equations. Nevertheless, when I repeat simulations in a serial transfer regime, results are qualitatively similar to those in a chemostat. Replicative transposons still invade, providing cost is not high, and conservative transposons do not. The exact equalities in equations 6 and 7 no longer hold, however.

As with Sawyer and Hartl's, our model was explicitly designed for direct testing with data from real populations. Laboratory populations of *E. coli*, the plasmid R100, and transposons Tn*3* and Tn*5* could be arranged to mimic the invasion-when-rare scenario illustrated in Figure 8–3 (Condit 1990). We found that the models quite successfully predicted changes in cell density caused by plasmid transfer, but because transposition rates were so low, we were unable to observe changes in frequency caused by transposition. That is, although the experimental data matched theoretical predictions, the tests of the model were not robust because transposition rate was never high enough to play a role in population dynamics.

8.2.3.4 Preliminary Models for the Dynamics of Beneficial Transposons. Initially, we felt that a situation where a transposon raises the fitness of its host was not particularly interesting. After all, any gene that raises host fitness will invade a population when rare, whether it transposes are not. If the fitness difference s (now s favors cells *with* a transposon) is much greater than I_0, then transposition would play a trivial role—invasion would be largely caused by differential fitness. On the other hand, if s is much lower than I_0, then the transposon is essentially neutral, and we return to the conclusions above.

Recently, however, we have reconsidered scenarios with beneficial transposons and decided they are not so trivial, providing we change the structure of the model. Much of what follows is based on the model presented in Condit and Levin (1990) whose purpose was different, but the structure of this model has relevance for modeling transposable elements.

The problem with my initial thinking about beneficial transposons was that it considered only whether genes within the transposable element were beneficial, not how the *act of transposition* itself could have positive fitness. If a transposon carries useful genes, then of course one anticipates its spread into a population, but once fixed, the ability to transpose should be lost while the genes within the element are maintained.

What is necessary for transposability to be maintained is a situation where new gene combinations have a selective advantage. We modeled one such scenario: two plasmids co-occur in a population of bacteria, each carrying a gene that can be beneficial to the bacteria; both of the genes are contained within transposable elements. In addition, the two plasmids are "incompatible," that is, single bacterial cells cannot maintain both for long (Novick 1987). This is modeled by having a high segregation rate, τ_p, among the cells N_{12}. If selection for both genes occurs simultaneously, so that cell type N_{12} has a selective advantage over N_1 and N_2, then a population of cells in which one of the transposons has moved between plasmids (or onto the chromosome) will invade. The reason is the instability of the plasmids. Cell type N_{12} will have the highest growth rate, but it will also lose plasmids at a high rate, creating progeny with low fitness. Cells with both transposons (and hence both useful genes) on the same plasmid do not suffer from this instability. We proved using simulations that the invasion rate of the new cell type is equal to the segregation rate of the two plasmids.

This situation leads to a selective advantage for *transposability* based on the unstable inheritance of separate pieces of DNA in the same cell. However, once the transposons find themselves in a new arrangement, the selective advantage to transposition is lost. The benefit to the transposon has only been temporary, unless new circumstances constantly adjust fitnesses and continually favor gene rearrangements. The problem of maintaining transposability thus appears to be the same as the general problem of the evolution of recombination and sex (Feldman et al. 1980). Clearly, further theoretical work is needed in this area.

8.2.3.5 Limitations of the Models Describing Dynamics. Some of the assumptions underlying the model have been mentioned already, but certain ones deserve particular attention before assessing the theoretical conclusions:

1. Populations were assumed to be at equilibrium before the transposon was introduced. If a plasmid carrying a transposon were introduced into a population formerly without that plasmid, the transposon could be swept to fixation by plasmid transfer alone. Perhaps nonequilibrium systems, with plasmids frequently invading and going extinct, offer a better avenue for invasion of selectively neutral transposons. In Condit et al. (1988), we argue that transposons are no more likely to invade nonequilibrium populations than equilibrium populations, but the idea has not been tested rigorously.
2. We did not consider whether transposons affect plasmids in any way. If a transposon is deleterious to its host, but somehow benefits the plasmid it is on (relative to other plasmids), its ability to invade would improve, but we have not quantitatively assessed this situation.
3. Our models explicitly model populations in liquid culture. Although I do not anticipate that surface populations would exhibit dramatically different behavior, the possibility should be considered further.

8.3 CONCLUSIONS

8.3.1 Explanations for the Invasion and Maintenance of Transposons

The models of Condit et al. (1988) and Condit and Levin (1990) represent standard population genetic approaches. We sought to understand the basic dynamics of genes that are able to transpose between pieces of DNA in simple model systems. Obviously, the model systems do not reflect reality exactly, but it is necessary to understand dynamics in simple systems before considering more complicated real populations. What do the models show?

It is plausible for transposons to be "parasitic DNA." This had been taken as a given in bacterial populations (Campbell 1981), but never proven. However, as Condit et al. (1988) argue, there are several reasons to doubt whether the parasitic DNA hypothesis can be generally applicable to bacterial transposons. The problem is twofold: first, bacteria are almost asexual, since the rate of plasmid turnover is usually quite low (Levin et al. 1979; Freter et al. 1983); second, transposition rate is extremely low for nearly all bacterial transposons (Foster et al. 1981a; Peterson et al. 1982; Iida et al. 1983; Meyer et al. 1983; Schmidt and Klopfer-Kaul 1984). Both rates are critical for invasion as a parasite, and since both are so low, it must be a rare event for a transposon to establish itself in a new bacterial strain without benefitting its host.

The above statements about magnitudes of transposition and plasmid turnover are based on very little data (at least for natural populations), and they should be considered provisional. Also, one should consider transduction as an alternative gene transfer mechanism, so our conclusion in the previous paragraph should be debated. Still, it does not seem plausible that transposons have become so widespread as parasitic DNA.

In contrast, Sawyer et al. (1987) concluded that their models suggest "... a moderate to strong detrimental effect of copy number [of transposable elements] on fitness...." for most insertion elements. But their group did not examine models for beneficial transposons, and as they readily acknowledge, their conclusions were not strong. Clearly, more data about the dynamics of transposable elements in natural bacterial populations are needed to establish whether some are acting as parasitic DNA.

Models that demonstrate how transposition can be selectively advantageous are available (Condit and Levin 1990), but they do not provide a compelling general argument for the maintenance of transposability. Although transposition can have a transient advantage, the theory ends there. How can the ability to move to new sites be continuously selected for? More theoretical work is needed in this area.

8.3.2 Applications of Models to Risk Assessment

On the basis of our study of the dynamics of the spread of transposons, and information about how low the critical rate parameters are, we predict that engineered genes released into bacterial communities will never become established in novel populations *as long as* they do not raise fitness of any strain they inhabit. Genes *will* transfer into new cells, but without a selective advantage these transfer events will be rare and isolated and will not lead to fixation of the gene at a new site.

Nevertheless, our models for beneficial transposons show that if a transposition event creates a new cell that has higher fitness than its progenitors, then a new strain carrying a rearrangement can become fixed very rapidly. This raises an important caution for risk assessment. No matter how rare genetic rearrangements might be, they can readily lead to new strains if the selective conditions are right.

It is debatable whether general predictions like this are useful for risk assessment. One should probably make assessments on a case-by-case basis, and it is possible to apply models to individual cases. In Condit (1988), I describe how mathematical models of bacterial populations might be used to make predictions about the fate of engineered bacteria released into the environment. Our models for the dynamics of transposon populations could be used to make such predictions. Given parameters for transposition, plasmid transfer, and selective coefficients, one could try to predict the fate of a particular engineered gene released into a particular bacterial population.

Although this might be a goal for modeling work in risk assessment, quantitative predictions based on detailed models are unlikely to be accurate enough as a basis for important decisions, at least in the near future. Instead, I see models as continuing to be a basic tool for evaluating the processes that control the abundance and distribution of transposable elements in natural populations of bacteria. This basic knowledge is crucial for making decisions about manipulating bacterial populations.

REFERENCES

Atwood, K.C., Schneider, L.K., and Ryan, F.J. (1951) *Cold Spring Harbor Symposia on Quantitative Biology* 16, 345–355.

Berg, D.E. (1977) in *DNA Insertion Elements, Plasmids, and Episomes* (Bukhari, A.I., Shapiro, J.A., and Adhya, S.L., eds.), pp. 185–203, Cold Spring Harbor Laboratory, New York.

Bottstein, D., and Kleckner, N. (1977) in *DNA Insertion Elements, Plasmids, and Episomes* (Bukhari, A.I., Shapiro, J.A., and Adhya, S.L., eds.), pp. 185–203, Cold Spring Harbor Laboratory, New York.

Campbell, A. (1981) *Ann. Rev. Microbiol.* 35, 55–83.

Campbell, A., Berg, D., Botstein, D., et al. (1977) in *DNA Insertion Elements, Plasmids, and Episomes* (Bukhari, A.I., Shapiro, J.A., and Adhya, S.L., eds.), pp. 15–22, Cold Spring Harbor Laboratory, New York.

Chao, L., Vargas, C., Spear, B.B., and Cox, E.C. (1983) *Nature* 303, 633–635.

Charlesworth, B., and Charlesworth, D. (1983) *Genet. Res. Camb.* 42, 1–27.

Cohen, S.N. (1976) *Nature* 263, 731–738.

Condit, R. (1988) in *Biotechnology: A Risk Assessment Framework* (Mittleman, A., ed.), pp. 88–157, prepared for the U.S. Environmental Protection Agency, Washington, D.C., by Technical Resources Inc., Washington, D.C.
Condit, R. (1990) *Evolution* 44, 347–359.
Condit, R., and Levin, B.R. (1990) *Amer. Natur.* 135, 573–596.
Condit, R., Stewart, F.M., and Levin, B.R. (1988) *Amer. Natur.* 132, 129–147.
Datta, N., and Hughes, V.N. (1983) *Nature* 306, 616–617.
Dykhuizen, D.E., and Hartl, D.L. (1983) *Microbiol. Rev.* 47, 150–168.
Doolittle, W.F., and Sapienza, C. (1980) *Nature* 284, 601–603.
Egner, C., and Berg, D.E. (1981) *Proc. Nat. Acad. Sci.* 78, 459–463.
Feldman, M.W., Christiansen, F.B., and Brooks, L.D. (1980) *Proc Nat. Acad. Sci.* 77, 4838–4841.
Foster, T.J., Davis, M.A., Roberts, D.E., Takeshita, K., and Kleckner, N. (1981a) *Cell* 23, 201–213.
Foster, T.J., Lundblad, V., Hanley-Way, S., Halling, S.M., and Kleckner, N. (1981b) *Cell* 23, 215–227.
Freter, R., Freter, R.F., and Brickner, H. (1983) *Infect. Immun.* 39, 60–84.
Gawron-Burke, C., and Clewell, D.B. (1982) *Nature* 300, 1–3.
Ginzburg, L.R., Bingham, P.M., and Yoo, S. (1984) *Genetics* 107, 331–341.
Hawkey, P.M., Bennett, P.M., and Hawkey, C.A. (1985) *J. Gen. Microbiol.* 131, 927–933.
Hickey, D.A. (1982) *Genetics* 101, 519–531.
Iida, S., Meyer, J., and Arber, W. (1983) in *Mobile Genetic Elements* (Shapiro, J.A., eds.), pp. 159–221, Academic Press, New York.
Klaer, R., Pfeifer, D., and Starlinger, P. (1980) *Molec. Gen. Genet.* 178, 281–284.
Kleckner, N. (1981) *Ann. Rev. Genet.* 15, 341–404.
Levin, B.R., Stewart, F.M., and Rice, V.A. (1979) *Plasmid*, 2, 247–260.
Levy, S.B., and Marshall, B.M. (1988) in *The Release of Genetically Engineered Micro-organisms* (Sussman, M., Collins, C.H., Skinner, F.A., and Stewart-Tull, D.E., eds.), pp. 61–76, Academic Press, London.
Meyer, J.F., Nies, B.A., and Wiedemann, B. (1983) *J. Bact.* 155, 755–760.
Monod, J. (1949) *Ann. Rev. Microbiol.* 2, 371–394.
Novick, A. (1955) *Ann. Rev. Microbiol.* 9, 97–110.
Novick, R.P. (1987) *Microbiol. Rev.* 51, 381–395.
Orgel, L.E., and Crick, F.H. (1980) *Nature* 284, 604–607.
Peterson, B.C., Hashimoto, H., and Rownd, R.H. (1982) *J. Bact.* 151, 1086–1094.
Read, H., Das Sarma, S., and Jaskunas, S.R. (1980) *Proc. Nat. Acad. Sci.* 77, 2514–2518.
Reynolds, A.E., Felton, J., and Wright, A. (1981) *Nature* 293, 625–629.
Sawyer, S., and Hartl, D. (1986) *Theor. Pop. Biol.* 30, 1–16.
Sawyer, S.A., Dykhuisen, D.E., Dubose, R.F., et al. (1987) *Genetics* 115, 51–63.
Schmidt, F., and Klopfer-Kaul, I. (1984) *Mol. Gen. Genet.* 197, 109–119.
Stewart, F.M., and Levin, B.R. (1977) *Genetics* 87, 209–228.
Tanaka, M., Yakamoto, T., and Sawai, T. (1983) *J. Bact.* 153, 1432–1438.

CHAPTER

9

Quantifying Fitness and Gene Stability in Microorganisms

Richard E. Lenski

Fitness represents the combined effects of all other phenotypic properties on the capacity for survival and reproduction by a particular genotype in a particular environment. Many genetically engineered microorganisms will be more fit than their wild-type counterparts under environmental conditions corresponding to their intended biotechnological application. For example, microorganisms engineered to degrade some environmental toxin may increase in frequency after their introduction into the contaminated environment if they can use that toxin as a growth substrate. The efficacy of the intended application may often be enhanced by the increased frequency of the engineered genotype.

On the other hand, genetically engineered microorganisms may often be *less* fit than their wild-type counterparts under other environmental conditions, owing to the "excess baggage" associated with carriage and expression of the recombinant genes (Regal 1986; Lenski and Nguyen 1988). If so, then microorganisms engineered for biodegradative functions, for example, will decline in frequency, and may eventually be lost (but see Chapter

I thank Toai Nguyen and Michael Rose for helpful discussions and comments. Preparation of this chapter was supported by cooperative research grants from the Gulf Breeze Laboratory of the U.S. Environmental Protection Agency (CR-815380 and CR-813401).

10) after they have detoxified a site or if they are transported to an uncontaminated site.

This "excess baggage" hypothesis has been frequently invoked as an argument for the safety of deliberate release of genetically engineered microorganisms, because it implies that unintended spread and any consequent adverse effects are unlikely (Brill 1985; Davis 1987). It is my opinion that this hypothesis will hold true in many cases, perhaps most, but that there may also be some exceptions (e.g., Bouma and Lenski 1988). Alternative points of view on this subject can be found elsewhere (Brill 1985; Colwell et al. 1985; Sharples 1987; Davis 1987; Lenski and Nguyen 1988; Regal 1988), and further discussion of this hypothesis is not the purpose of this chapter.

However, reports sponsored by the Ecological Society of America (Tiedje et al. 1989) and by the U.S. National Research Council (Committee on Scientific Evaluation of the Introduction of Genetically Modified Microorganisms and Plants into the Environment 1989) have emphasized the importance of confining a modified organism after its introduction into the environment in order to minimize the duration and scale of any potentially adverse ecological effects. Both reports also provide frameworks for assessing ecological risk or uncertainty, wherein the fitness of a genetically modified organism relative to its unmodified counterpart in the appropriate environment is an important criterion for evaluating confinement. In this chapter, I will present methods that can be employed to measure selection (i.e., a difference in fitness) that arises from the carriage and expression of engineered genes by microorganisms.

Selection is also important because its effects may be confounded with losses of an engineered gene due to the infidelity of replication or transmission of the recombinant DNA. Imagine, for example, that the level of expression of some engineered function diminishes with time in a population of recombinant microorganisms, thereby hindering the efficacy of the intended biotechnological application. If one assumes that this instability is due to genetic infidelity, and not to selection, then it would be reasonable to increase the number of copies of the recombinant gene in each individual. If, however, the primary cause of instability is selection against expression of the intended function, then an increase in copy number (and a concomitant increase in the level of expression) may actually aggravate the instability of the recombinant gene at the level of the population. In this chapter, I will also present methods that can be used to distinguish the effects of selection from the effects of genetic infidelity.

9.1 GENERAL PRINCIPLES

I will use the term segregation to refer to any losses of an engineered gene that are due to the infidelity of replication or transmission, including mutation. I will assume that these losses are irreversible, although models incorporating reversible events (e.g., back-mutation) can be readily developed. I will use the term selection to refer to changes in the relative abundance of two clones that result from genotypic properties that cause them to differ in their capacity for survival or reproduction.

I use the terms wild-type, parental, and segregant to refer to clones that lack the recombinant gene or genes. I assume that wild-type and engineered clones are isogenic except with respect to the recombinant gene or genes of direct interest. I will discuss exceptions to this assumption, both deliberate and unintentional, in Sections 9.3.1.3 and 9.4.1.

Dykhuizen and Hartl (1983) and Kubitschek (1970) provide excellent discussions of selection and mutation in microbial populations, which the reader is encouraged to consult.

9.1.1 Effect of Selection

Let p be the frequency of an engineered clone, and let $q = 1 - p$ be the frequency of the parental clone. The change in the frequency of the engineered clone is governed by the following differential equation (Kimura and Ohta 1971):

$$dp \,/\, dt = -s\, p\, q = -s\, p\, (1 - p) \tag{1}$$

where s is the selection coefficient, or the difference in fitness between the two clones, which we are assuming to be constant. If $s>0$, then selection favors the parental clone; if $s<0$, then selection favors the engineered clone. It is possible to integrate equation 1 and thereby obtain an expression for the frequency of the engineered clone:

$$p_t = \frac{p_0}{(1 - p_0)\, e^{st} + p_0} \tag{2}$$

where p_0 is the frequency of the engineered clone at time zero, and p_t is the frequency of that clone at some later time t. Equation 2 can be linearized, but one must first convert the *frequencies* of the two clones, p and q, to the *ratio* of one clone to the other: $R = p \,/\, q = p \,/\, (1 - p)$. With the relative abundance of the clones thus expressed, a natural logarithmic transformation yields (Dykhuizen and Hartl 1983):

$$\ln R_t = \ln R_0 - s\, t \tag{3}$$

where R_0 is the ratio of the engineered clone to the parental clone at time zero and R_t is that ratio at some later time t. In Section 9.3.1, I will discuss the use of linear regression to estimate the selection coefficient from empirical data.

9.1.2 Effect of Segregation

Segregation is typically an exponential decay process, whereby individuals with the engineered genotype are converted to segregants at a constant rate. The rate of change in the frequency of the engineered genotype is therefore described by the following differential equation:

$$dp / dt = - u p \tag{4}$$

where p is the frequency of the engineered genotype and u is the segregation rate. Integrating equation 4 yields:

$$p_t = \frac{p_0}{e^{ut}} \tag{5}$$

where p_0 is the frequency of the engineered genotype at time zero, and p_t is that frequency at some later time t. Equation 5 can be linearized by a natural logarithmic transformation of the frequencies of the engineered genotype:

$$\ln p_t = \ln p_0 - u t \tag{6}$$

In contrast, recall that under the model of selection, genotypic frequencies must first be converted to ratios before a logarithmic transformation linearizes the dynamics.

In Figure 9–1, the dynamics of an engineered genotype subject to *segregation* of the recombinant gene (dashed line) are contrasted with the dynamics of an engineered genotype subject to *selection* against carriage of the recombinant gene (solid line). With segregation, the rate of decline in the frequency of the engineered genotype is greatest when that frequency is near one, and it declines continuously as the frequency approaches zero. In contrast, the rate of change under selection is greatest when the frequency of

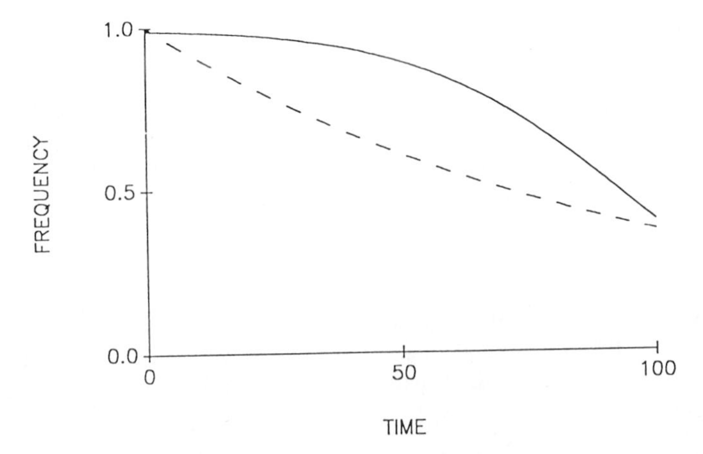

FIGURE 9–1 The dynamics of loss of the engineered genotype as a function of time. Solid line: model of loss of engineered genotype by selection (equation 1). Selection coefficient s is 0.05 per unit time; initial frequency p_0 is 0.99. Dashed line: exponential decay model of loss of engineered genotype by segregation (equation 4). Segregation rate u is 0.01 per unit time; initial frequency p_0 is 1.

the engineered clone is near 0.5, and is lowest when that frequency is near one (or near zero).

9.1.3 Combined Effects of Segregation and Selection

The joint effects of segregation and selection are described by the following differential equation:

$$dp \,/\, dt = -\,s\,p\,q - u\,p = -\,s\,p\,(1 - p) - u\,p \tag{7}$$

where p is the frequency of the engineered genotype, u is the rate of segregation, and s is the selection coefficient. I will now consider the case in which segregants are more fit (i.e., $s>0$), so that both segregation and selection drive the frequency of the engineered genotype downward.

It is possible to integrate equation 7 and thereby obtain an expression for the frequency of the engineered genotype remaining at time t (Charles et al. 1985; Lenski and Bouma 1987):

$$p_t = \frac{p_0\,(u + s)}{[u + s\,(1 - p_0)]\,e^{(u+s)t} + s\,p_0} \tag{8}$$

There is no transformation that can linearize equation 8 over the entire parameter space and range of frequencies (Cooper et al. 1987). However, one can use nonlinear regression to estimate simultaneously the segregation rate and the selection coefficient (Lenski and Bouma 1987), as I will show in Section 9.3.2.

9.2 METHODOLOGICAL ISSUES

9.2.1 Selection

Koch (1981) clearly presents various methods for measuring the growth rate of a microbial population in pure culture. In principle, one could determine the difference in fitness of two genotypes simply by measuring the growth rates of each in isolation, and then calculating the difference. In practice, however, it is often difficult or even impossible to estimate fitness differences in this way. For example, imagine two clones, each grown separately in a chemostat. Provided that each clone is capable of growing at a rate sufficient to offset washout, then each clone will become established. At equilibrium, each clone has a growth rate equal to the rate of flow through the culture vessel. Therefore, if both clones can become established, then they must have the same equilibrium growth rates when grown in isolation. Yet one clone may be more efficient than the other at exploiting some limiting resource, or one clone may produce a substance that inhibits the growth of the other. By reducing the level of resources, by the production of growth inhibitors, and so on, populations modify their environment. Thus, to compare the growth rates of two populations grown in isolation is to ignore the effects that each population would have on the other, as mediated by their effects on the environment.

There are other limitations to traditional measurements of growth rate obtained for pure populations. Most measurements of growth rate are made on exponentially growing populations. Starting from even a single cell, there are only a relatively few generations of exponential growth before nutrient limitation or metabolic by-products limit growth, and still fewer generations during which one can accurately quantify a population's density from measurements of light scattering using a spectrophotometer. If the difference in fitness between two clones is small, then it is unlikely that a comparison of exponential growth rates will reveal any significant difference. But small differences in fitness, when compounded over hundreds and thousands of generations, may be quite important. Also, exponential growth rates are typically measured only under conditions of superabundant resources, a situation that is ephemeral in nature.

Differences in fitness inferred from measurements of growth rates for pure populations may require a high degree of replication, especially if sampling error or variation among replicates is large. Direct measurements of selection, on the other hand, invariably have an internal control; whether a particular sample density is high or low, or whether a particular replicate is nutrient rich or poor, the same is true for both clones. This internal control makes direct measurements of differences in fitness more accurate than those calculated indirectly from separate measurements of growth rate for two populations, each of which inevitably contains its own independent error of estimation.

Dykhuizen and Hartl (1983) provide an excellent review of studies in which measurements of selection have played a central role; their review focuses on chemostat studies that were undertaken to test a variety of evolutionary hypotheses. Many of the studies reviewed by Dykhuizen and Hartl relate directly to the issue of whether or not there is measurable selection associated with expression of unused functions or associated with carriage of accessory genetic elements, including plasmids, prophages, and transposons (see also Cooper et al. 1987; Bouma and Lenski 1988; Lenski and Nguyen 1988; Nguyen et al. 1989).

It should be emphasized that measurements of selection can be made not only in highly simplified laboratory systems, such as chemostats, but also in more complex natural or semi-natural systems, such as microcosms (Pritchard and Bourquin 1984); all that is required is the ability to monitor the relative abundance of two clones sharing a common environment. Whether the total number of individuals in a population is increasing, constant, or decreasing, the relative frequencies of genotypes may be changing systematically as the result of differences in their fitness.

9.2.2 Gene Instability

The instability of a gene is often described in terms of the frequency of segregants in a population after a specified period of time. Implicit in this simple description is the assumption that the dynamics of gene loss are governed solely by an exponential decay process, so that the segregation rate can be calculated from equation 6. However, the dynamics of instability are altogether different if selection is acting in concert with segregation, as shown in Section 9.1.3. In fact, any number of combinations of segregation rate and selection coefficient can be chosen to pass through a particular pair of initial and final frequencies. Thus, it is essential to have a series of frequencies and to analyze the dynamics of instability using mathematical models that allow one to determine the effects of both segregation and selection.

Several studies have attempted to distinguish between the effects of segregation (including mutation) and selection on the dynamics of loss of some genotype from a population (Nordstrom et al. 1980; Charles et al. 1985; Boe et al. 1987; see also additional citations in Cooper et al. 1987; Lenski and Bouma 1987). In all of these studies, the frequency of individuals possessing the gene of interest was determined at several points in time, thereby permitting inferences regarding the importance of both segregation and selection.

While the authors of the papers cited above clearly recognized the significance of selection as well as segregation for the dynamics of instability, they did not present explicit procedures for estimating both the segregation rate and the selection coefficient from empirical data. Two recent papers (Cooper et al. 1987; Lenski and Bouma 1987) have filled this void, and although the details of their procedures differ, both permit an investigator to estimate simultaneously the segregation rate and the selection coefficient from a series of genotypic frequencies. Both procedures were first applied to experimental studies of bacterial plasmids, but they are equally applicable to any recombinant gene, whether chromosomal or extrachromosomal.

Cooper et al. (1987) identify alternative transformations that linearize genotype frequencies, at least over a portion of their range; the success of a particular transformation depends upon the relative magnitudes of the selection coefficient and the segregation rate. Standard linear regression techniques can be used to estimate the relevant parameters, but some of the experimental data invariably must be excluded from the analysis because frequencies fall outside the range over which a particular transformation successfully linearizes the dynamics.

Lenski and Bouma (1987) use nonlinear regression to estimate the segregation rate and selection coefficient; while nonlinear regression procedures are less familiar, they are widely available in statistical packages. One advantage of this approach is that it does not require identification of an appropriate transformation to linearize the dynamics, but instead utilizes a single mathematical expression to describe the dynamics expected for all parameter values and over the entire range of genotype frequencies. Another advantage of this approach is that it uses all of the experimental data. For these reasons, and because of my greater familiarity with my own work, I will present the approach used by Lenski and Bouma in Section 9.3.2.

9.3 SPECIFIC METHODS FOR ESTIMATING PARAMETERS

9.3.1 Measuring Selection

The basic logic of an experiment designed to estimate the difference in fitness between two clones is very simple. Two clones are mixed together in some initial ratio (usually about 1:1) in any environment of interest. At subsequent points in time, additional samples are obtained and the ratio of the two clones in each sample is determined. If the ratios shift upward or downward in a systematic fashion, then one can infer that one clone is more fit than the other in that particular environment; if the ratios remain essentially unchanged with time, then one must conclude that the two clones are equally fit, at least within the limits of one's statistical resolution.

The natural logarithm of the ratio of two clones is expected to exhibit a linear relationship with time (see equation 3). Therefore, the selection coefficient, s, can be estimated by the slope of the regression of the natural logarithms of these ratios against time and has units of inverse time. These units may be universal units, like hours, or they may require additional knowledge of the particular experimental system, like generations.

Data from one experiment are presented in Table 9–1. In that experiment, a clone of *Escherichia coli* resistant to the bacteriophage T4 was grown in mixed culture with its T4-sensitive progenitor (see Lenski and Levin 1985, for further details). A steady decline in the ratio of resistant to sensitive cells was observed, and a linear regression of the natural logarithm of the sample ratios against time yields a selection coefficient (here, equal to the negative value of the slope) of 0.15 per hr. The correlation coefficient gives a measure of the fit of the data to the regression line and is equal to 0.99 for this data set, which is significant at $p<0.01$ (Rohlf and Sokal 1981). Hence, one can conclude that there was selection against the T4-resistant clone.

In the sections that follow, I discuss the conditioning of clones prior to measurements of selection; several methods for enumerating clones, in-

TABLE 9–1 Effect of Selection on the Ratio of T4-Resistant to T4-Sensitive Genotypes of *Escherichia coli*

Time of Sample (hr)	R[1]	*ln* R[1]
0	1.6082	0.475
21.0	0.2972	−1.213
45.5	0.0040	−5.521
68.5	0.0001	−9.210

[1] *R*, ratio of resistant to sensitive cells; ln *R*, natural logarithm of *R*.
Source: Lenski and Levin 1985.

cluding the use of neutral genetic markers; and alternative approaches to making statistical inferences about estimates of selection coefficients.

9.3.1.1 Preconditioning. One important consideration for measuring the difference in fitness between two clones is the conditioning that both clones receive prior to their introduction into a common environment. In general, it is desirable to meet the following two conditions: (1) each clone should be preconditioned in the same environment as the other; and (2) the environment used for preconditioning should be the same as the environment in which the selection coefficient is to be estimated.

If (1) is not met, then subsequent changes in the ratio of two clones may reflect their physiological states rather than their genetic composition. For example, imagine that clone A possesses a recombinant gene that enables it to utilize some novel substrate for growth, while clone B is the parental wild-type. Let us say that A has been preconditioned on the novel substrate, while B has been preconditioned on the standard medium in which the two clones will compete. When the two clones are mixed together in the standard medium, it is observed that the ratio of A to B declines significantly. Does this indicate that carriage of the recombinant gene reduces fitness in the standard medium? It may, but it may indicate instead that there is a physiological effect of switching from one medium to another that has inhibited the growth of clone A for a few generations.

If (1) is met, but (2) is not, then significant changes in the ratio of two clones do reflect genotypic differences, as their physiological states should be identical. However, failure to meet (2) may alter one's interpretation of the environment in which selection was measured. For example, imagine that clones A and B have both been engineered to be resistant to some toxin; A expresses the resistance function constitutively, while B requires induction of the resistance function by the toxin. Both clones are preconditioned in the same toxin-free medium, so (1) is met. The selection coefficient is then estimated in medium supplemented with the toxin, so (2) is not met. It is observed that the ratio of A to B increases significantly. Does this indicate that A is more fit than B in medium supplemented with toxin? Once again, the interpretation is ambiguous. The data might indicate instead that A is able to respond more rapidly to the toxic challenge by virtue of its constitutive expression of the resistance function; had the two clones been preconditioned in the presence of toxin, thereby inducing the resistance function in B, there might have been no significant selection. Thus, when (2) is violated, one cannot infer whether a difference in fitness was characteristic of the experimental medium or whether the difference resulted from the *transition* between the preconditioning and experimental media.

In some instances, it may be difficult or impossible to ensure that (1) and (2) are met. If one clone has a highly unstable genotype, then it may be necessary to maintain that clone in a selective medium during the preconditioning step. If selection is to be measured in a natural or semi-natural environment, then it may be necessary to precondition clones in the laboratory. However, it is important to be aware of (1) and (2), and to interpret one's results with appropriate caution when either of these conditions has been violated. In general, violations of (1) or (2) are less likely to affect one's interpretation when the selection coefficient is estimated from a long-running experiment, because the alternative interpretations depend upon transitions between environments. However, the reader is cautioned that there is also a potentially serious complication that arises when these experiments are carried out too long. Longer experiments increase the likelihood that additional genetic changes will arise in one or both clones, confounding interpretation of the genetic basis of any observed difference in fitness (see Section 9.4.1).

There is one other important consideration regarding the treatment of clones prior to measurements of selection. It is essential that clones be stored properly, and that experiments be conducted with clones taken directly from this storage. If this is not done, then once again additional genetic changes may occur that confound interpretation of the genetic basis of differences in fitness (see Section 9.4.1). For example, imagine that one measures the difference in fitness between two clones, one carrying some recombinant gene and the other the parental clone lacking that gene. The parental clone has been stored in a low temperature freezer ever since its acquisition. After transformation with the recombinant DNA yielded an engineered clone of possible interest, a number of that clone's properties were determined, which required that it be propagated for several weeks in the standard laboratory medium. Only after these properties were determined was the engineered clone properly stored. Let us imagine that a subsequent experiment indicated that the engineered clone was significantly more fit than the parental clone when measured in the standard laboratory medium. Does this result mean that the engineered gene enhances fitness in the standard laboratory medium? It may, but it is also likely that the additional exposure of the engineered clone to the standard medium led to one or more genetic substitutions, unrelated to the engineered function, that rendered that clone better adapted to the medium. Evolution by natural selection does not stop when microorganisms are brought into the laboratory, and precautions must be taken to ensure that it does not confound interpretation of fitness differences.

In summary, the following steps are recommended prior to measuring the difference in fitness between two clones.

1. Clones should be properly stored (e.g., in a low temperature freezer) as soon as possible after their initial isolation. If a clone is subsequently determined to have inappropriate properties, then it can be discarded at that time.
2. Clones should be removed from storage and their relevant properties confirmed as directly as possible, preferably by a single round of selective plating.

3. The two clones should normally be preconditioned in the same medium as one another, and that medium should normally be the same as that in which the ratio of the clones will be monitored.

9.3.1.2 Enumeration. Clones cannot usually be distinguished by optical methods, such as microscopy or spectrophotometry, but their colonies can often be distinguished by some means on agar plates. (In the next section, I discuss the use of genetic markers to facilitate distinguishing colonies belonging to different clones.) There are three distinct plating procedures for determining the relative abundance of two clones during the measurement of selection, which I call *direct, replicative,* and *selective.* For all procedures, samples taken from the experimental environment should be processed immediately or stored in such a way that no further change occurs in the abundance of either clone.

For selective plating, a sample from the experimental environment, after appropriate dilutions, is plated on both selective and nonselective medium. The nonselective medium gives an estimate of the total density of both clones ($A + B$), while the selective medium gives the density of one of the clones (A). One calculates the ratio of A to B from the ratio of selective to (nonselective minus selective) plate counts. With selective plating, one must calibrate the plating efficiency of the selectable clone A on both selective and nonselective medium, and adjust the estimated ratios accordingly. Selective plating is quite useful when A is a small fraction of the total, but it is essentially useless when A is numerically dominant.

With both direct and replicative plating, a sample from the experimental environment, after appropriate dilution, is plated only on nonselective medium. In the case of direct plating, colonies are distinguished by their visual appearance, as for instance when a genetic marker causes differences in colony color (see Section 9.3.1.3). In the case of replicative plating, the relative abundance of the two clones is determined by testing all (or some randomly chosen subset) of the colonies isolated on the nonselective medium for some distinguishing genetic characteristic, usually by transferring the colonies to selective plates with toothpicks (colony by colony) or with velveteen (whole plate imprints). If only direct or replicative platings are used, then there is no need to determine relative plating efficiencies, as these efficiencies do not affect the estimate of a selection coefficient. This is so because a difference in plating efficiency between two clones will affect the intercept of the linear regression of the logarithmic ratios, but not the slope of that regression. Both direct and replicative platings are far preferable to selective plating when the selectable clone A is near fixation; both are also quite effective when the frequencies of the two clones are of a similar magnitude. However, if one clone is much rarer than the other, then selective plating is usually preferable.

9.3.1.3 Neutral Markers. I use the term marker to refer to a gene that facilitates identification of a particular clone, but is not itself of special interest. This marker should have little or no effect on fitness in the environment of interest; i.e., it should be effectively neutral. If the marker is not neutral, then it may obscure or confound interpretation of fitness effects associated with the genes of primary interest. Clearly, the putative neutrality of a marker must be established by a control experiment.

For example, imagine that one is interested in determining whether or not there is a difference in fitness between an engineered clone *A* and its parental clone *B*. If the two clones can be distinguished by selective or replicative plating, but not by direct plating, then one might use a marker to create an additional clone that permits direct plating, perhaps by causing differences in colony color on a certain nonselective medium. Alternatively, if clones *A* and *B* cannot be distinguished by replicative or selective plating, then any additional marker (e.g., antibiotic resistance) that permitted such platings would be quite useful. In either case, one would produce a third clone *C* that carried the marker but not the engineered gene. One then could estimate the selection coefficient acting on clone *B* relative to clone *C*, which indicates any effect of the marker, and the selection coefficient of clone *A* relative to clone *C*, which indicates the combined effects of the marker and the engineered gene. If the selection coefficient for *B* relative to *C* is not significantly different from zero, then one may conclude that the marker is neutral, within the limits of one's statistical resolution. If the selection coefficient of *A* relative to *C* is significantly different from the selection coefficient of *B* relative to *C*, then one must conclude that the engineered gene has a significant effect on fitness.

When using neutral markers, it may be necessary to determine that the marker and the engineered gene have retained their presumed linkage during the course of an experiment. If the engineered gene is on a plasmid, for example, it may segregate, thereby altering the linkage of the marker and the engineered gene. In such cases, it is important to confirm the presumed linkage by appropriate replicative or selective platings. If dissociation has occurred, then it may be necessary to use the methods presented in Section 9.3.2 to estimate both the segregation rate and the selection coefficient.

In *Escherichia coli*, genotypes with and without the ability to use certain sugars (including arabinose, lactose, and maltose) can be readily distinguished by their colony color on broth plates supplemented with the sugar and 2,3,5-triphenyltetrazolium chloride (Levin et al. 1977; Carlton and Brown 1981; Nguyen et al. 1989). In several studies, these markers have been shown to be neutral (or nearly so) in medium in which some other nutrient is limiting. Other types of markers include resistances to viruses and antibiotics, but many genes conferring resistance have large effects on fitness under "nonselective" conditions. In *E. coli*, resistance to virulent phage T5 is often neutral (or nearly so), whereas resistance to virulent phage T4 engenders a substantial reduction in fitness (Lenski and Levin 1985; see also Table 9–1). The neutrality of a particular marker in a particular clone in a particular environment is an empirical question that can be addressed only by an appropriate control.

9.3.1.4 Statistical Inference. The purpose of statistical inference is to formalize the degree of confidence (i.e., the significance level) in some conclusion. I wish to review briefly one important issue that relates to the statistical interpretation of selection coefficients. There are two distinct ways of assigning significance to an estimated selection coefficient. According to one, a single measurement of the selection coefficient is calculated from the slope of the linear regression of the natural logarithm of the ratio of two clones against time. One then asks: Is that slope significantly different from zero? The procedures for determining the answer to this question were briefly illustrated for the data in Table 9–1; further details can be found in almost any statistics book (e.g., Sokal and Rohlf, 1981) and need not concern us. Basically, the more sample points one has and the closer these points lie to the fitted line, the more confidence one has in the slope of that line. (Many statistics books also describe the calculations necessary to ask whether two slopes are significantly different from one another.) The resulting significance level is based upon the assumption that each point is a truly independent observation, which seems quite reasonable at first glance. But what if some chance event impinges upon our single measurement, and that event affects the relative fitness of the two clones? Imagine, for example, that a single colony used to found one of the two clones just happened to have contained a deleterious mutation; one would wrongly ascribe that clone's reduced fitness to its intended genotype.

An alternative approach is to obtain many estimates of the selection coefficient, each based upon an independent experimental replicate. A slope would be calculated for each experiment as before, but no significance level would be attached to any single slope. Instead, one would use a *t*-test to ask whether the *mean* of many slopes was significantly different from zero (e.g., Nguyen et al. 1989). (One could also use a *t*-test to ask whether two means differed significantly from each other.) This second approach is usually preferable, because statistical inferences based on many independent replicates are more robust than those based on a single experiment; chance events are not inadvertently ascribed a stronger association with the intended comparison than is appropriate. Hurlbert (1984) discusses the importance of proper replication in ecological experiments. Koch (1981) also illustrates the importance of independently replicated measurements in making statistical inferences.

9.3.2 Analyzing Instability

It is simple to estimate both the segregation rate and the selection coefficient, at least in principle. A population is founded in which the frequency of the recombinant gene is near one. The population is propagated and sampled at intervals, and the frequency of individuals possessing the recombinant gene is enumerated. From these dynamics, nonlinear regression can be used to estimate simultaneously the segregation rate and the selection coefficient, as illustrated in the next section. In practice, it may sometimes be difficult to distinguish between the effects of segregation alone and the combined effects of segregation and selection. In such cases, it is desirable to run a parallel experiment in which the population is deliberately "seeded" with segregants lacking the recombinant gene, as discussed in Section 9.3.2.2.

9.3.2.1 Nonlinear Regression. The precise mathematical methods used in nonlinear regression analysis need not concern us, as programs are widely available in statistical packages to perform the actual calculations (e.g., Dixon 1985). These programs can be used to find the intercept (p_0), segregation rate (u), and selection coefficient (s) in equation 8 that minimize the sum of the squared deviations about the fitted model for a set of frequencies of the engineered genotype obtained at different points in time (p_t's).

Table 9–2 presents a set of data corresponding to the dynamics of the loss of plasmid pACYC184 from *Escherichia coli* (Lenski and Bouma 1987); it includes the time at which the sample was taken, the number of individuals scored, and the frequency of the plasmid-bearing genotype. For each sample frequency, the approximate 95% confidence interval has been calculated from the binomial distribution, which is available in many statistical tables (e.g., Rohlf and Sokal, 1981). These confidence intervals are neither always symmetric about the sample frequency, nor are they always of equal breadth, despite the equal sampling effort at all points in time; this is characteristic of the binomial distribution.

The data from Table 9–2 are plotted in Figure 9–2, along with two fitted models. The solid line represents the fitted model when the intercept, the

TABLE 9–2 Dynamics of Instability of Plasmid pACYC184 in *Escherichia coli*

Time of Sample (hr)	*Number Scored*	*Frequency*[1]	*C.I.*[2]
0	300	0.963	0.94–0.98
4.5	300	0.810	0.76–0.85
10.5	300	0.560	0.50–0.62
16.5	300	0.133	0.10–0.18
22.5	300	0.020	0.01–0.04
28.5	300	0.000	0.00–0.01

[1] Frequency, p, of the plasmid-bearing genotype.

[2] C.I., the approximate 95% confidence interval for the sample frequency (from Rohlf and Sokal 1981).

Source: Lenski and Bouma 1987.

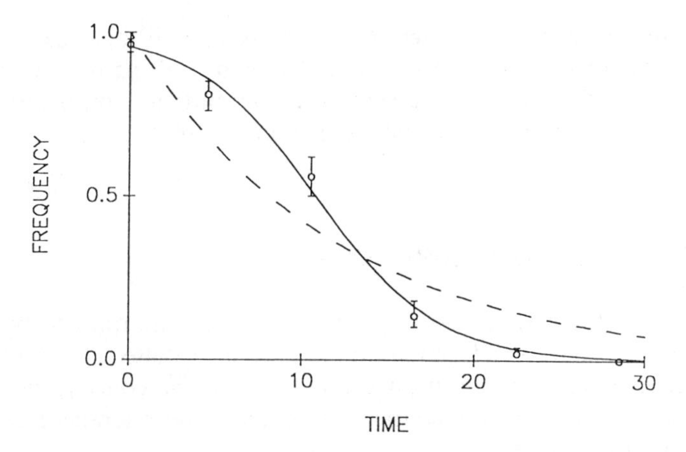

FIGURE 9–2 The dynamics of instability of plasmid pACYC184 in *Escherichia coli.* Raw data and 95% confidence intervals for each sample frequency are presented in Table 9–2. Solid line: model fit with both segregation and selection parameters. Dashed line: model fit with segregation parameter only. From Lenski and Bouma 1987.

segregation rate, and the selection coefficient are allowed to vary. The dashed line indicates the fitted model when the intercept and the segregation rate are allowed to vary, while the selection coefficient is held to zero. The dashed line, which ignores selection, is clearly inconsistent with the data; there are too few segregants in the early samples, and too many in the later samples, to be explained by segregation alone.

9.3.2.2 Manipulation of Initial Frequency. In some cases, it may be difficult to distinguish between the effect of segregation alone and the combined effects of segregation and selection. This is especially likely to be true when the frequency of segregants remains low over the entire course of an experiment. On the one hand, it is more parsimonious to accept the simpler model of segregation alone. On the other hand, it is possible to test more rigorously the null hypothesis that the selection coefficient is zero.

The most obvious solution to this problem is to continue the experiment for a longer period of time, thereby obtaining a greater range of sample frequencies. However, longer experiments may be complicated by additional genetic changes that go undetected, and that have effects on fitness that may be erroneously associated with the genotypic property of interest (see also Sections 9.3.1.1 and 9.4.1). A preferable approach is to perform a parallel experiment in which the frequency of segregants at the start of the experiment is deliberately increased by the inoculation of the appropriate wild-type clone, which is presumed to be identical to a segregant. This manipulation of the initial frequency of the wild-type will increase the range of frequencies subsequently observed, and thereby will permit better distinction between the effect of segregation alone and the combined effects of segregation plus selection (Lenski and Bouma 1987).

9.4 CAVEATS AND ASSUMPTIONS

No methods are without assumptions, and it is important to be aware of those situations in which assumptions may be violated. In this section, I will discuss several important assumptions and the consequences of their violation. The reader is encouraged to consult the references cited below for further discussion.

9.4.1 Fitness-Enhancing Mutations (Periodic Selection)

One of the most critical assumptions of the analyses presented in this chapter is that any relevant genetic differences between genotypes are known to the investigator. If this is not true, then these hidden genetic differences may confound one's interpretation of the effect on fitness caused by the known genetic factors.

Secondary mutations invariably arise such that any population actually contains many genetically distinct subclones. However, these secondary mutations are usually hidden to the investigator. Each subclone is perceived as belonging to one of the two intended clones, whose average fitness is thereby affected. Rare subclones have very little effect on the average fitness of the clone, and mutations that reduce fitness should remain rare. But if a mutation enhances fitness, then it will spread through the population, a process that has been termed *periodic selection* (Atwood et al. 1951; Kubitschek 1974; Dykhuizen and Hartl 1983). Neutral and even disadvantageous genes can be dragged along with the adaptive mutation, owing to the inevitable linkage that exists in clonal organisms (e.g., Helling et al. 1981).

The likelihood of confounding effects caused by periodic selection can be minimized by storing clones properly just after their isolation (see Section 9.3.1.1) and by limiting the length of experiments (see Sections 9.3.1.1 and 9.3.2.2). One can test whether anomalous results are due to the effects of periodic selection by isolating subclones derived from each original clone at the end of an experiment, and performing additional measurements of selection coefficients. If heritable changes in fitness have occurred, then the fitness of a derived subclone will have increased relative to its progenitor. Periodic selection is probably most frequent when a microorganism has been recently introduced into a novel environment, so that certain genetic variants, previously neutral or disadvantageous, are suddenly favored.

The likelihood that a fitness-enhancing mutation will enable an otherwise competitively inferior microorganism to persist after it has been introduced into the environment will be discussed in the next chapter.

9.4.2 Gene Transfer

The methods presented in this chapter assume that there is no gene transfer. The effects of gene transfer have been incorporated into mathematical models (Stewart and Levin 1977; Levin and Rice 1980), but I am unaware of any published papers that present explicit methods for simultaneously estimating the rate of gene transfer, the segregation rate, and the selection coefficient. I suspect that such methods can be developed, although this will not be trivial for the following reason. Segregation and selection can be distinguished because the rate of change due to the former depends on the frequency of the engineered genotype, whereas the rate of change due to the latter depends on the *product* of the frequencies of the engineered and segregant genotypes (see equation 7). The rate of change due to gene transfer (including conjugation, transduction, and transformation) depends on the product of two populations, and so it may be more difficult to distinguish its effects from the effects of selection. Of course, gene transfer can also break apart any presumed linkage of a marker and an engineered gene, as can segregation (see Section 9.3.1.3).

When a recombinant gene reduces fitness, segregation and selection act in concert to reduce the frequency of the engineered genotype. Under the continued action of these forces, the recombinant gene will eventually be lost completely from a population. However, transfer can maintain a gene or extrachromosomal element in a population despite the purging effects of segregation and selection (Stewart and Levin 1977; Levin 1980; Levin and Rice 1980; Levin and Lenski 1983).

9.4.3 Constancy of Parameters

Although perhaps obvious, it is important to emphasize that analyses of the models presented in this chapter are predicated on the assumption that the segregation rate and the selection coefficient are constants. However, there may be heterogeneity among cells even within a single genetically uniform subclone with respect to the effects of segregation or selection. For example, in continuous culture, cells attached to the walls of the vessel are subject to a reduced rate of turnover and consequently a reduced selection coefficient (Dykhuizen and Hartl 1983; Chao and Ramsdell 1985). An environmental refuge such as this may generate the appearance of an equilibrium, although in fact the rate of change due to selection may only have slowed.

In some cases, the selection coefficient may vary in direct response to the relative frequencies of the competing clones. If ecological feedbacks are such that one clone has a selective advantage when it is scarce, but it is selectively disadvantaged when common, then frequency-dependent selection can promote a stable equilibrium whereby the two clones coexist indefinitely (e.g., Lenski and Hattingh 1986).

9.5 SUMMARY AND CONCLUSIONS

Fitness represents the combined effects of all other phenotypic properties on the capacity for survival and reproduction by a particular genotype in a particular environment. For most environmental applications of genetically modified microorganisms, efficacy will be enhanced if the engineered genotype is more fit than its wild-type counterpart in the target environment. However, inadvertent spread of the engineered genotype will be less likely if it is less fit than the wild-type. Thus, the fate of a population of genetically engineered microorganisms, and the likelihood and magnitude of any environmental effects (whether beneficial or detrimental), will be strongly influenced by the relative fitnesses of modified and unmodified genotypes. In this chapter, I have presented theoretical principles and empirical methods for determining the relative fitnesses of engineered and wild-type clones. Selection coefficients were used to provide a quantitative measure of the difference in fitness between the two clones in a particular environment.

Many engineered genotypes are unstable, such that their frequencies decline with time. Instability may be caused by infidelity of replication or transmission of a particular gene (which is termed segregation), or it may be caused by a difference in the fitness of genotypes that retain or have lost that gene (selection). In this chapter, I have also presented theoretical principles and empirical methods for distinguishing the effects of selection and segregation.

Finally, it should be emphasized that selection coefficients and segregation rates can be estimated not only in highly simplified laboratory systems, but also in more complex natural or semi-natural systems, such as microcosms. All that is required is the ability to monitor the relative abundance of two clones (e.g., engineered and wild-type) that share a common environment.

REFERENCES

Atwood, K.C., Schneider, L.K., and Ryan, F.J. (1951) *Proc. Natl. Acad. Sci. USA* 37, 146–155.

Boe, L., Gerdes, K., and Molin, S. (1987) *J. Bacteriol.* 169, 4646–4650.

Bouma, J.E., and Lenski, R.E. (1988) *Nature* 335, 351–352.

Brill, W.J. (1985) *Science* 227, 381–384.

Carlton, B.C., and Brown, B.J. (1981) in *Manual of Methods for General Bacteriology* (Gerhardt, P., ed.), pp. 222–242, American Society for Microbiology, Washington, D.C.

Chao, L., and Ramsdell, G. (1985) *J. Gen. Microbiol.* 131, 1229–1236.
Charles, I.G., Harford, S., Brookfield, J.F.Y., and Shaw, W.V. (1985) *J. Bacteriol.* 164, 114–122.
Colwell, R.K., Norse, E.A., Pimentel, D., Sharples, F.E., and Simberloff, D. (1985) *Science* 229, 111–112.
Committee on Scientific Evaluation of the Introduction of Genetically Modified Microorganisms and Plants into the Environment (1989) *Field Testing Genetically Modified Organisms: Framework for Decisions*, National Academy Press, Washington, D.C.
Cooper, N.S., Brown, M.E., and Caulcott, C.A. (1987) *J. Gen. Microbiol.* 133, 1871–1880.
Davis, B.D. (1987) *Science* 235, 1329–1335.
Dixon, W.J., ed. (1985) *BMDP Statistical Software*, University of California Press, Berkeley.
Dykhuizen, D.E., and Hartl, D.L. (1983) *Microbiol. Rev.* 47, 150–168.
Helling, R.B., Kinney, T., and Adams, J. (1981) *J. Gen. Microbiol.* 123, 129–141.
Hurlbert, S.H. (1984) *Ecol. Monogr.* 54, 187–211.
Kimura, M., and Ohta, T. (1971) *Theoretical Aspects of Population Genetics*, Princeton University Press, Princeton, NJ.
Koch, A.L. (1981) in *Manual of Methods for General Bacteriology* (Gerhardt, P., ed.), pp. 179–207, American Society for Microbiology, Washington, D.C.
Kubitschek, H.E. (1970) *Introduction to Research with Continuous Cultures*, Prentice-Hall, Englewood Cliffs, NJ.
Kubitschek, H.E. (1974) *Symp. Soc. Gen. Microbiol.* 24, 105–130.
Lenski, R.E., and Bouma, J.E. (1987) *J. Bacteriol.* 169, 5314–5316.
Lenski, R.E., and Hattingh, S.E. (1986) *J. Theor. Biol.* 122, 83–93.
Lenski. R.E., and Levin, B.R. (1985) *Am. Nat.* 125, 585–602.
Lenski, R.E., and Nguyen, T.T. (1988) in *Planned Release of Genetically Engineered Organisms (Trends in Biotechnology/Trends in Ecology and Evolution Special Issue)* (Hodgson, J., and Sugden, A.M., eds.), pp. 18–20, Elsevier Publications, Cambridge.
Levin, B.R. (1980) in *Antibiotic Resistance: Transposition and Other Mechanisms* (Mitsuhashi, S., Rosival, L., and Krcmery, V., eds.), pp. 197–202, Springer-Verlag, Berlin.
Levin, B.R., and Lenski, R.E. (1983) in *Coevolution* (Futuyma, D.J., and Slatkin, M., eds.), pp. 99–127, Sinauer Associates, Sunderland, MA.
Levin, B.R., and Rice, V.A. (1980) *Genet. Res., Camb.* 35, 241–259.
Levin, B.R., Stewart, F.M., and Chao, L. (1977) *Am. Nat.* 111, 3–24.
Nguyen, T.N.M., Phan, Q.G., Duong, L.P., Bertrand, K.P., and Lenski, R.E. (1989) *Mol. Biol. Evol.* 6, 213–225.
Nordstrom, K., Molin, S., and Aagaard-Hansen, H. (1980) *Plasmid* 4, 215–227.
Pritchard, P.H., and Bourquin, A.W. (1984) *Adv. Microb. Ecol.* 7, 133–215.
Regal, P.J. (1986) in *Ecology of Biological Invasions of North America and Hawaii* (Mooney, H.A., and Drake, J.A., eds.), pp. 149–162, Springer-Verlag, New York.
Regal, P.J. (1988) in *Planned Release of Genetically Engineered Organisms (Trends in Biotechnology/Trends in Ecology and Evolution Special Issue)* (Hodgson, J., and Sugden, A.M., eds.), pp. 36–38, Elsevier Publications, Cambridge.

Rohlf, F.J., and Sokal, R.R. (1981) *Statistical Tables*, 2nd ed., W.H. Freeman, San Francisco.
Sharples, F.E. (1987) *Science* 235, 1329–1332.
Sokal, R.R., and Rohlf, F.J. (1981) *Biometry*, 2nd ed., W.H. Freeman, San Francisco.
Stewart, F.M., and Levin, B.R. (1977) *Genetics* 87, 209–228.
Tiedje, J.M., Colwell, R.K., Grossman, Y.L., et al. (1989) *Ecology* 70, 298–315.

CHAPTER 10

Quantifying the Risks of Invasion by Genetically Engineered Organisms

Junhyong Kim
Lev R. Ginzburg
Daniel E. Dykhuizen

The rate of evolutionary (genetic) change in most organisms is sufficiently slow so that it does not effect ecological dynamics over the time span that usually interests us. But, at certain times, the specific characteristics of the organisms under question require us to include or even to concentrate our efforts on their evolutionary properties in order to solve ecological problems. An important example is encountered in attempts to conserve rare species where, due to the very small size of the populations, the lack of genetic variability becomes an important issue, although the original problem was primarily demographical (Simberloff 1988; O'Brien and Evermann 1988). Interestingly, at the other end of the spectrum, when the population size is extremely large, new advantageous mutations may arise in the population every generation. If the generation time of such organisms is short, these advantageous mutants can be rapidly selected, and ecological parameters could change very fast.

Most genetically engineered organisms proposed for release into field are microorganisms. Microorganisms have precisely these characteristics of large population size and short generation time. They are also haploid and mostly clonal, enabling new mutations to be more easily fixed than in diploid organisms with same number of individuals. Thus, this potential for rapid evolutionary change makes it difficult to predict the ecological fate of microorganisms. The capability for producing novel mutants evokes popular fears of the creation of new pathogens, while demographic unpredictability would make the containment of such organisms difficult if they were ever to be released into nature. Yet genetically engineered microorganisms (GEMs) have the potential for large benefits, from controlling frost damage to cleaning oil spills (Lindow et al. 1989). We must be able to balance the benefits of a program for using GEMs against its possible risks.

Risk is defined roughly as the product of the probability of an event happening and the cost of the occurrence of the event. The same value of risk may be assigned both to a highly improbable event, such as the creation of a new pathogen, which has very high costs, and to the more probable escape of frost-control bacteria, which has much lower costs. However, since the exact costs of possible adverse events after the release of GEMs cannot be determined, we will consider the cost to be a constant and for our purposes define risk as just the probability of the occurrence of the adverse event alone. While some problems involved in estimating the probability of the evolution of a new pathogen are currently impossible to solve, other problems, given current ecological theory, can be addressed reasonably (e.g., the possibility of a released GEM invading the natural microbial community). Here, we will discuss one such problem and describe a model that may be used, coupled with experimentally measurable parameters, to obtain the risk of invasion of released GEMs in natural communities.

10.1 RISK ASSESSMENT USING EXTREME VALUE DISTRIBUTION

GEMs released in the field may either be competitively superior to the existing natural flora and thereby eventually displace the natural flora, or they may be competitively inferior and be displaced by the natural strains. In the former case, we cannot hope to contain GEMs unless there exists physical containment (probably impossible for microorganisms released in the field) or a method for selectively killing the GEMs when it becomes necessary to eliminate them (in this case, the original ecological properties of the released location may already have been altered through competitive displacement of the natural flora). In the latter case, the population of released GEMs will decline through time as they are displaced by the natural strains and eventually disappear, analogous to natural degradation of hazardous waste. Mutational derivatives of the released GEMs may either become immune to a proposed selective elimination or acquire higher competitive ability such that they evade competitive displacement. The probability of such a mutant occurring depends on the mutation rate and the total number of individuals that are subject to mutational events. For a given amount of time, we must consider the total number of individuals of GEMs that existed during the period. We will call this the *mutational exposure* of a given population. We will first consider a risk model that incorporates this notion of mutational exposure for the second case, i.e., the release of competitively inferior GEMs, after which we will discuss the problems associated with selective killing of GEMs.

10.1.1 Risk from Release of Competitively Inferior GEMs

As mentioned above, if released GEMs are less fit than at least one of the natural strains, the population of GEMs will go towards extinction more or less exponentially (see Appendix 10A; see also Chapter 9). During this period, both the GEM and the natural strain (termed the wild-type strain from here on) will generate an array of mutants. The risk that we are concerned with is the possibility that some mutant derived from the GEM may have higher fitness than the wild-type strain or any mutants derived from the wild type. As the population of GEMs decreases through time it will give off mutant derivatives of different fitnesses (we use fitness in a loose general sense) while the population of the wild type does the same. In the sample of different mutant derivatives of GEMs, there will be one strain that has the highest fitness. Likewise, such a "maximum-fitness" strain (termed the MF strain) will also exist for the wild-type strain. If the fitness of the MF strain derived from GEMs is higher than that derived from wild-type cells, we can consider the GEMs to have successfully invaded. We avoid the treatment of stochastic loss of advantageous mutants. We will model this process, making some simplifying assumptions, with incorporation of available experimental data.

To calculate the probability of such invasions, we must first obtain the mutational exposure of the GEMs, or the total number of individuals expected to be at risk for mutational events. We will determine this by assuming a two-species chemostat competition model. This model ignores both multi-species competition, which must be present in natural communities, as well as environmental complexity. We assume that before the introduction of GEMs, the wild-type strain is at equilibrium. As shown in Appendix 10A, if GEMs of lower fitness are added, the GEM population will decline more or less exponentially until it becomes extinct, while the wild-type strain will approximately maintain its equilibrium size. We can then write this in terms of the initial population sizes of the GEMs (N_0^{GEM}), the wild type (N_0^{WT}), and their relative fitness difference (Δf) measured as specific growth rates. Then the total time until extinction for the GEM is

$$t \approx \frac{\log N_0^{GEM}}{\Delta f} \tag{1}$$

This time can be used to determine the total number of GEMs and wild-type cells exposed to mutation during this period, which becomes

$$N_e^{GEM} \approx \frac{N_0^{GEM}}{\Delta f} \tag{2}$$

$$N_e^{WT} \approx N_0^{WT} \times \frac{\log N_0^{GEM}}{\Delta f} \tag{3}$$

We are, of course, ignoring the change in the dynamics that would occur as the fitness changes for each mutation. Even though the population size at any one time may be very large (say 10^{12}), the number of mutants of any given fitness will be quite small as the mutation rate to a *given* fitness will be very low. Thus, it is likely that there will only be a single new individual carrying the particular fitness change. If the change is to lower fitness, the single individual will immediately be eliminated from the population. Therefore, we will ignore any mutations to lower fitness or nonadvantageous mutations. The consequence of advantageous mutations that are not maximally fit will be discussed later.

Therefore, the risk that we desire to calculate is

$$\text{Risk} = \text{Prob}\{MAX[f_{mut}(N_e^{GEM})] > MAX[(f_{wt} \wedge f_{mut}(N_e^{WT})]\}$$

where f_{mut} is the fitness of mutants and f_{wt} is the fitness of wild type. Therefore risk is the probability that the maximum fitness of the mutants of N_e^{GEM} cells is greater than the maximum fitness of the mutants of N_e^{WT} cells and the original wild type. To determine this probability, we must first establish a probability distribution for the fitnesses of mutants. Unfortunately, this is a largely unstudied problem and little is known about the nature of such distributions, although they may be experimentally determinable. Advantageous mutations are likely to be very rare, and there must be a limit to increasing fitness indefinitely due to fundamental constraints. Lenski (1988) has shown that the fitness effects of advantageous mutations in strains of relatively lower fitness are different from those in strains of relatively higher fitness. The change in fitness was found to be larger when the mutants were derived from relatively lower fitness strains compared to the change in fitness of mutants derived from strains of higher fitness. This suggests that the variance of fitness distribution of mutants is inversely related to the fitness of the strains from which the mutants are derived.

Mutations of large advantageous fitness effects are observed to be rare, and therefore we will assume that the fitness mutants are distributed as an exponential distribution with the origin of the distribution at the current fitness of the GEM and the wild type (see Figure 10–1). The variance of the distribution is assumed to decline linearly with fitness. Linearity was assumed as a first approximation, since the shape of the function is unknown. The fitness probability density function of the GEM and wild type is then

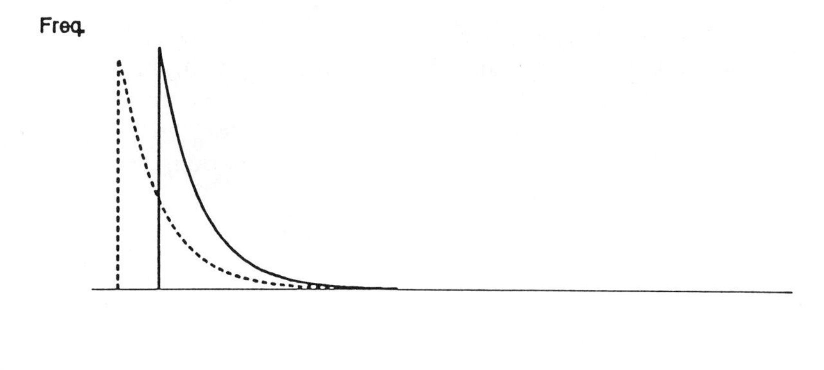

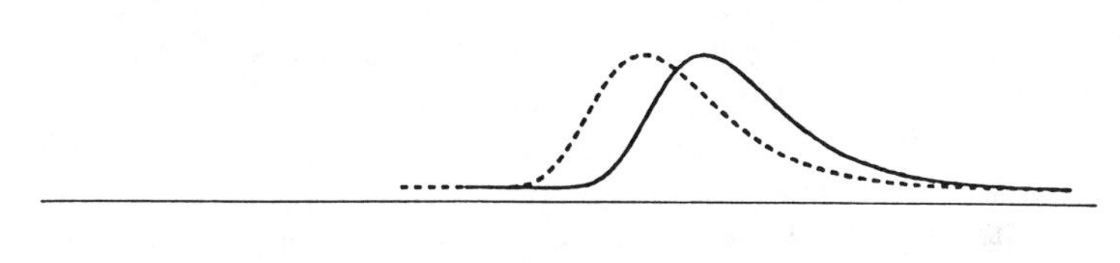

FIGURE 10–1 The plot of possible fitness mutations derived from strains of given fitness. Top: the exponential distributions of fitness mutations assumed to be underlying distribution for the distribution of the maximum fitness. The dashed line represents the GEMs, and the solid line represents the wild-type strains. Bottom: the corresponding maximum value distributions.

given by

$$f_{WT}(x) = \frac{1}{\gamma} e^{-\frac{(x-wf)}{\gamma}}, \qquad x \geq wf \tag{4}$$

$$f_{GEM}(x) = \frac{1}{\gamma'} e^{-\frac{(x-wf+\Delta f)}{\gamma'}}, \qquad x \geq wf - \Delta f \tag{5}$$

$$\gamma' = \gamma + \beta\Delta f \tag{6}$$

where x is fitness, wf is the fitness of the wild type, Δf is the difference in fitness between the wild type and GEM, and γ, γ' are variance parameters for the exponential distributions. The probability density functions are only defined for x values above wf and $wf - \Delta f$ respectively, and are defined as 0 elsewhere. Equation 6 represents the linear decline in variance of the exponential distribution with increase in fitness. The equation has been scaled such that γ represents the variance of the higher fitness strain (or the wild-type strain here) and Δf the absolute value of the fitness difference between the two; β represents the slope of the change in the variance with respect to fitness change. For both the GEM and wild type we sample from the two distributions $f_{WT}(x)$ and $f_{GEM}(x)$, n_1 and n_2 times respectively, and we desire to compare the maximum value from the multiple sampling. The number of times sampled for each strain during a given period is the total number of individuals that existed during that period, or the mutational exposure, multiplied by the rate of advantageous mutations. From equations 2 and 3, this is

$$N_s^{GEM} = N_e^{GEM} \times P_{adv}^{GEM} \quad (7)$$

$$N_s^{WT} = N_e^{WT} \times P_{adv}^{WT} \quad (8)$$

$$P_{adv}^{GEM} = P_{adv}^{WT} + \rho\Delta f \quad (9)$$

where P_{adv}^{GEM}, P_{adv}^{WT} is the mutational rate to advantageous mutants for GEMs and wild-type strains, respectively. As with equation 6, we assume that the proportion of advantageous mutations declines linearly with respect to absolute fitness. This also implies that there is a fitness at which no more advantageous mutations can occur. Equation 9 has been scaled in a similar way to equation 6, such that P_{adv}^{WT} represents the proportion of advantageous mutations of the organism that is at the higher fitness. The parameter ρ, like β, represents the slope of the linear decline.

From these considerations we want to obtain the probability distribution of the *maximum* value of taking n samples from an exponential distribution since we are concerned with the fitness of the "highest fitness" strain. For any distribution with the cumulative distribution function $F(x)$, the cumulative distribution function of the maximum of taking n samples from this distribution is simply $F^n(x)$, that is, the probability that all n values are less than x. Although this is algebraically simple, it becomes impossible to calculate if n is large. We solve this problem by using an approximation given by extreme value theory, as presented in Appendix 10B. When the underlying distribution is exponential, the extreme value distribution of taking n samples is given by,

$$\Phi_n(x) = e^{-e^{-y}} \quad (10)$$

$$\phi_n(x) = \alpha_n e^{-y-e^{-y}} \quad (11)$$

$$y = \alpha_n(x - u_n) \quad (12)$$

where $\Phi_n(x)$ is the cumulative distribution function and $\phi_n(x)$ is the probability density function, while y, α_n, and u_n are parameters described below. The right hand curves in Figure 10–1, shows the extreme value distribution for the corresponding exponential distribution shown as the curves on the left. The distribution is characterized by extremely fast decline and skewness to the left. The parameters of the extreme value distribution with the exponential distribution as the underlying distribution are given by

$$\alpha_n = \frac{1}{\gamma} \quad (13)$$

$$u_n = \gamma \log n \tag{14}$$

Risk is then shown by the equation

$$\text{Risk} = \int_0^\infty (1 - e^{-e^{-y'}})\alpha_n e^{-y-e^{-y}} dx \tag{15}$$

where y is the parameter for the wild-type strain and y' is the parameter for the GEM, and the integral describes the probability of maximum fitness of the GEM being larger than that of the wild-type strain taken over 0 to ∞. The parameters are given by

$$y = \alpha_n(x - u_n) \tag{16}$$

$$u_n = \gamma \log N_s^{WT} + wf \tag{17}$$

$$\alpha_n = \frac{1}{\gamma} \tag{18}$$

$$y' = \alpha'_n(\mathrm{x} - u'_n) \tag{19}$$

$$u'_n = \gamma' \log N_s^{GEM} + wf - \Delta f \tag{20}$$

$$\alpha_n = \frac{1}{\gamma'} \tag{21}$$

The terms wf and $wf - \Delta f$ in equations 17 and 20 represent the location of the fitness of the wild type and the GEM on the fitness scale. We will estimate the above parameters and then calculate the risk as a function of the initial population sizes of the two strains and their fitness differential.

10.1.2 Estimation of Parameters and Results

Few empirical studies have been done specifically to estimate the fitness effects of random mutations. It is doubtful that even with sufficient data we will find that there are universal values to the particular parameters used in our model. Different species in the same environment and the same species in different environments are likely to have different values for these parameters. However, for a given strain of microorganisms, it should be possible to design experiments that would permit us to obtain roughly the parameters of the model. Since we do not have such data on hand we will attempt to determine some reasonable values from chemostat experiments using *Escherichia coli*. In substrate-limited chemostats, there is constant selection of mutations that are better able to utilize the substrate (Dykhuizen and Hartl 1981). Since we have modeled growth the same way it is modeled in chemostats, chemostat experiments can be used to estimate the parameters. However, since the experiments we use below were designed for other purposes we have made additional assumptions and the values should be viewed only as rough estimates.

Helling et al. (1987) conducted a very long term chemostat experiment using derivatives of *E. coli* K12 strains. A single clone was used to inoculate a chemostat under glucose limitation, and the subsequent population fluctuations were observed for approximately 1,800 generations. In one example, four different clones were isolated from the culture after approximately 800 generations. During this period, eight episodes of periodic selection (Atwood et al. 1951) (where the population of the culture is replaced by "fitter" mutants) were observed. The initial specific growth rate of the clone that was used to inoculate the chemostat was 0.44 hr^{-1}. The highest growth rate in the clones isolated after 800 generations was 0.60 hr^{-1}. Under the traditional model of chemostat competition (Monod 1950; Kubitschek 1970), both the specific growth rate and the maximum saturation coefficient (the ability of the organism to sequester the nutrient) determine the outcome of the competition. In fact, Helling et al. found that the change in the saturation coefficient was more important in competition between the isolated strains. However, since the measurements of the saturation coefficients were not reported, we will ignore this for our calculations and use a highly schematic outline of this experiment for our estimations.

For our purposes we assume that in the chemostat the existing population was displaced at equal intervals of 100 generations by more fit mutants and this occurred eight times, during which the specific growth rate changed from 0.44 hr^{-1} to 0.60 hr^{-1}. The fitness difference at each displacement event is assumed to have been equal, and therefore each displacement event resulted in $(0.60 - 0.44)/8 = 0.02$ hr^{-1} increase of growth rate. This value, 0.02 hr^{-1}, is assumed to represent the modal value of the extreme value distribution, that is, the most probable value. Then from equation 17

$$0.02 + wf = \gamma \log N_s + wf \tag{22}$$

where N_s represents the mutational exposure, $N_e \times P_{adv}$. In this equation there are two unknowns, P_{adv} and γ. If direct measurements had been made for γ we would have been able to obtain the other quantity, P_{adv}. However, in this case we cannot simultaneously obtain both values. Chao and Cox (1983) conducted chemostat experiments using a strain that produced higher rate of mutations than normal (*mutT* strains). Using their results, they calculated that normal strains of *E. coli* have mutation rates of 5.0×10^{-9} per generation for advantageous mutations. This calculation assumes that a mutation to higher fitness is only a single event. We may also try to approach this in a different way: in *E. coli* there are approximately 4.5×10^6 bp (base pairs)/genome, and the observed mutation rate is approximately 2×10^{-10}/bp-generation. Under conditions where one nutrient is limiting, only a small number of genes can increase fitness, and only a few of the many possible mutations in these few genes will be advantageous. If the number of genes is 10 and only 10 changes in each gene are advantageous, then only 100 possible changes are advantageous. This is about $100 \times 2 \times 10^{-10}$ or 2×10^{-8} cell/generation. This value is only fourfold higher than the value calculated by Chao and Cox, suggesting their estimate is reasonable. We will therefore use this as our estimate of P_{adv}. Then from equation 22

$$\gamma = 0.02/(\log N_e + \log P_{adv})$$

In the experiments of Helling et al. (1987), the chemostat had a total population size of 1.7×10^{10} to 2.4×10^{10} cells. If we take the lesser value and multiply by 100 generations, total mutational exposure N_e was 1.7×10^{12}, and, therefore, $\gamma = 0.00221$.

In our model, we have assumed that the variance γ of the underlying exponential distribution changes as a function of current fitness (see equation 6). We now assume that when current fitness is 0.52 (or the mid-value between 0.44 and 0.60), the variance has the value 0.00221. We will make another assumption in order to estimate the slope coefficient β. We consider that there exists a mutational state (a strain) at which the fitness is maximum. This may be due to either physiological or physical constraints irrespective of the genetics of the organism. At this point, the variance is expected to be 0, since there is not further higher fitness state. Under optimal growth conditions, *E. coli* has a generation time of about 20 min. We will assume that this is the highest fitness state possible (which is a generous value) and, therefore, the maximum point for fitness is 3.0 hr^{-1}. Using these two points, and substituting them into equation 6, we find that $\beta = 0.000891$ and that when fitness is zero (or at zero growth rate), $\gamma = 0.00267$.

Likewise we can estimate ρ for the linear equation describing P_{adv}. Chao and Cox (1983) made their calculations based on strains that had average growth rate of 0.41 hr^{-1}. We can again assume that when a strain is at maximum possible fitness (3.0 hr^{-1} in our example) that the proportion of advantageous mutations should be zero. Using these two points, $\rho = 1.931 \times 10^{-9}$ and when the growth rate is zero, $P_{adv} = 5.792 \times 10^{-9}$.

Using the parameters calculated above, we can compute the risk given by equation 15. The question that we are most likely to encounter is, given a GEM strain, what would the risk be if we inoculated a field with n number of cells? To use the model, we would first need to do a survey of the existing microflora of the field in which the GEMs were to be released. Determination of Δf between the GEM and the wild representations of this species (or other bacteria that could competitively exclude the GEM) is required. The reliability and precision of various laboratory techniques for measuring Δf are currently unknown. While it is assumed that Δf will be a complex function of the environmental and genetic characteristics of competition, this assumption could be unfounded, given that little systematic investigation of relative fitnesses has ever been done. Thus as a first approximation, Δf can be determined using chemostats as microcosms. If a pilot experiment can be done in quasi-natural conditions (see Chapter 9), Δf can be estimated from the initial decline in the numbers of genetically marked GEMs and

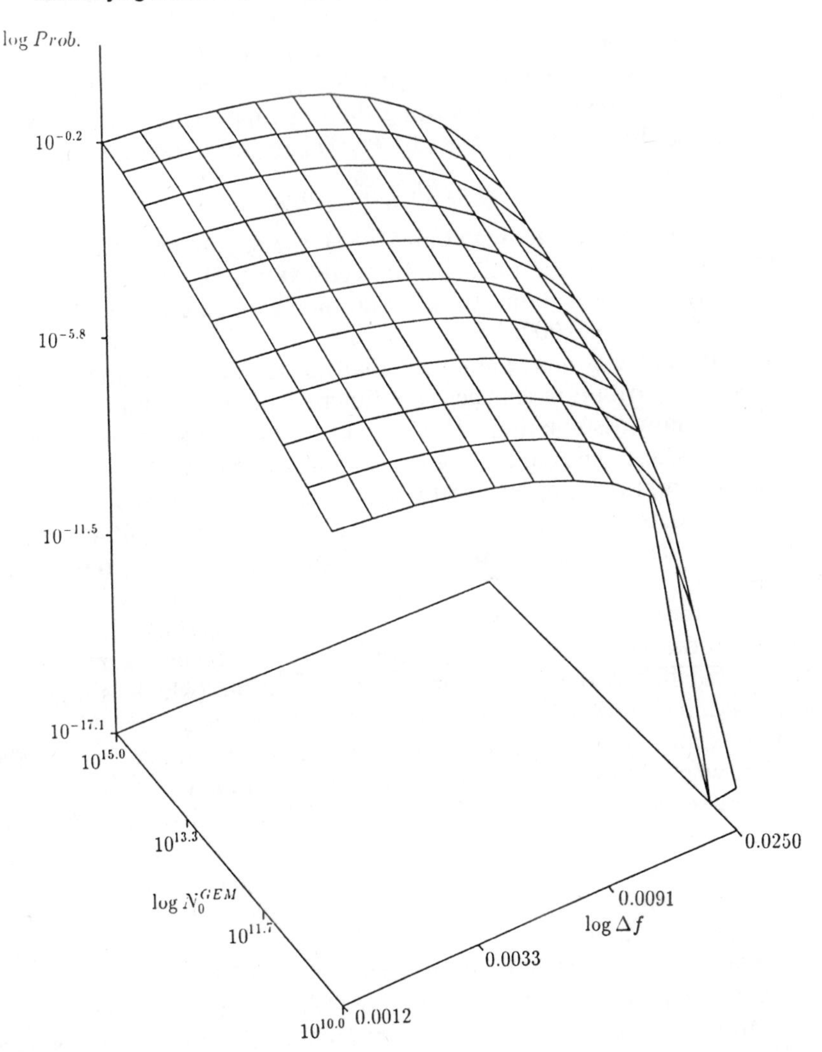

FIGURE 10–2 Three-dimensional plot of the probability (risk) surface in terms of N_0^{GEM}, the initial population size of the GEMs, and Δf, the relative fitness between the GEMs and the wild-type strains. All axes are logarithmic but the labels are in original scale for clarity. All other parameters are assumed to be fixed, using the values given in the text.

wild-type strains after inoculation of both. We also need to assay the population size of the wild-type strain in its natural environment. For many strains, this can be done using standard techniques (Atlas and Bartha 1987). The estimated values of Δf and respective population sizes can be substituted into equation 15 with the parameters determined as in the above example, and the risk can be computed. Since we may manipulate the ge-

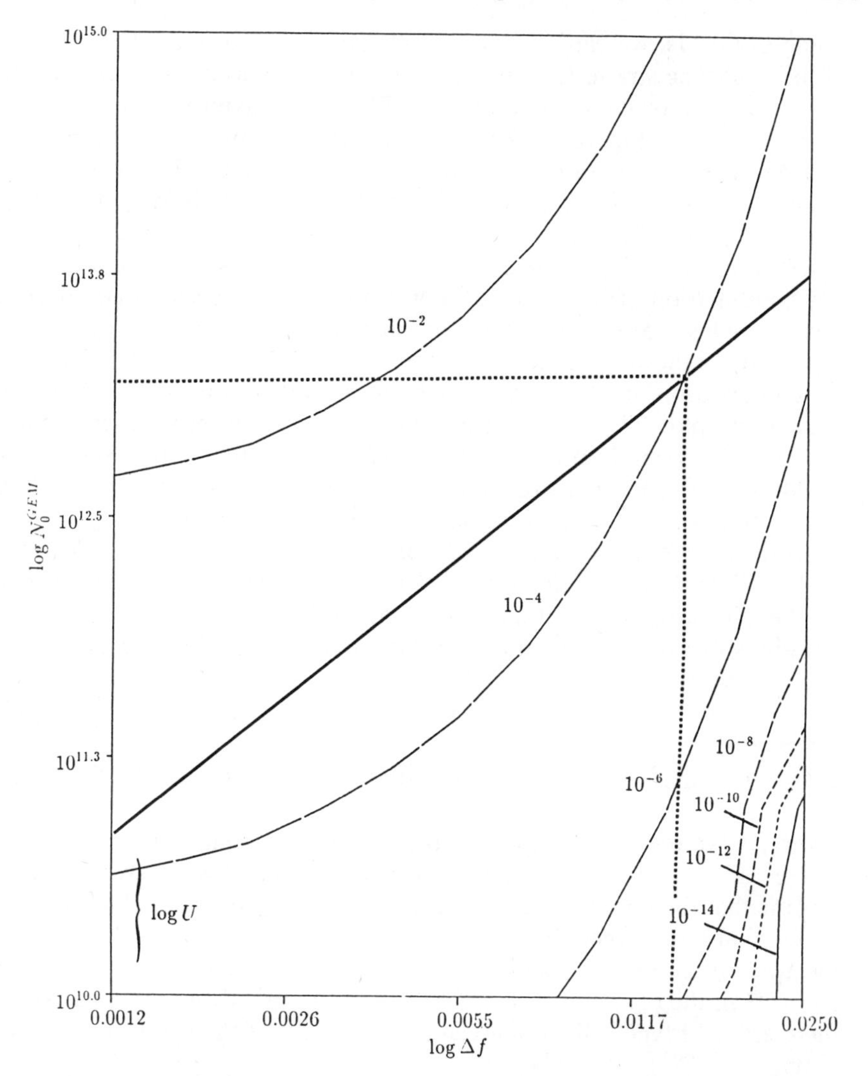

FIGURE 10–3 Contour plot of the risk surface from Figure 10–2. N_0^{GEM} is the initial population size of the GEMs, while Δf is the relative fitness differential between the GEMs and the wild-type strains. The linear line represents the utility function, and the intercept of the function is the log of utility. The intersection of the utility function with the 10^{-4} contour represents the N_0^{GEM} and the Δf values that need to be used in order to realize the risk level at the given value of utility. All axes are logarithmic but the labels are in original scale for clarity. All other parameters are assumed to be fixed, using the values given in the text.

nome of GEMs, we can control the relative fitness of the GEM with that of the wild-type strain. It is then useful to plot risk as a function of Δf and N_0^{GEM} (initial inoculation size of the GEM) for a given population size of wild-type strain. Figure 10–2 shows this plot for different values of N_0^{GEM} and Δf, using the parameters determined above and N_0^{WT} (total existing population size of the wild type before introduction of GEMs) equal to 10^{12} cells. The plot is given as a "log-log-log" plot for clarity. This can be plotted as a contour plot (Figure 10–3) and used to graphically evaluate the risk for any combination of N_0^{GEM} and Δf. For example, if we decide that 10^{-6} is an acceptable level of risk, and there are approximately 10^{12} cells of the wild-type strain in the area where we want to release the GEMs, we can determine the combination of N_0^{GEM} and Δf that corresponds to this level of risk from the contour plot. Therefore, if 10^{12} cells of GEM are to be released, we need to insure that there is at least $10^{-1.93} = 0.012$ fitness differential between the GEM and the wild-type strains. It can be seen from both Figure 10–2 and Figure 10–3 that the change in the risk surface is much faster with respect to Δf than N_0^{GEM}. *This implies that it is much more important to determine and regulate the fitness differential than the total amount of cells released.*

For the simple case of $\Delta f = 0$, or neutral relative fitness, we can analytically solve the risk equation. In this case equation 15 reduces to

$$\text{Risk} = \frac{N_s^{GEM}}{N_s^{GEM} + N_s^{WT}} \tag{23}$$

This is just the ratio between the mutational exposure of the GEMs and the wild-type strains to advantageous mutations. Since the fitness differential is zero, we cannot use equations 2 and 3 to determine mutational exposure. There will not be any exponential decay of the GEMs. However, we can assume that the total mutational exposure will be proportional to the initial population sizes. Since the risk in this case is only sensitive to the ratio of the two mutational exposures, we can substitute N_0^{GEM} and N_0^{WT} for N_s^{GEM} and N_s^{WT} in the equation above. Therefore, if more GEMs are released than there are wild-type cells in a given area, the risk of invasion can approach unity.

10.1.3 Applications of Extreme Value Risk Analysis to the Suicide-Gene Model

If the GEMs could be selectively killed without disturbing the natural flora, then much of the risk would be eliminated. One way of achieving selective death is by inserting genes into the GEMs that are capable of killing the host cell when triggered by an external cue. Such genes have been dubbed "suicide genes." Bej et al. (1988) constructed a model suicide vector for *E. coli* that contained the *hok* gene, which codes for a lethal polypeptide under the control of the *lac* promoter. The *hok* gene could be induced by the addition of isopropyl-β-**D**-thiogalactopyranoside, which induces the *lac* promoter. This suicide vector also carries a gene for resistance to the antibiotic carbenicillin, which, when added to the growth medium, provided strong selection against losing the suicide vector. However, tests showed that even though the induction of the *hok* gene initially reduced the cell numbers, some cells survived and increased in numbers after the induction. The surviving cells had intact *hok* genes but seemed to have mutations elsewhere that made them immune to its action. As exemplified by this study, suicide genes do not completely eliminate the risk, and we must carefully analyze the factors affecting the fate of GEMs with suicide genes.

The suicide-gene model may be viewed as a process that is comprised of two steps—before and after the activation of the gene by the trigger. During the first step, the GEMs carrying the suicide gene will interact with the existing flora. As outlined in the risk model above, the GEMs and the wild-type strains will both give off advantageous mutants, and there will be competitive interaction between the two. Therefore, at the end of the first step, the GEMs and its mutant derivatives may either have competitively excluded the existing wild-type strains, or they themselves may have been excluded, or they may be in some intermediate state with the wild-type strains. At the same time, however, mutations may also occur in the suicide gene such that the GEM may no longer be killed by the external trigger (these are termed immune mutations). The mutations to immunity may also have pleiotropic effect with respect to fitness such that fitness may be either increased or decreased as a result of immune mutations. In the second step, the external trigger will be applied, and all GEMs containing functional suicide genes will be eliminated. The population of GEMs will then be reduced to that of mutants with inactive suicide genes. After this reduction event, we will again have competitive interaction among the GEMs, the wild types, and their mutant derivatives.

The dynamics of what happens is complex. We must consider the population dynamics of the interaction among the GEMs and the wild-type strains and their fitness derivatives, as well as the probability of mutations in the suicide genes. The difficulty in the analysis arises mainly due to the fact that the induction of the suicide gene introduces an abrupt discontinuity in the dynamics. We will call the time of the triggering t_c, for critical time. Because of this discontinuity, the risk model presented above can only be applied to the case of suicide genes during two periods separately, i.e., before and after t_c, but not simultaneously. Therefore, we need to be able to estimate the parameters of the model, Δf, N_0^{GEM}, and N_0^{WT} for the time point t_c, from the values of the parameters at time $t = 0$. We can do this rather simply for two extreme cases: first, if t_c is very large (i.e., we induce the suicide gene after a long period), we can ignore the presence of the suicide gene, and the risk can be calculated with respect to only the initial fitness difference, Δf, and the respective initial population sizes, N_0^{GEM}, N_0^{WT}. Second, if t_c is small enough, such that there is not enough time for more than one mutation to occur, we may assume that there are no additional fitness mutations in the immune mutants. In this case, the risk model may be used with the parameters Δf, N_0^{WT}, and $N_0^{GEM} \times$ (probability of mutation to immunity). The last term, $N_0^{GEM} \ldots$, must be taken over the time period between release and t_c. For intermediate cases, we must be able to estimate two quantities: the absolute numbers of mutants immune to the actions of the suicide gene, and the relative fitness of these mutants with respect to the wild-type strain. This implies that there are two different dynamics that must be considered—the "within-population" dynamics of change in frequency of the immune mutants within the GEM population, and the "between-population" dynamics of competition between the GEMs and wild-type strains. The combination of the two will determine the absolute numbers of immune mutants that are expected at time t_c.

If the immune mutations have no effects on the fitness (i.e., are neutral), the mutants will accumulate at the neutral drift rate of haploid organisms. However, if there are mutations elsewhere in the genome that cause fitness changes, the frequency of immune mutants will also change as a result of linkage. Change in the competition dynamics of the GEMs and wild types will also cause concomitant change in the absolute numbers of immune mutants. The dynamics in this case will be similar to that of neutral mutations observed in periodic selection in chemostats (Dykhuizen and Hartl 1983). It may also be possible that the immune mutations themselves have pleiotropic effects on the fitness. In fact, in the example of Bej et al. (1988), the immune mutation caused changes in the cell membrane resulting in selection against the immune mutants when the suicide gene was not induced. In this case, the immune mutants may achieve mutation-selection balance and maintain an equilibrium frequency. In the case of positive effect, mutations will increase in frequency much faster than the neutral rate (Moser 1958). It may be possible to determine the pleiotropic effects of immune mutations through experimental work. Although we would not be able to observe every possible mutation that leads to immunity, it is likely that a given suicide-gene system will pose strong constraints on possible immune mutations. Although these cases describe the dynamics of the frequency change, it is very difficult to follow the associated change in relative fitness. This is especially true since the fitness of the wild type will also be changing through mutations. We will, however, defer discussion of this problem at this time and approach it in future papers.

Reducing the population size of GEMs can greatly reduce the risk of invasion. However, if the goal of the release of the GEMs is to perform some specific function, we would want to increase the population size as much as possible to increase the chances of success. Suicide genes greatly facilitate this by reducing the population size by several orders of magnitude when the suicide trigger is applied. Under the potential selection pressure of the triggering event, the cells carrying the suicide gene must first become immune in order to affect the fate of the GEMs. If such immune mutations carry fitness costs, as in the *hok* gene example, suicide genes may further reduce risk by constraining the fitness of the GEMs. However, we cannot assume that these genes will eliminate all risk by completely eliminating the GEMs, since the risk of mutations to immunity increases with time and population size. This must be carefully considered in further analyses.

10.2 DISCUSSION

There are actually two different types of events that need to be avoided during the release of GEMs. First, there is the possibility of the released organisms causing an ecological imbalance by displacing the existing flora. Therefore, although the GEMs may not actively cause any harm, they may have unknown consequences for the ecosystem. One can imagine, for example, that some organisms engineered to prevent root fungal disease in crops could eliminate the nitrogen-fixing bacteria in the soil. Second, there is the possible harm caused by the unknown side effects of the action of the released organisms. Engineering the strains to prevent ice nucleation, for example, may also cause it to produce some undesirable by-products in certain environments. Even if we are able to determine the actions of the engineered organisms in the laboratory, further mutational changes may cause the strain to produce undesirable products.

It seems obvious from considerations of our risk model that no GEMs should ever be released that are competitively superior to existing flora. From an overall ecological perspective, three different risky events may happen to competitively *inferior* GEMs released into nature. First, the GEMs may acquire advantageous mutations and increase in numbers for a while but eventually be eliminated by even-better wild-type mutant strains. Second, the GEMs may become established in some area where it is very difficult to dislodge them, even if they are competitively inferior (competition in structured habitats). Third, the GEMs may become competitively superior and completely displace the wild-type flora. If we are worried about the possible ecological imbalance caused by the GEMs, even the first type of event may be undesirable. Ecological disturbance may be caused not only through complete elimination of the existing flora, but also if the population of GEMs reaches some critical size. In this case, we would need to determine the undesirable critical population level and then follow the dynamics of all advantageous mutations of the GEMs. The second type of event would also be important if we are worried about the unknown side effects of the gene products of the GEMs. Unfortunately, the dynamics of organisms in structured environments has not been well studied, especially in terms of persistence. However, the risk of persistence of GEMs will certainly be much higher than the risk of competitive exclusion of wild-type strains. This will no doubt prove to be a fertile field of research in the future.

The last type of event probably has the highest costs associated with it, and it is the kind of event most commonly considered when we talk about the risk of invasion of genetically engineered organisms. This is also the kind of risk that we have considered in our risk model (see Section 10.1). Therefore, we have evaded the complex dynamics of intermediate advantageous mutations and concentrated on the question of the probability of complete exclusion of wild-type strains by the GEMs. We ask a highly simplified question: "Given x numbers of GEMs and y numbers of wild-type cells, what is the probability that one of the GEMs will be competitively superior than any of the wild type strains?" Our simplification is in the way we calculated the quantities x and y. We assumed that they could be calculated from fixed, unchanging, time-dependent functions, i.e., the chemostat equations. In fact, any advantageous mutations in either the wild-type strains or the GEMs will cause this function to change. An advantageous GEM mutant will cause the number of GEMs to increase for a while even if it is eventually displaced by a more fit wild-type mutant arising later. Therefore, the equations for the quantity x that we give may be an underestimate. On the other hand, advantageous mutations in the wild type could cause GEMs to be eliminated faster than expected and result in an overestimate. Analogous cases could also be found for quantity y that would result either in its underestimate or overestimate. In reality, both quantities x and y are fitness dependent. While we are concerned with the risk from fitness changes of the strains, we assume that x and y are fitness *independent*. This may seem contradictory, but the assumption was made in order to make the problem analytically tractable. We propose this as a first approximation to the problem. More realistic approaches may have to be developed through computer simulations. However, if the degree of underestimation of both x and y are similar, from consideration of equation 23, our approximation may prove to be robust, and the gross observations from the model should be fair.

Using our model we may make some additional interesting observations. As with most chemical agents, greater quantities of the organisms will probably make it more efficient in performing the desired function. Therefore, the utility of a GEM will directly depend on the population size of the GEMs and the duration of their action. In other words, the utility of the organisms will depend on the same factors that determine its risk. *The higher the utility, the more risk will be taken.* It will be desirable to find a way of maximizing the utility while keeping risk at acceptable levels. Figure 10–3 shows the contour plot of risk against the log-log axis of initial population size and Δf. The utility of a GEM can be expressed as the total number of individuals expected perform the desirable action, or N_e^{GEM} given by equation 2. If we express this as U for utility and take the log on both sides of the equation, it becomes

$$\log U = \log N_0^{GEM} - \log \Delta f \tag{24}$$

This can be represented by a line with a slope of 45 degrees and with log U as the intercept in Figure 10–3. Therefore, we can maximize utility for given level of risk by taking the intersection of the risk contour and the line represented by equation 24. The intersection of the utility line with a given risk level would show what the values of N_0^{GEM} and Δf must be to realize that risk and utility. Obviously to maximize utility, we would want the intercept of the utility line to be as high as possible, while maintaining the intersection with the desired risk contour. As can be seen in the figure, we may increase utility indefinitely, as long as we reduce fitness. This is due to the fact that the risk surface is much more dependent on fitness differential than on initial population size. *Thus, it will be desirable to use large inoculations of the GEMs while keeping the fitness low, thereby reducing the residence time.*

The parameters of the model described above are all determinable using existing experimental systems but have not been studied enough to obtain the appropriate values. We propose that the relative fitness can roughly be measured using chemostats, or similar devices such as gradiostats (Wimpenny 1982). Studies of relative fitness under various conditions—temperature, energy or nitrogen limitation, etc.—in microcosms could provide an understanding of the relative competitiveness. This approach has been used successfully to predict which species of algae would be selected during various periods of the year in a shallow lake (Zevenboom 1980). The distribution of advantageous mutations may be determined in several ways. One would be to induce random mutations using mutagens and then to measure fitness using chemostats. The fitness of strains before and after the application of mutagens can be used to assess both the distribution of random fitness mutations and their dependence on initial fitness. With more understanding of the sources of fitness and mutational variation, a guideline could be constructed where a set of required experiments are specified from which a predetermined set of parameters must be measured. Therefore, comprehensive risk management may involve the selection of a suitable mathematical risk model with the incorporation of appropriate experimental work.

Engineering can never completely prevent disasters. Even for physical problems such as the design of nuclear reactors, risk is quantified through the use of probabilistic models. Biological organisms, however, are much more volatile than physical materials or mechanical constructs. But just as material failure rates can be determined by engineering tests, biological "failure rates" may be estimated using proper experiments. Such experiments must be designed keeping in mind suitable quantitative risk models in order to determine what it is that needs to be measured.

APPENDIX 10A APPROXIMATE MUTATIONAL EXPOSURE OF GEMs

We assume that the engineered strain is less fit than the wild-type strain, and we want to estimate total exposure of the introduced strain to mutations, N_e^{GEM}, defined as

$$N_e^{GEM} = \int_0^\infty N^{GEM}(t)dt \qquad (25)$$

We assume the competition process to be governed by the following chemostat-like equations:

$$\frac{dN^{WT}}{dt} = N^{WT}[\mu_{WT}(S) - D]$$

$$\frac{dN^{GEM}}{dt} = N^{GEM}[\mu_{GEM}(S) - D]$$

$$\frac{dS}{dt} = (S_0 - S)D - K_{WT}\mu_{WT}(S)N^{WT} - K_{GEM}\mu_{GEM}(S)N^{GEM}$$

Where N^{WT}, N^{GEM} is the abundance of the wild strain and the engineered strain respectively, D is the dilution rate, S_0 is the concentration of the incoming substrate, S is the concentration of the substrate in the media, K_{WT}, K_{GEM} are efficiency constants, $\mu_{WT}(S)$, $\mu_{GEM}(S)$ are the functions relating the growth rate of both strains to the substrate concentration.

Functions of μ_{WT} and μ_{GEM} are usually assumed to be of the Michaelis-Menten type and have the general form

$$\mu(S) = \frac{\mu_{max}S}{S + K_s} \tag{26}$$

where μ_{max} is the maximum specific growth rate and K_s is the saturation coefficient, or the efficiency coefficient. Assume that prior to the release, the wild-type strain is at equilibrium, that is,

$$\mu_{WT}(S) = D, S = S^* \tag{27}$$

Immediately after the release of the engineered strains, the substrate concentration in the medium will decline causing a decline in N^{WT} followed by the asymptotic recovery of N^{WT} to its equilibrium level. N^{GEM} will be declining through time, first somewhat faster, then asymptotically, at the rate of $\mu_{GEM}(S^*) - D = \mu_{GEM}(S^*) - \mu_{WT}(S^*)$. We will denote this value as $-\Delta f$, the deficiency in fitness of the engineered strain relative to the wild strain. It will be conservative to assume that the rate of decline of N^{GEM} is equal to its asymptotic rate all the time. This will only lead to the overestimation of the N^{GEM} (effective). Under this assumption,

$$N^{GEM}(t) \approx N_0^{GEM}e^{-\Delta ft} \tag{28}$$

and

$$N_e^{GEM} \approx \frac{N_0^{GEM}}{\Delta f} \tag{29}$$

The degree of conservatism of this estimate depends on how quickly the substrate concentration recovers towards its equilibrium value of S^*. If we make the reasonable assumption that $S \rightarrow S^*$ much faster than $N^{GEM}(t)$

$\rightarrow 0$, the approximation will be quite accurate. In any case, we are operating here with orders of magnitude rather than with precise values since only $\log(N_e^{GEM})$ will participate in our final estimates. For this purpose equation 29 can be viewed as sufficiently accurate.

Certainly, when $\Delta f = 0$, two strains are selectively neutral with respect to each other and the engineered strain will have infinite exposure. With the selective differential $\Delta f = 10^{-2}$, the engineered strain will effectively expose to mutations $10^2 \times N_0^{GEM}$ cells for the overall period of time until it reaches extinction.

APPENDIX 10B ASYMPTOTIC DISTRIBUTION OF THE LARGEST VALUE IN A SAMPLE

The following derivation of the asymptotic extreme value distribution is for cases where the underlying probability distribution is unbounded to the right and decreases exponentially. Most useful distributions, such as exponential, normal, and gamma distributions, fall into this category. The derivation is adapted from Gumbel (1954).

We desire to obtain the distribution of the random variable that represents the largest value of n independent samples taken from a known underlying distribution. Since each of the n observations are independent, the exact form of the c.d.f. (cumulative distribution function) and the p.d.f. (probability density function) is simply

$$\Phi_n(x) = F^n(x) \tag{30}$$

$$\phi_n(x) = nF^{n-1}(x)f(x) \tag{31}$$

where $F(x)$ and $f(x)$ are the c.d.f. and p.d.f. of the underlying distribution from which the samples are obtained, that is, the probability that all n observations are less than or equal to x. Although this form is analytically simple, it is often numerically difficult to compute when n becomes large. Therefore, an asymptotic distribution is derived that is valid for large n and large values of x.

First, two quantities are defined that will be used later. The parameter u_n is defined as the value satisfying the equation

$$F(u_n) = 1 - 1/n \tag{32}$$

that is, the quantile at which the underlying c.d.f. equals $1 - 1/n$. This value is called the *characteristic largest value* (Gumbel 1954) and represents the quantile at which the expected number of observations in a sample of n exceeding this value is 1. The second parameter, α_n, is called the *extremal intensity* and is defined as

$$\alpha_n = \frac{f(u_n)}{1 - F(u_n)} = \frac{f(u_n)}{1 - 1 + 1/n} = nf(u_n) \tag{33}$$

Differentiation of equation 32 with respect to n gives

$$f(u_n)\frac{du_n}{dn} = \frac{1}{n^2} \tag{34}$$

and from equation 33,

$$\frac{du_n}{d \ln n} = \frac{1}{\alpha_n} \tag{35}$$

We will first find the mode of the largest value distribution and derive the asymptotic distribution around the mode. Differentiation of $\phi_n(x)$ leads to

$$\frac{d\phi_n}{dx} = n[(n-1)F(x)^{n-2}f(x)^2 + F(x)^{n-1}f'(x)] \tag{36}$$

and setting the right-hand side to 0,

$$(n-1)\frac{f(x)}{F(x)} = \frac{-f'(x)}{f(x)} \tag{37}$$

The mode is x that satisfies this equation.

Since we are concerned with large values of x, we use the fact that as x becomes large ($x \to \infty$), $f(x) \to 0$ and $1 - F(x) \to 0$. We assume that this is sufficiently fast that we can use L'Hôpital's rule and set

$$\frac{f(x)}{1 - F(x)} \sim \frac{f(x)'}{-f(x)} \tag{38}$$

From equations 37 and 38 it can be seen that the mode goes to u_n as x becomes large.

We now expand the underlying distribution in Taylor's series around the mode u_n, and write

$$F(x) = F(u_n) + F'(u_n)(x - u_n) + 1/2F''(u_n)(x - u_n)^2 + \dots + 1/n!F^{(n)}(u_n)(x - u_n)^n + \delta \tag{39}$$

We assume each derivative of the underlying p.d.f. exists and approaches 0 for the kinds of distributions mentioned above, and we will again use L'Hôpital's rule to obtain approximate relations for each derivative of $F(x)$:

$$F(u_n) = 1 - \frac{1}{n}$$

$$F'(u_n) = f(u_n) = \frac{\alpha_n}{n}$$

$$F''(u_n) = f'(u_n) \approx -f(u_n)\alpha_n = -\frac{-\alpha_n^2}{n}$$

$$F'''(u_n) = f''(u_n) \approx \frac{f'(x)^2}{f(x)} = \frac{\alpha_n^3}{n}$$

$$\vdots$$

and

$$F(x) = 1 - \frac{1}{n} + \frac{\alpha_n}{n}(x - u_n) - \frac{1}{2}\frac{\alpha_n^2}{n}(x - u_n)^2 + \ldots$$
$$- \frac{1}{2k!}\frac{\alpha_n^{2k}}{n}(x - u_n)^{2k} + \frac{1}{2k+1!}\frac{\alpha_n^{2k+1}}{n}(x - u_n)^{2k+1} + \delta \quad (40)$$

Therefore,

$$F(x) \approx 1 - \frac{1}{n} e^{-\alpha_n(x - u_n)} \quad (41)$$

and the desired equation for the largest value distribution is

$$\Phi^n(x) = F^n(x) = (1 - \frac{1}{n} e^{-\alpha_n(x - u_n)})^n \approx \exp(-e^{-\alpha_n(x - u_n)}) \quad (42)$$

and the p.d.f. is

$$\phi^n(x) = \alpha_n \exp[-\alpha_n(x - u_n) - e^{-\alpha_n(x - u_n)}] \quad (43)$$

The conditions for this approximation to hold depend on the use of L'Hôpital's rule for the convergence of the Taylor series. It can be shown that distributions with exponential decay and with the existence of all moments converge to this asymptote as n becomes large. Exact conditions can be found in Gnedenko (1943).

REFERENCES

Atlas, R.M., and Bartha, R. (1987) in *Microbial Ecology*, 2nd ed., pp. 195–223, Benjamin/Cummings Co., Reading, MA.
Atwood, K.C., Schneider, L.K., and Ryan, F.J. (1951) *Proc. Natl. Acad. Sci.* 37, 146–155.
Bej, A.K., Perlin, M.H., and Atlas, R.M. (1988) *App. Environ. Microbiol.* 54, 2472–2477.
Chao, L., and Cox, E.C. (1983) *Evolution* 37, 125–134.
Dykhuizen, D., and Hartl, D. (1981) *Evolution* 35, 581–594.
Dykhuizen, D.E., and Hartl, D.L. (1983) *Microbiol. Rev.* 47, 150–168.
Gnedenko, B. (1943) *Ann. Math.* 44, 423–453.
Gumbel, E.J. (1954) *Statistical Theory of Extreme Values and Some Practical Applications*, Appl. Math. Ser. 33, U.S. Dept. Commerce, Washington, D.C.
Helling, R.B., Vargas, C.N., and Adams, J. (1987) *Genetics* 116, 349–358.

Kubitschek, H.E. (1970) *Introduction to Research with Continuous Cultures*, Prentice Hall, Inc., Englewood Cliffs, NJ.
Lenski, R.E. (1988) *Evolution* 42, 433–440.
Lindow, S.E., Panopoulos, N.J., and McFarland, B.L. (1989) *Science* 244, 1300–1307.
Monod, J. (1950) *Ann. Inst. Pasteur Paris* 79, 390–410.
Moser, H. (1958) *The Dynamics of Bacterial Populations Maintained in the Chemostat*, Carnegie Institution of Washington, Washington, DC.
O'Brien, S.J., and Evermann, J.F. (1988) *TREE* 3, 254–259.
Simberloff, D. (1988) *Ann. Rev. Ecol. Syst.* 19, 473–511.
Wimpenny, J.W.T. (1982) *Phil. Trans. R. Soc. Lond. B* 297, 497–515.
Zevenboom, W. (1980) *Growth and Nutrient Uptake Kinetics of Oscillatoria Agardii.* Krips Repro Meppel, Amsterdam.

CHAPTER 11

Quantifying the Spread of Recombinant Genes and Organisms

Robin Manasse
Peter Kareiva

Before anything can be said about the risk of releasing genetically engineered organisms, one needs to know the *rate* at which these organisms are likely to spread (U.S. Congress 1988). By this we mean the rate at which populations (or genotypes) grow in number and come to occupy larger and larger areas. Spread is thus the combined result of survival, reproduction, and dispersal. All else being equal, it is clear that an organism whose rate of spread is minimal poses negligible risk compared to an organism that can multiply its population and rapidly expand its range. Indeed, it is the prospect of uncontained spread that underlies many of the worries environmentalists express regarding biotechnology.

We thank Steven Lindow of the University of California–Berkeley, and Xavier Delannay and Sally Metz of Monsanto Co. for access to unpublished information. We could not have written this paper without a correspondence course on spread models given to us by Hans Weinberger. Bill Morris contributed insightful comments on the manuscript, and Selina Gleason prepared the figures and helped summarize the results of simulations. This material is based on work supported by the Cooperative State Research Service, USDA (Biotic Stress, Weed Science Program); any findings, conclusions, or recommendations expressed in this publication are those of the authors and do not necessarily reflect the view of the USDA.

Despite theoretical explorations of problems that might be encountered upon field releases of genetically engineered organisms, ecologists have offered little practical advice about how to use data from field trials to anticipate the spread of these organisms (Tiedje et al. 1989). The absence of such advice is surprising because there is a rich theoretical literature that might be mined for insight regarding the risk of spread (Diekmann 1978; Fisher 1937; Mollison 1977; Murray et al. 1986; Okubo 1980; Okubo et al. 1990; Shigesada et al. 1986; Skellam 1951; Thieme 1979; van den Bosch et al. 1988; Weinberger 1978 and 1982). Drawing on this existing theory, we aim to prompt population biologists to collaborate with those performing field releases in order to develop predictive spread models.

We begin by reviewing the mathematical models that have been used to address questions of spread in a variety of evolutionary and ecological contexts. These are the models that should provide a jumping-off point for any particular "real-world" spread assessment. Then we describe some simulation results that highlight the spread one is likely to observe for advantageous transgenic genes that escape (through pollen movement and hybridization) from a crop into neighboring weed populations. We conclude by examining the results of two experimental field releases of transgenic organisms (rapeseed and "ice-minus" bacteria) in the context of spread models. We use these case studies to emphasize that our discussion is not about purely abstract matters, and to suggest that the data from field releases might be improved by consideration of theoretical approaches prior to the design of sampling schemes.

11.1 REVIEW OF SPREAD MODELS

Theoreticians have examined the spread of organisms or genes by developing models that combine the effects of dispersal and population growth. The standard modeling tools are partial differential equations (e.g., Fisher 1937; Skellam 1951), and more recently, difference equations (Weinberger 1978 and 1982), and stochastic point processes (Mollison 1977). Although the details vary, all of these approaches use mathematics to portray changes in population densities or gene frequencies in time and space. After such a model has been formulated and the initial conditions specified (i.e., the initial mutation, invasion, or release described in mathematical terms), spread can be investigated by calculating the population densities, or gene frequencies, over a grid of spatial coordinates in successive time intervals. Because such predictions are tedious and difficult to develop, one usually seeks more concise descriptions of spread. An especially useful summary statistic that serves this purpose is called the "asymptotic rate of spread" (ARS hereafter). ARS, measured in units of distance/time, is the rate at which an organism's or gene's aerial range increases in radius. (Asymptotic refers to the fact that this rate of range expansion often starts off slowly and accelerates towards a constant that is determined by the life history traits of the organism or gene in question.) Although ARS is an upper bound on the velocity of range expansion, it is a bound that may be attained within a modest number of generations when one considers actual reproductive and dispersal rates (e.g., Lubina and Levin 1988; Andow et al. 1990).

Much of the mathematical literature on spread is concerned with proving that an ARS exists for different models; since an ARS need not exist in principle, this is an important exercise. Of course, those concerned primarily with the real-world problem of risk assessment are probably not going to worry about proving the existence of an ARS. Instead, more practical paths would be: (1) to explore the literature in search of models that approximate the biology of interest and for which an ARS has been calculated, (2) to run computer simulations that yield actual rates of spread, or (3) to enlist the aid of applied mathematicians who might be able to generate analytical results for a custom-designed model. Whatever route is taken, we encourage attempts to quantify the rate of spread, because they provide a common currency for comparing rates of spread among different types of transgenic organisms, ecological conditions, or release protocols. Without such a quantitative estimate of spread, risk analysis is doomed to anecdotal status.

11.1.1 The Spread of Genes in Idealized Populations

Over 50 years ago, R.A. Fisher asked how rapidly a mutant gene initially appearing in a single location would spread through a spatially distributed population (Fisher 1937). The model he used dealt with spread in a one-dimensional environment (i.e., a line), but can easily be extended to two dimensions. He addressed the problem by assuming a continuously breeding diploid population of constant density that was subject to weak selection, and in which gene flow could be described as a diffusion process:

$$\frac{\partial p(x)}{\partial t} = D\frac{\partial^2 p(x)}{\partial x^2} + mp(x)[1 - p(x)] \qquad (1)$$

where $p(x)$ is the frequency of the new mutant at position x, D is the diffusion coefficient for genes, and m is the intensity of selection in favor of the mutant. Fisher conjectured (and Kolmogorov et al. [1937] rigorously established) that this particular "diffusion and selection" equation predicts an asymptotic spread of the introduced gene in the form of a traveling wave with an ARS equal to $2\sqrt{Dm}$. Although it represents a pioneering effort, Fisher's model should be used cautiously because implicit in its derivation is a simplifying step that cannot be rigorously justified. In particular, Fisher assumed a Hardy-Weinberg equilibrium in order to reduce information on the relative frequency of three genotypes to a statement about the frequency of one allele (p in equation 1). If there were no (or only extremely weak) selection, this would not be a problem; but when there is selection and when one is dealing with continuously breeding (or reproducing) organisms (which is implicit in the partial differential equation formulation), the simplification used by Fisher is not appropriate (the reasons for this can be found in Weinberger [1978]). For continuously breeding organisms, it is only under very special circumstances that one can reduce the spatio-temporal dynamics of three genotypes to one equation (Weinberger 1978). But while Fisher's original theoretical treatment may be on shaky ground, subsequent models have been improved with respect to both biological realism and mathematical rigor (e.g., Aronson and Weinberger 1975).

One of the most important recent contributions to the theory of gene spread is an analysis by Weinberger (1978) that is tailored to populations that have discrete generations (such as annual plants), experience strong selection, and for which gene flow can be represented as virtually any continuous function (Weinberger 1978 and 1982). To illustrate Weinberger's approach, let us consider a concrete example that entails the spread of a mutant gene through an annual plant population with the following life cycle: selection—migration—random mating—production of the next generation of adult plants—and then selection again. We begin by defining $u_N(x)$ as the frequency of a dominant mutant, denoted A, in the Nth generation at position x (we treat the problem in a one-dimensional environment for convenience). The recursion equation below iterates the $u(x)$ spatial profile of gene frequency from one generation to the next:

$$u_{N+1}(x) = \int v(x-y)\, g[u_N(y)]\, dy \tag{2a}$$

where, dropping the generation subscript for convenience,

$$g[u(y)] = \frac{W_{AA}\, u(y)^2 + W_{Aa}\, u(y)\, [1 - u(y)]}{W_{AA}\, u(y)^2 + 2W_{Aa}\, u(y)[1 - u(y)] + W_{aa}[1 - u(y)]^2} \tag{2b}$$

Here, W_{AA}, W_{Aa}, and W_{aa} are the relative fitnesses of the AA, Aa, and aa genotypes, respectively; $v(x-y)$ is the probability of migration from position y to x; and g(u) is a function that takes u from one generation to the next as a result of selection and random mating. Whereas equation 1 assumes that migration is diffusive, the model above allows $v(x)$ to be a probability density function of any shape. The probability density could thus be chosen to match the leptokurtic distributions of dispersal distances that have repeatedly been found in empirical studies of seed or pollen movement (Bateman 1947; Ellstrand and Foster 1983; Handel 1982, 1983, and 1985; Levin and Kerster 1974; Nieuwhof 1963; Schaal 1980). The convolution integral in equation 2a sums up over the entire line all the genes that arrive at each point x, following the events summarized by $g(u)$, with v weighting the contribution from each point depending on how far away it is. Weinberger has shown that the ARS for this model is:

$$\text{ARS} = \min_{u>0} \{\frac{1}{u}\log[g'(0)\int \exp^{ux} v(x)dx]\} \tag{3}$$

Once the function v and the relative fitnesses of genotypes are known, it is often possible to compute an ARS in terms of a few biologically measurable parameters (an example is given later). The way one uses the above seemingly abstract expression is to substitute in $v(x)$ and $g'(0)$, and then set the derivative of equation 3 equal to zero as a means of identifying the minimizing condition. Thus, although equation 3 lacks the simplicity of $2\sqrt{Dm}$, it is no less valuable a result.

The models described above may be too simplistic to describe the spread of particular recombinant genes. However, they do illustrate the two major types of models one would want to pursue when examining gene spread: partial differential equations if the organisms were continuously reproducing microbes, or recursion equations for discrete breeders such as transgenic plants.

11.1.2 The Spread of Populations in Idealized Environments

Fisher's model for the spread of genes can easily be redefined to describe the spread of organisms. Moreover, the problems associated with collapsing three genotype equations into one gene frequency equation disappear if we are concerned only with the density of organisms. In fact, mathematical analyses of models such as Fisher's have concentrated much more on ecological as opposed to genetic questions, and ecological theories of spread have become quite general. For instance, if we imagine a continuously breeding organism whose dispersal can be approximated by a diffusion process, then that organism's spread can be analyzed using the following model:

$$\frac{\partial N(x,y,t)}{\partial t} = D\left(\frac{\partial^2 N}{\partial x^2} + \frac{\partial^2 N}{\partial y^2}\right) + F(N) \tag{4}$$

where $N(x,y,t)$ is the density of organisms at position x,y and time t, $F(N)$ is the rate of population growth (which is assumed to be a function of local density), and D is the diffusion coefficient for the organism. For most plausible population growth functions, a point inoculum of organisms governed by equation 4 will attain an ARS of:

$$\text{ARS} = 2\sqrt{F'(0)D} \tag{5}$$

where $F'(0)$ is the per capita rate of population growth when population densities are at zero. Thus, if $F(N)$ were a logistic equation, $F'(0)$ would be the intrinsic rate of increase for the species in question. One advantage of the above formula is that it requires knowledge of only two aspects of a species' life history: the average dispersal distance and the rate of population growth at low densities. Although this simple diffusion model might seem hopelessly unrealistic, equation 5 has actually been used to predict geographical rates of range expansion as recorded over a timespan of decades (e.g., Andow et al. 1990). Of course, there are counterexamples for which diffusion models totally fail; but the point is not that these models are right, but that they are eminently testable. It is also worth noting that many of the less-palatable assumptions underlying equation 4 (e.g., a homogeneous environment) can be relaxed, without sacrificing the ability to generate predictions about an ARS (Bramson 1983, Diekmann 1978, Shigesada et al. 1986, Thieme 1979, van den Bosch et al. 1988). For instance, if we consider organisms whose demographic and dispersal rates change as a function of their habitat, and the habitat varies in some regular spatial pattern, then it is still possible to calculate an ARS. Moreover, the result looks remarkably like equation 5, with the predicted ARS given by the expression:

$$\text{ARS} = 2\sqrt{\langle D\rangle_H \langle r\rangle_A} \tag{6}$$

where $\langle D\rangle_H$ is the harmonic mean of the spatially varying diffusion coefficient, and $\langle r\rangle_A$ is the arithmetic mean for the spatially varying per capita rate of population growth at low densities (Shigesada et al. 1986). Other extensions of the theory include density-dependent dispersal and age-structured populations (Alt 1985, Hastings 1990, Turchin 1989).

One can also relax the assumption of simple diffusion and continuous reproduction by modifying Weinberger's (1978 and 1982) genetic models to address ecological questions. For instance, the spread of an annual plant whose per capita seed production declines with increased crowding by conspecifics can be described by:

$$N_{t+1}(x) = r \int v(x-\xi)\, N_t(\xi) \exp[-sN_t(\xi)]\, d\xi \tag{7}$$

where $N_t(x)$ is the population density at position x in a one-dimensional environment in year t; r is the maximum reproduction rate, which is reached as population density tends to zero; and s is a parameter representing the extent to which increases in plant density reduce per capita reproductive rates. Seed dispersal enters the model through the term $v(x-\xi)$, which is the probability that a seed from a plant at position "$x-\xi$" will disperse to position x. It can be shown that the ARS for such a plant upon invading an empty habitat is (Andersen 1987):

$$\text{ARS} = \min_{u>0} \left\{ \frac{1}{u} \log[r \int \exp^{ux} v(x) dx] \right\} \tag{8}$$

which is the ecological parallel to the ARS for gene spread in annual plants (i.e., equation 3). Notice that the information necessary for predicting the speed with which plants spread into vacant habitats can be easily obtained from field experiments—$v(x)$ could be estimated from seed dispersal experiments (e.g., McEvoy and Cox 1987)—and r is simply the rate of population growth at low densities.

11.1.3 The Spread of an Invader that Strongly Interacts with a Resident Species

A final suite of potentially useful models involves pairs of partial differential equations representing two species that interact with one another and then disperse, such as a host and a pathogen (e.g., Murray et al. 1986), or two competitors (e.g., Okubo et al. 1990). For example, the spread and population growth of an invader that competes with closely related residents can be examined by analyzing standard Lotka-Volterra equations to which a dispersal term has been added:

$$\frac{\partial N_1(x,t)}{\partial t} = r_1 N_1 \left[1 - \frac{N_1}{K_1} - \frac{a_{12}N_2}{K_1}\right] + D_1 \frac{\partial^2 N_1}{\partial x^2} \quad (9a)$$

$$\frac{\partial N_2(x,t)}{\partial t} = r_2 N_2 \left[1 - \frac{N_2}{K_2} - \frac{a_{21}N_1}{K_2}\right] + D_2 \frac{\partial^2 N_2}{\partial x^2} \quad (9b)$$

Here $N_1(x,t)$ is the density of species 1 at position x and time t; $N_2(x,t)$ is the corresponding density of species 2; D_1 is the diffusion coefficient for species 1; K_1 is the carrying capacity for species 1; r_1 is the intrinsic rate of increase for species 1; a_{12} is the competitive effect of species 2 on species 1; and so forth. Given the above model, we can ask at what rate should we expect species 1 to spread when introduced into an environment filled with species 2. For certain special cases, it is possible to compute an ARS for the invader in terms of r_1, D_1, K_1 and a_{12} (assuming the invader is species 1). Surprisingly, the resulting ARS is not very different from the ARS given by equation 5 (Okubo et al. 1990). In fact, although equations 9a and 9b represent an enormous simplification of interspecific competition and dispersal, they may nonetheless capture enough of the essentials to truly describe the spread of real organisms. For instance, they do a good job predicting the spread of the grey squirrel, which was introduced into Britain at the turn of the 20th century and subsequently has replaced the native red squirrel (Okubo et al. 1990).

11.2 USING FIELD DATA TO SIMULATE SHORT-TERM SPREAD

Ultimately, we seek to use data from field releases (such as those described below) to predict a range of plausible spread rates. We are particularly interested in examining spread over the short term under a wide variety of ecological circumstances, and in comparing our predictions with both field experiments and less-detailed, but more general, analytical results (e.g., equations 3, 5, and 6). To address these issues, we have developed a simulation model that yields predictions of spread for a recombinant gene that enters an annual weed population. By presenting this simulation we hope to make clear the data one requires to predict spread, and to illustrate the extent to which a biologically detailed simulation model meshes with the asymptotic calculations of Weinberger (1978 and 1982).

For convenience, we imagine a population of annual plants that occupies a linear habitat 660 m in length, and that is subdivided into 10-m sections. If spread is not directionally dependent, a linear habitat can be viewed as a transect through a two-dimensional habitat. We imagine a dominant, selectively advantageous recombinant gene that appears at one end of this linear habitat at an initial frequency of 1.0 (as homozygotes). For a variety of selective regimes and dispersal rates, we want to know how rapidly that gene will spread outward. The population of plants is assumed to be at a stable equilibrium density, and to initially include homozygous recessives for the wild-type gene at every position except where the recombinant gene first appears. Pollen is assumed to disperse according to an exponential distribution, and seed dispersal is discounted as negligible compared to pollen movement. We assumed that the recombinant gene is dominant, which is the usual case for "engineered" traits such as herbicide resistance. We simulated spread according to the algorithm summarized in Figure 11–1, and varied the following traits: the magnitude of pollen dispersal (by varying the term B in the exponential distribution $p(x) = B \exp^{-Bx}$); the relative fitness of plants carrying the recombinant gene compared to wild type; and the "patchiness" of the plant population (which was assumed to be either evenly distributed along the entire line or to be restricted to three patches, with 250 m between each patch).

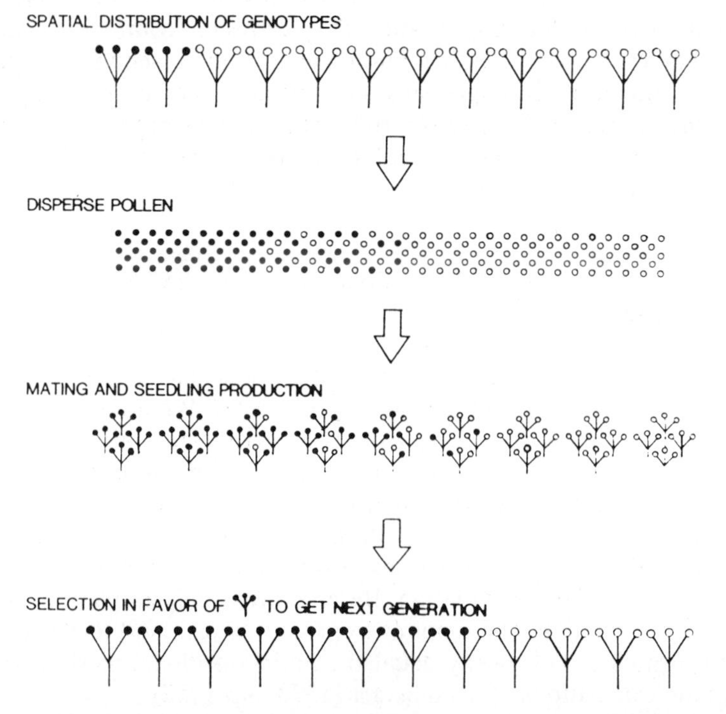

FIGURE 11–1 The sequence of events simulated in our gene-spread model for annual plants, with solid circles representing favored gametes or seeds. The solid circles spread and increase in frequency as a result of selection and dispersal.

We examined the spread of recombinant genes corresponding to selective coefficients against the wild type of 0.2, 0.1, and 0.025. These selection coefficients are severe underestimates if we assume herbicide is sprayed broadly and frequently, but may approximate selection given sporadic herbicide usage. We used mean pollen dispersal rates of 10 m or 20 m per generation, which represent values for *B* in the above exponential distributions of 0.1 and 0.05. These pollen dispersal rates are well within the range of reported pollen dispersal rates for the cultivar *Brassica oleracea* (Nieuwhof 1963). (More detailed analysis will appear elsewhere, in conjunction with an exploration of alternative sampling designs.) In all simulations involving the homogeneous habitat, a "wavefront" of the recombinant gene developed rapidly and moved along the linear habitat (Figure 11-2), as was predicted by classical models, such as Fisher's. We quantified spread rate in these cases simply as the number of meters the wave advanced in each generation. In agreement with theory and common sense, we found that when all else was equal, either increasing the relative fitness of the recombinant gene or increasing the dispersal rate enhanced the rate of spread. A more challenging question is exactly how do relative fitness and dispersal rate combine to generate rates of spread? Since we ran our simulations for only 50 generations and used a relatively small habitat, there

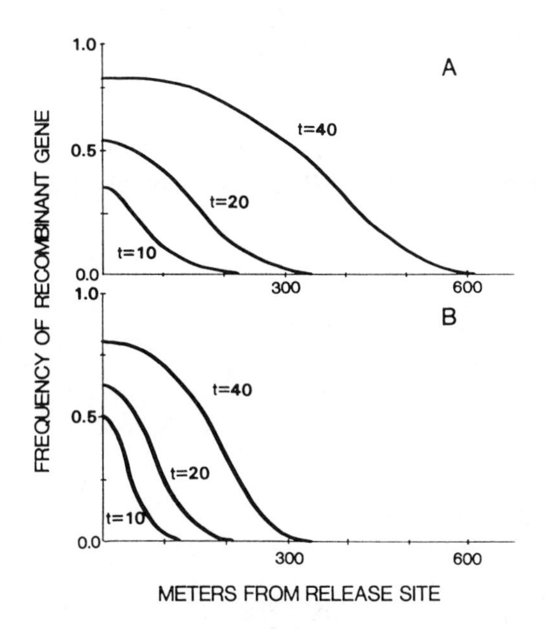

FIGURE 11-2 Two simulations of the travelling waves of the spread of recombinant genes at 10, 20, and 40 generations after the recombinant gene was first released, with a relative fitness for the "new" homozygote or heterozygote equal to 1.25 and a mean pollen dispersal rate of (A) 20 m per generation and (B) 10 m per generation.

is no guarantee we would observe anything like the asymptotic rate of spread. Nonetheless, we can use Weinberger's equation, our equation 3, to calculate the ARS for this model and ask how well the theoretically expected ARS matches up with the short-term-realized rate of spread. For our model, the ARS is given by:

$$\text{ARS} = \frac{1}{B}\sqrt{\frac{2s}{1-s}} \tag{10}$$

where $1/B$ is the mean associated with our exponential distribution of dispersal distances, and s is the coefficient of selection against the nonengineered homozygote recessive—so that, referring to equation 2b, $W_{aa} = (1-s)$, and $W_{AA} = W_{Aa} = 1.0$. We were amazed to see that the above formula predicted our observed rates of spread extraordinarily well (Figure 11–3), with an error that averaged less than 10%. It will be interesting to see whether theory maintains such success over a wider range of selection and dispersal parameters, or when spread is allowed to continue for even-briefer time periods before its velocity is calculated. Finally, we can use our simulation to investigate the influence of "isolation corridors" on spread, with the 250-m gaps between plant populations representing these zones. The effect of such isolation (or patchiness in plant populations) is dramatic; for example, a favored gene's invasion can be delayed by 20 to 60 generations simply because of a 250-m stretch of hostile habitat (Figure 11–4).

The merits of modeling spread in an explicit way are that one can easily examine the consequences of different selective regimes or habitat structures for a variety of dispersal processes. Since it is impossible to do field release experiments under all possible conditions, such models are useful in exploring opportunities for spread if, for example, the mean dispersal rate is

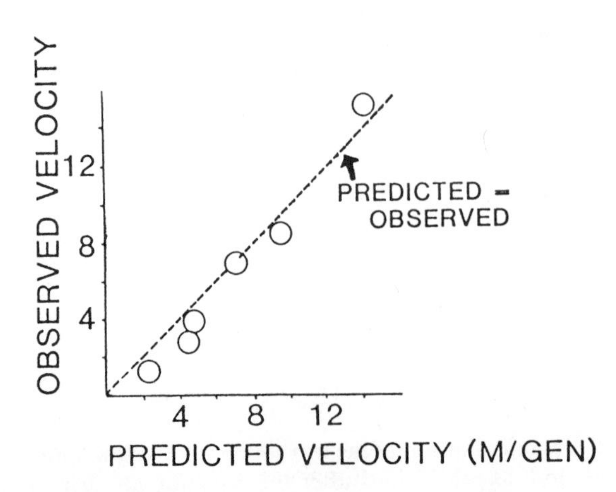

FIGURE 11–3 The relationship between simulated short-term rates of spread and the ARS predicted by equation 10.

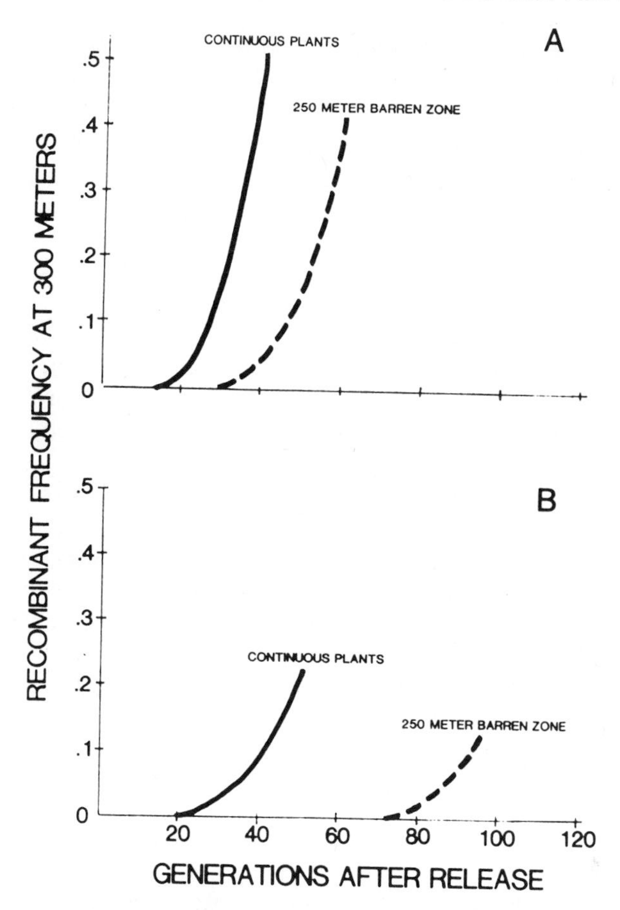

FIGURE 11–4 Two simulations showing the effect of a zone barren of plants on the frequency of increase of a highly favored recombinant gene in a stand of non-transgenic plants 300 m away from a transgenic plot. For these simulations, the relative fitness of the transgenic genotypes was 1.25, and the mean pollen dispersal rate was either (A) 20 m per generation or (B) 10 m per generation. Note that the delay due to a barren zone is reduced when the pollen dispersal rate increases (as it is in part A compared to part B).

doubled, or if selection is imposed as a spatial patchwork of no selection (i.e., some areas not sprayed with herbicide) with intense selection (i.e., some areas are sprayed with herbicide). The data needed for these models are empirical measures of gene flow and some estimate of selection coefficients. It will be particularly important to quantify the relative fitnesses of transgenic plants and recombinant genes under environments for which they were not initially designed. For instance, in order to anticipate the spread of a gene for herbicide resistance, it is essential that we know the fitness of that gene in the absence of herbicide spray.

11.3 SOME GENERAL LESSONS FROM EXISTING THEORY ON SPREAD

For a wide variety of ecological and genetic processes, there is a rich body of theory concerning the spread of genes or organisms. If only an approximate answer is needed, many of the existing simple models may suffice for estimating the likelihood of spread. If more exact answers are desired, one can turn to computer simulations that use models such as equations 1 through 10 as their starting point.

In spite of its limitations, the theory to date suggests three trends that may turn out to be quite general: (1) The velocity of spread consistently appears to be proportional to an organism's or gene's net reproductive rate, measured when the organism is at low density and unhindered by the effects of crowding. The fact that net reproduction may only need to be measured at low densities could greatly simplify the challenge of predicting spread. (2) Theory also indicates that to predict spread, we need to characterize the frequency distribution of dispersal distances. Indeed, although the measurement of dispersal under field conditions will always be difficult, it should be clear that there can be no risk assessment without such measurements. (3) Virtually all spread models predict that at first the observed rate of spread will be slow and will gradually accelerate to some asymptotic maximum rate. Thus, without some form of quantitative analysis, it is not wise to conclude spread will be negligible simply because after one generation, most genetically engineered organisms or genes are recovered close to their initial point of release.

11.4 THE NATURE OF THE DATA AVAILABLE FROM FIELD RELEASES

11.4.1 The Field Release of Transgenic Strains of *Brassica*

The movement of genes from transgenic canola into nontransgenic canola (*Brassica napus*, Westar variety) has been studied in a field experiment conducted by the Monsanto Co. and Agriculture Canada on an experimental farm in Scott, Saskatchewan, Canada. The design of this experiment used a central series of plots containing transgenic canola, and also five peripheral blocks of nontransgenic canola at distances of 400, 180, 100, and 50 m away from the transgenic plants (see Figure 11–5A for a detailed map). The transgenic plots were surrounded by mowed barley, whereas the nontransgenic blocks were planted in a large field of unmowed wheat.

Most plants within the central plots had at least one insertion for a transgenic gene that confers resistance to the herbicide Roundup (glyphosate) and is expressed in a dominant manner. Nontransgenic plants that are sprayed with glyphosate die shortly thereafter, long before maturation. Thus, in an environment where Roundup was heavily applied, selection in favor

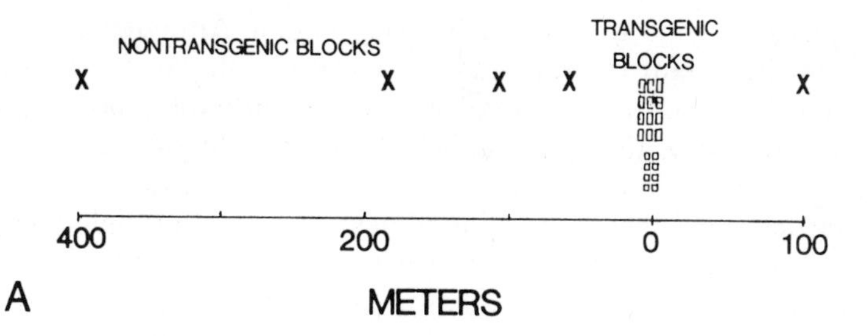

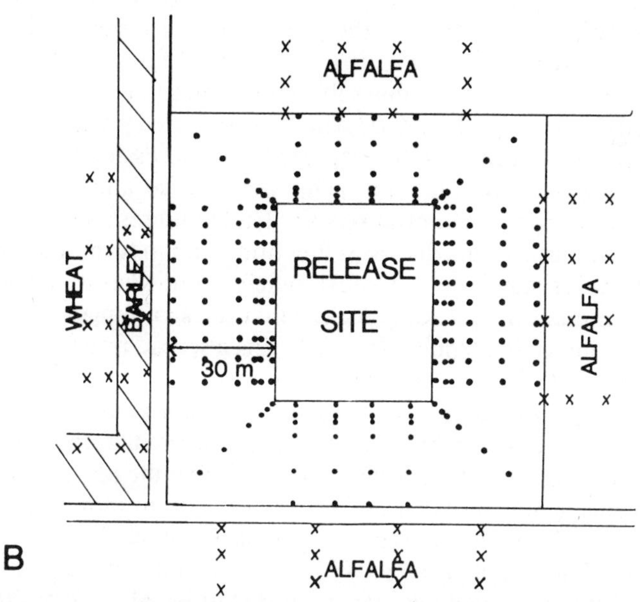

FIGURE 11–5 Sampling designs for two early field releases of transgenic organisms. (A) Blocks of transgenic rapeseed and the transects of nontransgenic plots used to measure gene movement (the nontransgenic plots are denoted with X's). Gene movement was assessed by collecting seeds from the nontransgenic plots and screening these seeds for the presence of herbicide resistance (which was the gene inserted into the transgenic rapeseed). (B) Release site for transgenic ice-minus bacteria and surrounding sample stations; the dots represent petri dishes or bean plant traps, and the X's indicate locations from which either alfalfa, barley, or wheat vegetation was cultured for the presence of ice-minus bacteria.

of the transgenic gene would be extremely strong. Although such selection might seem irrelevant to an annually harvested cultivar like canola, if the gene were to "escape" and become incorporated into the genome of a weedy conspecific *Brassica*, then natural selection could favor the spread of the transgenic gene through populations of these weeds. Monsanto conducted this experiment to assess the likelihood of such escape through pollen movement (and to evaluate the degree of herbicide resistance under field conditions). Nontransgenic canola growing in blocks outside the central set of plots were not sprayed with herbicide and were allowed to produce seeds, which were sampled and checked for herbicide resistance. In particular, 10,000 seeds were collected from each of the blocks planted with nontransgenic canola, and germinated on medium containing 2.5 mM glyphosate. The presence of a foreign, transgenic gene was scored when seedlings survived past germination. Although the gene dispersal observed in this experiment was between cultivars, it is a good model for spread into weedy populations, because canola is remarkably similar to some cruciferous weeds (such as weedy *Brassica campestris*, which crosses freely with canola bred from *B. campestris*). Monsanto has reported an outcrossing rate of 0.022% in the block lying 50 m from the transgenic gene source, and 0.011% in each block lying 100 m from the central blocks. These outcrossing rates provide a baseline for estimating gene dispersal functions, and in turn, rates of spread for the herbicide resistance gene. Unfortunately, from what we have already learned about models of spread, it is clear that the sampling design used by Monsanto is probably inadequate for predicting a spread rate for glyphosate-resistance genes; this is because the detection of outcrossing at only two distances from the transgenic source-plants does not allow one to characterize the dispersal function that is key to spread predictions [i.e., $v(x)$ in equation 2]. A better sampling design would involve the placement of nontransgenic plants at several positions between the 50-m plot and the transgenic block, as well as nontransgenic transects in other directions in case of directionally biased pollen flow. With better spatial coverage in its sampling design, Monsanto could estimate the $v(x)$ in equation 2, and then use a model such as the simulation we describe in Section 11.2 to anticipate spread under a variety of selective regimes or degrees of isolation.

11.4.2 The Field Release of Ice-Minus *Pseudomonas syringae*

In 1987 and 1988, Lindow et al. (1989) conducted large-scale field experiments in which ice-minus *Pseudomonas syringae* strains were released into plots of potatoes. By sampling the surrounding area for the ice-minus strains, Lindow was able to quantify the degree of spread for this "engineered invader." The spread detected represented the outcome of both airborne transport and population growth. Because the ice-minus strain is likely to compete with the nearly identical ice-plus strain, equations 9a and 9b may provide a good model for interpreting the results of this field experiment. In particular, one would analyze equations 9a and 9b, with the ice-minus strain representing the invader, and the ice-plus *P. syringae* assumed to be established as the resident population. One value of analyzing field releases using explicit reaction-diffusion models is that it forces us to be more precise in the inferences we draw from the data. For instance, the spatial distribution of transgenic organisms following their release is clearly a result not just of their dispersal (i.e., their diffusion), but also of the competition they experience due to wild-type microbes. Thus we might detect negligible spread because of competition, even though dispersal rates were high. These sorts of distinctions are crucial when exploring the risks associated with different ecological scenarios.

Here we simply describe the sampling protocol adopted by Lindow et al. (1988) because we believe it represents the best experiment to date on spread of an engineered organism. Unlike the Monsanto field release, Lindow's sampling was sufficiently stratified in space to provide a good portrait of spread. For convenience we discuss only the experiment reported in Lindow et al. (1988), in which ice-minus *P. syringae* strains were applied to potato tubers prior to planting or sprayed onto the potato foliage after the plants were well established. The distribution and density of the ice-minus strains were estimated by culturing bacteria sampled from petri dishes with agar solution that had been placed on the ground, or from potted bean plants that were deployed as "traps," on media containing the antibiotic rifampin. The ice-minus organisms were marked with a mutation for resistance to this antibiotic. The petri dish and bean plant sampling stations were distributed in a regular grid (the dots in Figure 11–5B) to ensure thorough sampling of the bare ground surrounding the potato crop that had been treated with ice-minus bacteria. In addition, in order to sample longer-range dispersal, Lindow's team also examined bacterial cultures from samples of vegetation collected from alfalfa, wheat, and barley in fields 30 or more meters away from the sprayed potato plots (the *X*'s in Figure 11–5B). Not only was the spread of ice-minus bacteria thoroughly sampled in space, it was also sampled on several different dates so that a preliminary indication of the velocity of spread might be obtained from this field trial. In general, this experiment is distinguished by its sophisticated attention to population level phenomena, such as spread, survival, and reproduction, and should be a standard for subsequent microbial work. However, because the data have not been synthesized into a model, even this excellent experiment has not been exploited to the fullest extent possible.

11.5 FINAL REMARKS

We have introduced a variety of theoretical perspectives on spread in order to encourage less anecdotal and more quantitative discussion of the risk of spread. Spread has the units length/time, and should always be reported as such. Thus outcrossing rates for plants do not correspond to measurements of gene spread—they are but one ingredient of spread and need to be supplemented with information regarding relative fitnesses and the probability densities of gene flow.

A pessimist might point out that dispersal and fitness are sure to vary in time and space in ways that cannot be accurately described, and that therefore predicting spread is hopeless. We disagree. One can make substantial progress simply by obtaining estimates of the ranges over which dispersal rates and fitnesses vary. Furthermore, models can inform us of the consequences of isolation zones or of spatially variable habitats. Perhaps most importantly, modeling forces us to collect and report data on spread in a unified context. For the spread of engineered microbes, that context includes an estimation of diffusion rates and population dynamics when at low densities, but in a variety of competitive environments. For the spread of genes, that context includes an estimate of selective coefficients and how they vary, and estimates of gene flow probability densities. Only when experimentalists tailor their field releases to provide the information required by "spread models," will risk analysis move beyond the blind guesswork stage.

REFERENCES

Alt, W. (1985) *Lect. Notes in Biomathematics* 57, 33–38.

Andersen, M. (1987) *The spread of plant populations into empty habitats: the theory of wind dispersal.* M.Sc. thesis, Biomathematics Program, University of Washington, Seattle, WA.

Andow, D., Kareiva, P., Levin, S., and Okubo, A. (1990) *Landscape Ecology* (in press).

Aronson, D., and Weinberger, H. (1975) *Lect. Notes in Math.* 446, 5–49.

Bateman, A. (1947) *Heredity* 1, 303–336.

Bramson, M. (1983) *Memoirs Amer. Math. Soc.* 44(285), 1–190.

Diekmann, O. (1978) *J. Math. Biol.* 6, 109–130.

Ellstrand, N. (1988) *Trends Ecol. Evol.* 6, S30–S32.

Ellstrand, N., and Foster, K. (1983) *Theor. Appl. Genet.* 66, 323–327.

Fisher, R. (1937) *Ann. Eugenics* 7, 355–369.

Handel, S. (1982) *Amer. J. Bot.* 69, 1538–1546.

Handel, S. (1983) *Evol.* 37, 760–771.

Handel, S. (1985) in *Structure and Function of Plant Populations* (Haeck, J., and Woldendorp, J., eds.) pp. 251–265, North-Holland Publishing, Amsterdam.

Hastings, A. (1990). *Theor. Pop. Biol.* (in press).

Kolmogorov, A., Petrovskii, I., and Piskunov, N. (1937) reprinted in *Applicable Mathematics of Non-Physical Phenomena* (Oliveira-Pinto, F., and Connolly, B., eds.), 1982, John Wiley & Sons, New York.

Levin, D., and Kerster H. (1974) *Evol. Biology* 7, 139–220.

Lindow, S.E., Knudsen, G.R., Seidler, R.J., et al. (1988) *Appl. Environmental Microbiol.* 54, 1557–1563.

Lindow, S., Panopoulos, N., and McFarland, B. (1989) *Science* 244, 1300–1307.
Lubina, J., and Levin, S. (1988) *Amer. Nat.* 131, 526–543.
McEvoy, P., and Cox, C. (1987) *Ecology* 68, 2006–2015.
Mollison, D. (1977) *J. Roy. Stat. Soc.* B 39, 283–326.
Murray, J., Stanley, E., and Brown, D. (1986) *Proc. Roy. Soc. Lond.* B 229, 111–150.
Nieuwhof, M. (1963) *Euphytica* 12, 17–26.
Okubo, A. (1980) *Diffusion and Ecological Problems: Mathematical Models*, Springer-Verlag, New York.
Okubo, A., Maini, P., Williamson, M., and Murray, J. (1990) *Proc. Roy. Soc. London. B.*, 238, 113–125.
Schaal, B. (1980) *Nature* 284, 450–451.
Shigesada, N., Kawasaki, K., and Teramoto, E. (1986) *Theor. Pop. Biol.* 30, 143–160.
Skellam, J. (1951) *Biometrika* 38, 196–218.
Thieme, H. (1979) *J. Math. Biology* 8, 173–187.
Tiedje, J.M., Colwell, R.K., Grossman, Y.L., et al. (1989) *Ecology* 70, 298–315.
Turchin, P. (1989) *J. Anim. Ecology* 58, 75–100.
U.S. Congress, Office of Technology Assessment (1988) *New Developments in Biotechnology: Field-Testing Engineered Organisms: Genetic and Ecological Issues*, OTA-BA-350, Washington, DC.
van den Bosch, F., Zadoks, J., and Metz, J. (1988) *Phytopathology* 78, 59–64.
Weinberger, H. (1978) *Lect. Notes Biomath* 648, 47–96.
Weinberger, H. (1982) *SIAM J. Math. Anal.* 13, 353–396.

PART IV

Regulation, Research Trends, and Social Values

CHAPTER

12

Regulation of Biotechnology by the Environmental Protection Agency

Marvin Rogul
Morris Levin

Before the age of genetic engineering and new wave biotechnology, introductions of microbes, such as viruses, bacteria, fungi, and protozoans into the environment were of little concern to the U.S. Environmental Protection Agency (EPA). Until the 1980s, review and registration was only required of microorganisms that were used as pesticides. And even when these live microbial pesticides were randomly mutated or otherwise genetically manipulated there was no great concern or alarm. There was, in fact, in the mid-1970s an attempt to shorten the review process for these products, which were then known as "biorational pesticides," because they were considered less dangerous to the environment than chemical pesticides. This complacency was disturbed by the discovery of new ways to easily and specifically mutate and manipulate microbes. These new techniques rapidly

The material in this chapter is based on a series of reports by the EPA and on discussions with agency personnel. Although the initial reports were reviewed by the agency, the opinions herein are those of the authors and do not constitute official policy of the U.S. Environmental Protection Agency.

and easily combined DNA from sources that were previously believed to be genetically incompatible.

The discovery of restriction enzymes and of the ability to make specific genetic constructs was accompanied by a rhetoric and hyperbole that described a new control over the natural order. These products were designed to easily carry out syntheses and functions that previously required the expenditure of great energy and resources. Consulting firms, financial analysts, and industry leaders forecasted increases and new environmental applications of commercial biotechnologies. They predicted large increases in the numbers of new microbes released that would have beneficial effects on the environment.

In response to the belief that genetically engineered microorganisms (GEMs) would become major factors in the environment, the EPA examined its governing statutes and concluded that the Agency could and should regulate certain constructs of GEMs and their products under preexisting legislation. The two statutes deemed most appropriate for the task were the Toxic Substances Control Act (TSCA) and the Federal Insecticide, Fungicide, and Rodenticide Act (FIFRA).

From the inception of FIFRA in 1947, any chemicals used as pesticides were reviewed and registered under FIFRA. In 1948, the first microbial pesticides, *Bacillus popilliae* and *B. lentimorbus*, were registered by the U.S. Department of Agriculture. By 1970, the diverse authorities for pesticide registration were transferred to the newly formed EPA.

In 1974, EPA announced the intent to develop special policies and testing guidelines to regulate microbial pest-control agents. The agency explained that biological-control agents were much different from chemical pesticides and should be tested and treated differently (Federal Register 1974). In 1982, EPA announced that GEMs were included in this policy (Federal Register 1982). This was followed by the recombinant DNA-testing guidelines, which were published in 1984 (Federal Register 1984).

In contrast, GEMs not used as or in pesticides, foods, food additives, cosmetics, drugs, and medical devices were considered for possible regulation under the TSCA. One of the major features of TSCA is that it is invoked for substances that are interpreted as being "new." Prior to the publication of the "Coordinated Framework for Regulation of Biotechnology; Announcement of Policy and Notice for Public Comment" (Federal Register 1986) the EPA paid little attention to the release of nonpesticide microorganisms, regardless of their lineage or how they were produced. Now, with the exceptions described above, all microorganisms produced for environmental, industrial, or consumer uses are potentially regulatable under TSCA (Federal Register 1986). It was not until intergeneric constructs of GEMs were interpreted as "new" that EPA pursued the possible notification requirement and regulation of microorganisms (Federal Register 1986).

As of yet, only limited experimental applications of GEMs have been allowed in the environment. The opportunities to examine ecological risks have been minimal. Historically, only the introduction of naturally occurring or conventionally manipulated microorganisms can be used as paradigms for the new GEM constructs. In this regard, the most notable ecological alteration caused by a microbe was due to the introduction of myxoma virus in Australia as a control agent against the pest rabbit *Oryctolagus cuniculus* (Fenner and Ratcliffe 1965). The sudden decimation of the huge rabbit population caused ecological and economic havoc. No similar pathogens of mammals have been submitted for registration at the EPA. Consequently, any risk assessments for this kind of product and application are based on limited data.

A more recent human ecology problem has been described in the USSR, where industrial production of single-cell protein (SCP) in over 100 factory facilities has come to an abrupt halt. This low-level biotechnology (as far as we know it is not recombinant DNA technology) appears to be the cause of atmospheric protein-dust pollution, bronchial asthma, allergic reactions, and a generalized reduction in immunity to disease. In addition, this report noted claims that thousands of farm animals died after being fed SCP produced by one such factory (Rimington 1989).

Whether risk assessments in the laboratory are translatable to the environment has to be judged on a case-by-case basis. Nester et al. (1985) recently reported an unexpected host-range change for a strain of *Agrobacterium tumefaciens* whose tumor-inducing virulence gene had been inactivated by mutation. The wild type functional *vir*C gene caused protective hypersensitive reactions in resistant plants. However, if this gene was inactivated, a previously resistant plant did not respond with a protective hypersensitive reaction and was subsequently susceptible to tumor induction by the mutant *A. tumefaciens* (Nester et al. 1985).

One of the most frequent points made by the EPA is that only the recombinant DNA products are considered for regulation, not the process used to produce them. Developers of these products argue that recombinant DNA products are indeed regulated differently. To the manufacturers it appears that it is the process that is being singled out for special attention, since similar products were not previously considered for regulation.

This chapter will examine the rationale for reviewing recombinant DNA products and GEMs, focusing on how and why the reviews are conducted and when the reviews are initiated.

12.1 BACKGROUND

Shortly after the major recombinant DNA discoveries were announced, scientists and businessmen became especially active in commercializing and selling the new genetic technologies and their powerful uses. To raise financial and public support, start-up companies heralded astounding new products that would have profound effects on practices in areas such as medicine, agriculture, the environment, and energy obtainment and use.

Public and governmental responses were not always as positive as proponents wished, because many proposed uses of this new technology in the environment also invoked the fears of possible ill effects, perhaps irrevocable ones in some cases. Still, there was no movement to regulate until the early 1980s.

A group of researchers at the Kingston, R.I., Laboratory of the EPA first foresaw the need for research programs at the EPA that would be oriented to productive uses and potential regulation of this new technology and its products. EPA scientists at this laboratory envisioned a time when the processes and products of genetic engineering might be used in the environment and the safety of these processes and products might be called into question.

In the summer of 1978, the laboratory convened the first of three biotechnology workshops to explore the EPA's relationship with biotechnology. This first workshop, held for the purpose of defining the EPA's interests in biotechnology, was an attempt to distinguish between regulatory needs and opportunity for environmental enhancement.

The sense of the 1978 workshop was that the first genetically engineered organisms to be released would probably be microbes; and that they would probably enter the environment through hazardous waste detoxification processes, municipal waste treatment, or escapes from commercial fermentation processes. In 1979, the second workshop focused on environmental risks; and in 1980, the third and last workshop dwelt mainly on harnessing biotechnology for pollution control, and environmental enhancement. The combined effects of these workshops resulted in the development of a pilot research program and agenda for studying the problems of release, genetic exchange, survival, colonization, and dissemination that might occur from the use of GEMs or eucaryotic recombinant DNA plants. The research program was first presented publicly at a conference on plasmid ecology (Levin et al. 1983).

The workshop participants believed that water quality was at risk and that hazardous waste sites were potentially dangerous breeding grounds, because GEMs and their products could find their way into these substrates and could volatilize or leach the products into the surroundings. In addition, commercial fermentation facilities were of concern because they might release recombinant DNA or GEMs into the environment.

If these conclusions were correct, then it was logical to assume that EPA would initially regulate genetically engineered products on the basis of the media involved, such as air, water, or soil, and consequently invoke the Clean Water Act (CWA); the Clean Air Act; the Resource Conservation and Recovery Act (or RCRA); and the Comprehensive Environmental Response, Compensation and Liability Act (CERCLA, or the Superfund Act).

12.1.1 Regulatory Background

Most of the laws governing the EPA are abatement laws that deal with the removal, destruction, or detoxification of existing pollutants, and also increases in pollution. Most of these laws regulate in specific media, such as drinking water, fresh and marine waters, land in the form of abandoned or future waste disposal sites, and air.

In contrast to the workshop conclusions, EPA initially regulated biotechnology products through the Toxic Substances Control Act (TSCA) in the Office of Toxic Substances (OTS) and the Federal Insecticide Fungicide and Rodenticide Act (FIFRA) in the Office of Pesticide Programs (OPP). These two acts are usually described as "gateway" legislation. Their approaches are not primarily aimed at abatement, but rather, are prophylactic and predictive. They are invoked before application of a new product to a medium in hopes of preventing pollution. The concept of newness invokes TSCA or FIFRA. Traditionally, new pesticides and chemicals or new uses for old ones are the trigger mechanisms that invoke TSCA or FIFRA. The most comprehensive description of this policy of regulation was published in the document entitled "Coordinated Framework for Regulation of Biotechnology" (Federal Register 1986).

12.2 STATUTORY ARGUMENTS

12.2.1 The Toxic Substance Control Act

In the mid-1970s, one of the first tasks of the OTS was to develop an inventory list of chemical products that would be exempt from TSCA regulations because of prior use in commerce (consequently they were exempted from regulation). A number of microorganisms were listed on the TSCA inventory. These were naturally occurring microorganisms used in commerce and listed by their genus or species names. Nevertheless, there is nothing in the legislative history that clearly stated that the intent of the TSCA statute included microbes. An interpretation was required to determine the regulatory status of recombinant DNA life forms.

In response to informal inquiries, a memorandum entitled "Status of Recombinant DNA and New Life Forms Under TSCA," dated March 14, 1983, was written by the EPA Associate General Counsel, Stanley H. Abramson, to Dr. John A. Todhunter, Assistant Administrator for Pesticides and Toxic Substances (Staff Report 1984). The general counsel's office concluded that TSCA had authority to regulate recombinant DNA and new life forms. Their reasoning was that TSCA generally grants authority to regulate chemical substances and that (1) DNA is a chemical substance; (2) life forms are comprised of "chemical substances" as defined in TSCA; and, moreover, (3) life forms are themselves "chemical substances" under TSCA because they are combinations of organic substances. In addition to the statutory argument, the associate general counsel submitted a policy argument for invoking TSCA authority. His reasoning was based on the concept that Congress intended TSCA to be a "gap-filler." Since no other federal regulatory scheme currently regulated life forms on a wide comprehensive basis, involving TSCA jurisdiction over life forms was thought to be consistent with the gap-filling function.

In support of regulation by TSCA, the associate general counsel stated that (Staff Report 1984):

> [4] It is difficult to think of any way that a DNA molecule would not be a "chemical substance" under the TSCA definition. However, a DNA molecule per se does not have any use aside from the life form of which it is a part. . . .
>
> [5] In the past the Agency has taken the position that DNA and life forms are chemical substances under TSCA, in documents such as the Inventory reporting instructions and in testimony during Congressional hearing on recombinant DNA research. . . .

The inference was that it did not make sense to regulate new forms of DNA without regulating the life forms that carried them.

12.2.2 Federal Insecticide, Fungicide, and Rodenticide Act

FIFRA was enacted in 1947 to regulate chemical pesticides. Microbes acting as pest-control agents were first registered in 1948. When recombinant organisms came into being, this established convention also extended to their regulation. Interestingly, the term "pesticide" as defined by the FIFRA also includes plant growth regulators that can be used to control plants considered to be weed pests (FIFRA Amendment P.L. 86–139, 1959).

In November 1982, the OPP published "Subdivision M of the Pesticide Assessment Guidelines for Biorational Pesticides." These guidelines were for the development of data required by 40 Code of Federal Regulations (CFR) Part 158 for determining the fate of biorational (living microorganisms) pesticides in the environment and their adverse effects on humans and other nontarget organisms. GEMs were marked for special consideration in these guidelines (Federal Register 1984a). In 1984, EPA again singled out GEMs in a *Federal Register* notice that explained EPA's policy on testing these microorganisms in the environment. A revised version of the Pesticide Guidelines was published in 1989 (Subdivision M of the Pesticide Testing Guidelines 1989). This time, however, GEMs were not especially singled out for additional testing.

Although EPA has other legal authorities, such as the Clean Water Act, the Resources Conservation and Recovery Act, etc., the agency has not found it necessary to extend their application to other genetically engineered products at the present time.

12.3 SPECIFIC EPA CHANGES FOR REGULATING BIOTECHNOLOGY

Eventually, laboratory constructs of genetically engineered organisms matured to the point where they needed testing in greenhouses or in test plots in the open environment. Since genetically engineered microorganisms are not conventional chemicals, the EPA required new skills and expertise for their reviews. Therefore, new units specifically oriented to the development and coordination of policy and review procedures for recombinant DNA microorganisms were created.

12.3.1 EPA Units and Functions

In 1984, the Office of Pesticides and Toxic Substances (OPTS) instituted the EPA Biotechnology Advisory Committee (BAC). One of the Committee's first changes was to determine what legislation was appropriate for the regulation of recombinant DNA products. It was clear that the regulation of recombinant, or biorational, microbial pesticides would probably proceed much along the lines of other pesticides that were also microorganisms, whereas regulation under TSCA was a different matter. Since there were no TSCA-type biotechnology products to regulate at the time and recombinant DNA research was not being funded by the EPA, there was little for the BAC to do.

The only other visible sign of regulatory activity at that time was the creation of an EPA liaison representative to the National Institute of Health (NIH) Recombinant DNA Advisory Committee (RAC) and a commitment from the agency to abide by NIH RAC decisions on scientific experiments and issues, and large-scale fermentations.

12.3.1.1 Special Assistant for Biotechnology. A small office and the position of special assistant (SA) for biotechnology were created in 1987 to aid and coordinate biotechnology activities of the assistant administrator of the OPTS. The Office of the Special Assistant deals with biotechnology at policy levels. The SA serves as a facilitator to unify diverse elements into a consensus policy. The office is too small, however, to handle detailed reviews. An example of the SA's role was in the development of the agreeable terms in the consent order that was arranged between Monsanto and OPTS for field testing a strain of *Pseudomonas fluorescens* containing the *Escherichia coli lac*ZY gene. Because there was some possible environmental impact, the SA arranged intergovernmental cooperation among OPTS, the U.S. Department of Agriculture Animal and Plant Health Inspection Service (APHIS), the Food and Drug Administration (FDA), and water quality experts at EPA.

The most significant point in this process was that the consent order was the result of review and regulation of a commercial product at the experimental research level. Prior to this time, there was always an exemption for experimental testing of chemicals. The reasons for this early intervention were based on the apprehension that living microorganisms might spread beyond the test plots.

12.3.1.2 Biotechnology Science Advisory Committee. The SA is also the executive secretary to the EPA Biotechnology Science Advisory Committee (BSAC). This 11-person committee was formed in 1986 to advise the administrator of EPA. The BSAC creates subcommittees that review risk assessments produced by the Office of Toxic Substances (OTS) and Office of Pesticide Programs (OPP). The committee also forms expert panels to investigate and review science and policy issues. The annual reports of the BSAC are available from the office of the executive secretary.

12.3.1.3 Office of Pesticides and Toxic Substances. At the EPA, the umbrella Office of Pesticides and Toxic Substances (OPTS) contains the two regulatory offices, OTS and OPP, that conduct risk assessments (RA) for microbes that are destined for release to the environment. As stated before, a new position of Special Assistant on Biotechnology to the Assistant Administrator of OPTS was created to develop uniformity and consistency in policy and actions. One of the duties of the Special Assistant is to coordinate activities and policies between the OPP and the OTS.

12.3.1.4 Office of Pesticide Programs. The OPP is mandated to implement the Federal Insecticide, Fungicide and Rodenticide Act (Public Law 92-516) (or FIFRA). According to this legislative act, the OPP has the authority to regulate the registration and use of microbial pest-control agents (MPCA; microbes that are used for controlling pests and weeds). FIFRA mandates the registration of pest-control products and "economic poisons" prior to production and sale. In order to register a product, an applicant must submit complete data as provided in the statute (7 USC 136c). The OPP also establishes tolerance levels for consumption of treated food products. The FDA has the responsibility for enforcing the levels. Civil penalties for violation of FIFRA vary according to the nature of the offense, whether it is a repeated offense, and the type of applicant.

During the course of its product evaluations, the OPP conducts reviews or risk assessments that are the functional equivalent of an environmental assessment (EA) or an Environmental Impact Statement (EIS). From 1948 to 1988, 17 MPCAs were registered with the USDA or EPA (Table 12–1). Before 1984, none of these were reviewed at the research level, notification

TABLE 12–1 Registration of Microbial Pesticides (MPCAs) by the EPA

Microbial Pesticide	*Year Registered*	*Pest Controlled*
Bacteria		
Bacillus popilliae	1948	Japanese beetle larvae
B. lentimorbus	1948	Japanese beetle larvae
B. thuringiensis berliner	1961	Moth larvae
Agrobacterium radiobacter	1979	Crown gall (*A. tumefaciens*)
B. thuringiensis israeliensis	1981	Mosquito larvae
B. thuringiensis aizawai	1981	Wax moth larvae
Pseudomonas fluorescens	1988	*Pythium, Rhizoctonia*
B. thuringiensis San Diego	1988	Coleopterans
B. thuringiensis tenebrionis	1988	Coleopterans
Viruses		
Heliothis nuclear polyhedrosis virus (NPV)	1975	Cotton bollworms, budworms
Tussock moth NPV	1976	Douglas fir tussock moth larvae
Gypsy moth NPV	1978	Gypsy moth larvae
Pine sawfly NPV	1983	Pine sawfly larvae
Fungi		
Hirsutella thompsonii	1981	Mites
Phytophthora palmivora	1981	Citrus strangler vine
Colletotrichum gloeosporioides	1982	Northern joint vetch
Protozoa		
Nosema locustae	1980	Grasshoppers

was not needed for field testing in plots that were under 10 acres in size, and no Experimental Use Permits (EUP) were required. Starting in 1984, notification of intent to field test GEMs and nonengineered nonindigenous pathogenic and nonpathogenic microorganisms was required, regardless of testing site size. An EUP may now be required if the agency deems it necessary. Table 12–2 lists the current notification and review requirements of the OPP.

Since 1984, the OPP has received 21 notifications and 9 EUP requests for genetically engineered or nonindigenous MPCAs (Table 12–3). This table shows that EUPs were not required for small-scale testing of certain types of microbes. For example, Monsanto was not required to obtain a EUP for testing a killed genetically engineered pseudomonad containing the *Bacillus thuringiensis* (BT) toxin. Instead, this material went through the usual testing procedures for inanimate pesticides. No EUPs were required for undirected mutagenesis, certain transconjugants, or for plasmid-cured strains. The strain of *Pseudomonas aureofaciens* containing an *Escherichia coli*

TABLE 12–2 Prior Notification and Review Requirements of FIFRA and TSCA

	Subject to Review under FIFRA		*Subject to Review under TSCA*	
Microbial Product	*<10 Acres*	*>10 Acres*	*<10 Acres*	*>10 Acres*
Genetically engineered:				
Intergeneric microorganism	Yes	Yes	Yes	Yes
Intrageneric microorganism				
Pathogen	Yes	Yes	Yes	Yes
Nonpathogen	Abbreviated	Yes	Abbreviated	Abbreviated
Nongenetically engineered:				
Nonindigenous[1] pathogen	Yes	Yes	Abbreviated	Yes
Nonindigenous[1] nonpathogen	Abbreviated	Yes	Abbreviated	Abbreviated
Indigenous pathogen	No	Yes	Abbreviated	Yes
Indigenous nonpathogen	No	Yes	Abbreviated	Abbreviated

[1] Not native to the continental United States (including Alaska) and the immediately adjoining countries (i.e., Canada and Mexico).

Source: excerpted from Federal Register (1986); *Fed. Regist.* 51, 23319.

TABLE 12–3 Chronological Summary of Notifications and EUPs for GEMs under FIFRA

Applicant (Type of Submission)	*Nature or Purpose of Application*	*Chronology of Events*	*Major Regulatory Issues and Possible Impact*
Advanced Genetic Sciences (AGS) notification	Recombinant DNA deletion product Yielding bacteria for frost protection *Pseudomonas syringae* and *P. fluorescens*	November 1984; Notification received February 1985; Letter requesting additional information and an EUP application	Potential adverse effect on plants, insects, weather patterns Competitiveness
University of California notification	Recombinant DNA deletion product *P. syringae* to be used for frost protection for plants	December 1984; Notification received February 1985; SAP comments March 1985; request for additional data and an EUP June 1985; EPA, NIH, and Cabinet Council discuss both agency's reviews	Potential adverse effects on plants, weather patterns Competitiveness of ice-minus strains may displace ice-plus bacteria
Monsanto notification	Recombinant DNA addition product *B. thuringiensis* toxin gene inserted into *P. fluorescens* Insect control	January 1985; Notification received March 1985; SAP meets April 1985; request for additional information and an EUP	Could toxin gene insertion mechanism cause environmental problems? Possible adverse effects on nontarget species
USDA notification	Undirected mutagenesis product: two species of fungi to be used to control pest fungi *Trichoderma viride* and *T. harzianum*	January 1985; Notification received April 1985; 90-day clock suspended; EPA requests additional information May 1985; USDA withdraws notification	Relative level of risk (undirected mutagenesis vs. other altered products) Extent of data needed (and practical to obtain on these products).

(*continued*)

TABLE 12–3 Chronological Summary of Notifications and EUPs for GEMs under FIFRA (continued)

Applicant (Type of Submission)	*Nature or Purpose of Application*	*Chronology of Events*	*Major Regulatory Issues and Possible Impact*
Mycogen (informal review)	Recombinant DNA addition product *B. thuringiensis* toxin gene inserted into *Pseudomonas* for insect control; Cells killed	April 1985; Notification received June 1985; Mycogen informed that Interim Policy does not apply to their killed product	Applicability of Interim Policy Mycogen provides data to substantiate nonviability of their product
McNeese State University (USDA) notification	Nonindigenous protozoan to control mosquitos	May 1985; Notification received September 1985; EPA meets with USDA to discuss adequacy of data November 1987; EPA review suspended pending receipt of additional information	Amount of data needed (and practical to obtain) October 1985; EPA informs McNeese data inadequate; also asked to comply with USDA/NIH
Advanced Genetic Sciences EUP	Recombinant DNA deletion product yielding bacteria for frost protection of plants *P. syringae* and *P. fluorescens*	July 1985; EUP application received November 1985; approval announced December 1986; AGS submits information on proposed sites February 1987; EPA approves test sites, reinstates EUPs April 24, 1987; AGS tests ice-minus bacteria, EPA monitors for aerial dispersion	Additional data requested on competitiveness, plant pathogenicity, colonization, identification, detection, and risk level November 1985; Agency sued February 1986; outdoor testing of recombinant DNA deletion product disclosed March 1986; EPA audits AGS; suspends EUPs; issues civil complaint July 1986; final order signed: penalty—$13,000

Monsanto EUP	Recombinant DNA addition product *B. thuringiensis* toxin gene inserted into *P. fluorescens* Insect control	October 1985; EUP application received May 1986; additional data requested September 1986; protocols submitted for data development reviewed November 1987; EUP review suspended pending receipt of additional data	Agency reviews company data on major scientific issues Colonization and effects on nontargets not fully addressed by data. Determination of confidential business information content
University of California EUP	Recombinant DNA deletion product yielding bacteria for frost protection of plants *P. syringae* *P. fluorescens*	December 1985; EUP application received June 1986; EUP granted April 1987; treated potatoes planted January 1989; request for EUP amendment April 1989 approved for EUP amendment	August 1986; University of California and state sued: trial postponed April 1987; go ahead for experiment issued
Ecogen notification	Plasmid-cured and transconjugant strains of *Bacillus thuringiensis* var. *kurstaki* and *azawai* to control lepidopteran pests	May 1986; Notification received June 1986; Decision: no EUP required	Applicability of Interim Policy Strains may occur naturally Relevance of existing data base on *B. thuringiensis*
Sands notification	Undirected mutagenesis (UV) product *Sclerotinia sclerotiorum* Control plant pests	May 1986; Notification received August 1986; Decision: no EUP required	Relative risk compared to parental strain Contingency plans for test termination

(*continued*)

TABLE 12–3 Chronological Summary of Notifications and EUPs for GEMs under FIFRA (continued)

Applicant (Type of Submission)	*Nature or Purpose of Application*	*Chronology of Events*	*Major Regulatory Issues and Possible Impact*
Harman notification	Protoplast fusion product *Trichoderma harzianum* Control pest fungi	June 1986: Notification received July 1986; SAP and other agencies review September 1986; EUP necessary	Relative risk compared to parental strains Competitiveness and effects on beneficial fungi Amount of data and information needed
Ecogen notification	Plasmid-cured and transconjugant strains of *B. thuringiensis* Use new strain inactive against beetles	March 1987; Notification received April 17, 1987; Decision: no EUP required	Expedited review Risk issues similar to submission 5/86 Additional host range
Ecogen notification	*B. thuringiensis* strains similar to submission of 5/86 Test on cotton and vegetables	March 20, 1987; Notification received May 12, 1987; Decision: no EUP required	Expedited review Risk issues similar to submission of 5/86
Ecogen notification	*B. thuringiensis* strains similar to submission of 5/86 Test on forest trees Use strains active against Lepidoptera and Coleoptera	April 10, 1987; Notification received April 27, 1987; Decision: no EUP required.	Expedited review Risk issues similar to submission of 5/86 Additional exposure of forest nontarget organisms

University of Arkansas notification	Chemically induced mutant *Colletotrichum gloeosporides* (EPA-registered product) Use on rice and soybeans	April 1987; Notification received June 1987; USDA/APHIS comments July 1987; Decision: no EUP required	Expedited review Possible host-range changes
Montana State notification	Transposon-mediated mutagenesis of *P. syringae*	June 22, 1987; received notification	Initiated testing before approval August 1987; letter of reprimand to MSU president
Ecogen notification	*B. thuringiensis* strain similar to Ecogen submission of 5/86 Test on cotton and vegetables	July 10, 1987; received notification Risk assessment completed September 1987; Decision: no EUP	Uses active Lepidoptera and Coleoptera strains Similar to 5/86 submission
Ecogen notification	Similar to 5/86	July 10, 1987; received notification Generic risk assessment of *B. thuringiensis* conjugation and/or plasmid-curing products. Initiated by HED September 1987 Decision: no EUP	Exemption from requirement for notification for *B. thuringiensis* strains derived by plasmid-curing and transconjugation Includes a description of areas of endangered species to be avoided
AGS amendment to EUP		September 1987; AGS requests EUP amendment to conduct 2nd test at same site November 1987; EPA amends EUP	Field trial slightly expanded Reduction in protective clothing proposed

(*continued*)

TABLE 12–3 Chronological Summary of Notifications and EUPs for GEMs under FIFRA (continued)

Applicant (Type of Submission)	*Nature or Purpose of Application*	*Chronology of Events*	*Major Regulatory Issues and Possible Impact*
Ecogen (4 EUPs)	Various *B. thuringiensis* transconjugants	January/February 1988; applications received Under review	
Crop Genetics (CGI) EUP	Recombinant DNA addition product to control corn borer *B. thuringiensis* delta toxin gene added to *Clavibacter xyli*	December 28, 1987; application received May 1988; EPA grants EUP June 1988; CGI initiates field studies	Mechanisms of plant to plant transmission of *C. xyli* Gene loss competition, reversion
Harmon notification	Protoplast fusion products of *Trichoderma harzianum* Control fungi	January 1988; Notification received; expansion of testing conducted in 1987 April 1988; Decision: no EUP	None identified at this time
Sandoz notification	Transconjugant *B. thuringiensis* strains Microbial insecticide	March 1988; notification received April 1988; EPA requests additional information	Product identity
Inotec Industries Ltd. notification	UV-induced mutant of *T. harzianum* for fungicidal use	May 1988; notification received June 1988; Decision: no EUP	Same strain as the parental strain used by Dr. Harman; field tested previously
Monsanto notification	Recombinant DNA product *E. coli lac*ZY gene as marker *P. aureofaciens*	July 1988; Notification received October 1988; Decision: no EUP	None

Crop Genetics International EUP	Recombinant DNA addition product Plant endophytic bacteria engineered to product toxin gene of *B. thuringiensis* *Clavibacter xyli*	December 1988; EUP received April 1989; EUP approved	Proposed test in Minnesota and Illinois outside of known range of *C. xyli*
Eastman Kodak notification	Protoplast fusion products Hybrid strains of *Trichoderma harzianum* to control pest fungi	January 1989; Notification received	Additional data requested
Boyce Thompson Institute notification	Recombinant DNA deletion product Deletion in polyhedrin gene To demonstrate that mutant is nonpersistent in field *Autographa californica* nuclear polyhedrin virus	January 1989; received notification	

*lac*ZY gene was routed through the TSCA testing procedure. In contrast, EUPs were required of all live recombinant DNA GEMs.

Risk Assessment Procedures. Procedures for Risk Assessments are initiated when applicants notify OPP that they are ready to field test a potential pesticide. In the June 26, 1986, Coordinated Framework document (Federal Register 1986a), the EPA announced a two-tiered review system (levels one and two) for risk assessment of microbial pesticides with special consideration given to nonindigenous and GEMs. According to OPTS, genetically altered or genetically engineered is defined as follows (*Fed Regist.* 51, 23330): "Any human intervention beyond removal from the environment and selection for the desired variant populations . . . should be considered to result in an engineered microorganism."

Some genetically engineered nonindigenous organisms that are not intergeneric constructs or constructed from certain pathogens may require only the Level I reporting necessary for small-scale field testing. Level I review requires a minimal amount of information and allows the agency up to 30 days for review. If not contacted within 30 days, the producer can proceed to small-scale testing. If the agency is concerned about the safety of a particular pesticide, the agency will require more information in a Level II full notification with a review period of up to 90 days. If the agency is still concerned, they may require more information or they may ask the producer to apply for an EUP. The countdown is halted during periods when the applicant has been requested to provide additional information. At present, the agency has suspended the small-scale field test exemption and is issuing EUPs. As is done with chemical pesticides, an EUP is always required for large-scale microbial pesticide testing on more than 10 acres of land or more than 1 acre of water.

On February 15, 1989, in its notice in the *Federal Register* requesting comments on EPA's proposed amendment of EUPs (Federal Register 1989), the agency requested comments on the following issues: (1) the scope of genetically modified microbial pesticides subject to notification; (2) the review of nonindigenous microbial pesticides at small-scale levels; and (3) establishment of independent expert review groups (environmental biosafety committees) patterned after NIH's Institutional Biosafety Committees.

In some cases, such as the applications for plasmid-cured strains and undirected mutagenesis (see Table 12–2), initial and follow-up submissions were so similar and uncomplicated that the OPP was able to treat the second and subsequent reviews in a generic fashion. The same can be said for most repetitive notifications.

Notifications of intent, correspondence among the agencies and the submitter, and the risk assessments are all on file at the OPP Docket Room. Confidential business information and other public documents are kept at the Document Center.

The Conduct of Risk Assessments. The risk assessments are conducted as part of the pesticide registration process. The purpose is to protect human health and the environment from damage by pesticides by conducting and documenting a thorough review of potential adverse impacts. A major departure from the EIS procedure is that the risk assessments do not start out with a public scoping session, but rather ask for comments at the end of their assessments. However, the courts have decided that pesticide registration reviews are the functional equivalents of Environmental Impact Statements (EIS).

The OPP conducts risk assessments within the Hazard Evaluation Division (HED). Risk assessments are conducted by a team of appropriate experts, with one individual from each branch of the HED. There is one representative from the Science Integration Staff (acting as team leader) and one person each from the Ecological Effects Branch, the Exposure Assessment Branch (responsible for fate and transport), the Residue Chemistry Branch (responsible for product components and identity), and the Toxicology Branch. Thus, risk assessments are not always conducted by the same team of experts.

Comments are also solicited when needed from other units in the EPA, USDA, FDA, NIH Recombinant DNA Advisory Committee (NIH RAC), National Science Foundation (NSF), and the Department of the Interior (DOI; for consideration of endangered species). Teams have also sought the advice of outside experts convened as OPP Science Advisory Panels (SAP). The EPA's Biotechnology Science Advisory Committee (BSAC) reviews the final product.

Individual experts from academia, EPA laboratories, other agencies, and the industry also serve as sources of information. OPP does not use consulting firms for these assessments. The internal and external reviewers and experts change with each assessment.

The risk assessments of genetically altered microbial pesticides are usually conducted on a case-by-case basis. The exception is when variations on the particular pesticide and/or its use are so minimal that the OPP feels it can employ a generic risk assessment. This has been done for some small-scale field test applications. The OPP has reserved judgement on whether generic risk assessments will be appropriate for full-scale registration.

As of yet, there are no descriptive or predictive risk assessment models for recombinant DNA pesticides, but the EPA laboratories have been working on models that might become useful. The OPP is receptive to the idea of using a computerized expert system if an adequate one can be developed.

Scientific Disciplines Used to Develop Risk Assessments. Risk assessments are developed in accordance with the specific pesticide and its application. In general, risk assessments for genetically altered microbial pesticides are conducted by molecular biologists, microbiologists, plant and insect pathologists, environmental toxicologists, biochemists, and other appropriate experts on an as needed basis. An example of adding an expert was the addition of a meteorologist to a SAP to advise the EPA on the possible effects that the large-scale use of ice-minus strains of *Pseudomonas syringae* might have on the weather.

The kinds of data required for risk assessments of genetically engineered microbial pesticides are listed in Part 158 subdivision M and the *Federal Register* (1986) Coordinated Framework document of June 26, 1986.

As of 1988, in conjunction with USDA, EPA has conducted 28 informal reviews of transgenic tomato, tobacco, and cotton plants with genetically engineered pesticidal qualities. These allelophyte plants were endowed with genes for tobacco mosaic virus (TMV), *Bacillus thuringiensis* (BT) toxin, and chitinase, among others. Table 12–4 summarizes the applications for permission to test these transgenic plants in the environment. The chronology of events in the review process and the issues involved are also presented. These reviews were conducted with USDA as the lead agency.

12.3.1.5 Office of Toxic Substances. The functions of the Office of Toxic Substances (OTS) are mandated by 15 *United States Code* 2601-2929; also known as the Toxic Substances Control Act (TSCA) of 1976 (Public Law 94-469). This Act is frequently referred to as gap-filling legislation, which means that chemicals not covered by other laws are covered under TSCA. TSCA requires manufacturers to notify OTS through a formal Premanufacturing Notice (PMN) 90 days before they are going to manufacture and distribute new chemical substances for commercial purposes. At this point EPA has the right either to ask for additional data, to allow limited production and use within stipulated limits which are agreed to by the manufacturer (called a consent order or agreement), or to ban the product. After 90 days, if the OTS does not take regulatory action, the notifier must inform OTS of intention to manufacture. The chemical is then placed on the TSCA inventory, and the manufacturer is allowed to proceed without any more PMN requirements.

OTS Policy. As previously discussed, OTS considers recombinant DNA and RNA products to be new chemicals not previously listed on the TSCA inventory and therefore subject to PMN. When nucleic acids from different microbial genera are hybridized, the EPA defines the resulting recombinant microorganism as "new" for regulatory purposes. The same is true for microorganisms constructed by cell fusion. When manufacturers develop and manufacture these "new" microbes they must report them to the OTS even if they are to be used only in closed systems.

In the June 26, 1986 *Federal Register* notice, the agency considered the following approaches that would affect the scope of TSCA coverage: the use of a list mechanism; of categories of microorganisms (e.g., pathogens); and the type of genetic exchange, as well as whether it was an intra- or inter-

TABLE 12–4 Applications for Testing Transgenic Plants that Produce Pesticidal Chemicals and Their Review by EPA

Applicant (Type of Submission)	*Nature or Purpose of Application*	*Chronology of Events*	*Major Issues*
Rohm and Haas (informal review)	Tobacco plant with *B. thuringiensis* toxin gene	August 1986; Summary of experiment received September 1986; EPA informed Rohm and Haas that it foresaw no unreasonable risks; EPA provided additional suggestions as to ensure minimal risk	Transfer of *B. thuringiensis* toxin gene to other organisms, inactivation of vector EPA regulatory responsibility for plants engineered to contain pesticide producing genes Transfer of introduced genes to other plants Crop destruction provision avoids tolerance
Monsanto (review of submission made to USDA)	Tomato plant with *B. thuringiensis* toxin gene Tomato plant with tobacco mosaic virus (TMV) coat-protein gene	May 1987; USDA/APHIS received May 1987; EPA informs USDA/APHIS that no unreasonable risks foreseen from field test; EPA provides additional suggestion to minimize risk of gene transfer	None
Monsanto (informal review)	Tomato plant with *B. thuringiensis* toxin gene Tobacco plant with TMV coat protein gene	March 1988; submission received and reviewed; comments provided to USDA/APHIS by EPA	None
Agrigenetics (informal review)	Tomato plant with *B. thuringiensis* toxin gene Tobacco plant with TMV coat	March 1988; submission received from APHIS and reviewed	None

(*continued*)

TABLE 12–4 Applications for Testing Transgenic Plants that Produce Pesticidal Chemicals and Their Review by EPA (continued)

Applicant (Type of Submission)	*Nature or Purpose of Application*	*Chronology of Events*	*Major Issues*
Agrigenetics (continued)	protein gene	March 1988; comments provided to USDA/APHIS	
Monsanto (informal review)	Tobacco plant with TMV coat protein gene	March 1988; submission received from APHIS and reviewed March 1988; comments provided to USDA/APHIS	None
Sandoz/Rohm & Haas (informal review)	Tomato plant with *B. thuringiensis* toxin gene	March 1988; submission received for APHIS and reviewed March 1988; comments provided to USDA/APHIS	None
DuPont (informal review)	Tobacco plant with chitinase gene	July 1988; submission received and under review at EPA	None
Monsanto (informal review)	Tomato plant with *B. thuringiensis* toxin gene	January 1989; submission received from APHIS and reviewed March 1989; comments provided to USDA/APHIS	Test site in Florida
Monsanto (informal review)	Tomato plant with *B. thuringiensis* toxin gene	February 1989; submission received from APHIS	Test site in California
Rohm & Haas (informal review)	Tomato plant with *B. thuringiensis* toxin gene	January 1989; submission received from APHIS	None
Agracetus (informal review)	Cotton plant with *B. thuringiensis* toxin gene	January 1989; submission received from APHIS	None

generic exchange. The present policy was adopted after convening expert committees and considering their reports. GEMs constructed using well-characterized intergeneric nucleic acid segments of noncoding regulatory regions are exempt from this policy. The creation and use of this kind of new recombinant DNA does not require a PMN.

On the basis of function, if a product (microbe) is to be employed for significant new uses, the OTS may invoke the Significant New Use Regulation (SNUR), and the product then becomes subject to regulation. For example, strains of *Pseudomonas syringae* are used to enhance snow making on ski slopes. If the same microorganisms were to be applied to enhance rainfall in cloud seeding operations, a SNUR might be issued based on the need to assess the risk of an increased exposure of different populations.

Generally, microorganisms will not be reviewed under TSCA when used to produce foods, food additives, drugs (including vaccines), cosmetics, or medical devices, if they are regulated by the Food and Drug Administration or the USDA. But microorganisms used in the production of a chemical end-product will be subject to TSCA if the chemical end-product is any chemical substance used for a purpose other than the aforementioned exempted uses. For example, microorganisms that are used in the production of pesticides, fuels, solvents, dyes, or chemical intermediates—even if they are only used in closed systems—are subject to TSCA authority and PMN requirements if the chemicals or microorganisms are new. As with OPP, OTS reviews are considered functional equivalents of National Environmental Policy Act (NEPA) reviews.

In the *Federal Register* (1986) Coordinated Framework document, the EPA explained that the agency would leave unanswered for a time the question of whether microorganisms containing genetic material from other microorganisms in the same genus (i.e., products of deliberate intra-generic combinations) and those which are developed from a single source microorganism (e.g., products of undirected mutagenesis, microorganisms with deletions) should also be considered "new." The document concluded that it is possible that, in the future, EPA will decide that such microorganisms are "new," but for now they are not subject to PMN requirements.

The OTS has a TSCA Assistance Office at EPA headquarters, which maintains and circulates information concerning current and past submissions and notification procedures.

Submissions. PMNs have been received from BioTechnica and Monsanto for environmental releases of recombinant DNA microorganisms. Table 12-5 summarizes the PMN applications that have been reviewed under TSCA for testing GEMs in closed systems and in the environment. The history of the reviews and the major issues for each application are included in the table. Consent orders and negotiated agreements were the mechanisms whereby the two companies were allowed to release their microbes on test plots in the open environment; the BioTechnica submissions were for en-

TABLE 12–5 PMN Applications for Microorganisms Covered under TSCA and Their Review by EPA

Applicant	*Nature or Purpose of Application*	*Chronology of Events*	*Major Issues*
Testing in the environment			
BioTechnica International, Inc.	Recombinant DNA addition products 2 strains[1] of nitrogen-fixing *Rhizobium meliloti* Commercial product: bacteria to enhance yield of alfalfa	February 1987; submission received Review suspended as company develops monitoring data March 1988; consent order issued to ensure proper conduct of proposed field test	Consent order ensures use of necessary constraints determined by agency and data reporting No yield increase
Monsanto	Recombinant DNA addition product *Pseudomonas aureofaciens* with genes from *E. coli* Commercial product: product to monitor microorganisms in field	June 1987; submission received October 1987; consent order issued March 1988; requested to permit distribution to noncommercial institutions August 1988; requested modification of monitoring protocols September 1988; modification granted	Information required on economic impact on water quality testing and dairy product shelf-life is needed to evaluate *widespread* releases Consent order ensures use of necessary constraints determined by agency and ensures data reporting
BioTechnica Agriculture, Inc.	Recombinant DNA addition product 8 strains[1] of genetically engineered *Rhizobium meliloti* Test to evaluate competitiveness and method of application	March 1988; submission received BTA withdraws several PMNs August 1988; consent order sent to BTA	Genetic stability; transposon (Tn*5*) was not inactivated Host range Long-term survival Data on antibiotic resistance at site November 1988; change requested in test site

BioTechnica Agriculture, Inc.	Recombinant DNA addition products 4 strains[1] of genetically engineered *Bradyrhizobium japonicum*	May 1988; submission received August 1988; BTA conducts tests	Soybean yield Antibiotic resistance in field plot Need for controls in monitoring protocol
BioTechnica International, Inc.	Recombinant DNA addition product Commercial product: strain to enhance nitrogen fixation *Rhizobium meliloti*	January 1989; submission received	New test site (Wisconsin)
BioTechnica International, Inc.	Recombinant DNA addition product Commercial product: strains[1] to enhance nitrogen fixation[1] *Bradyrhizobium japonicum*	February 1989; submission received	New test site (Louisiana)
Use in closed systems			
International Minerals and Chemical Co.	Recombinant DNA addition products *E. coli* engineered to contain human gene for insulin-like growth hormone Commercial product: component in cell culture media	February 1987; submission received May 1987; review completed, production and disposal as described in submission judged to be of low risk, no regulatory action taken	None identified at this time
(Company name confidential)	Addition product *Bacillus subtilis* engineered for enhanced production of a protease Commercial product: (kind and name of product confidential)	May 1987; submission received July 1987; review completed, production and disposal as described in submission judged to be of low risk, no regulatory action taken	None identified at this time

(continued)

TABLE 12–5 PMN Applications for Microorganisms Covered under TSCA and Their Review by EPA (continued)

Applicant	*Nature or Purpose of Application*	*Chronology of Events*	*Major Issues*
(Company name confidential)	Addition product *Bacillus* strain engineered for enhanced production of hydrolase Commercial product: (kind and name of product confidential)	July 1987; submission received October 1987; review completed, production and disposal as described in submission judged to be of low risk, no regulatory action taken	None identified at this time
(Company name confidential)	*Staphylococcus aureus* and *Bacillus alcalophilus* engineered for enhanced production of protease Self-cloned to contain multiple copies using plasmid vector from Commercial product: (kind and name of product confidential)	April 1988; submission received July 1988; review completed, production and disposal as described in submission judged to be of low risk, no regulatory action taken.	None identified at this time
Novo Biochemical Ind. Ltd.	Recombinant DNA addition product Lipase enzyme *Aspergillus oryzae*	November 1988; submission received February 1989; review completed; no regulatory action taken	None identified at this time
Enzyme Bio-Systems	Recombinant DNA addition product Alpha amylase *Bacillus subtilis*	December 1988; submission received March 1989; end of 90-day review period	None identified at this time

[1] Each strain requires a separate application.

hanced nitrogen-fixing bacteria and the Monsanto submission was for testing a new recombinant DNA microbial monitoring system in the environment. OTS has also reviewed PMNs from six companies for GEMs to be used for varied purposes in closed systems (see Table 12–5). No regulatory action has been taken on these closed systems yet.

Because of statutory limitations, noncommercial and purely academic researchers who have: (1) constructed intergeneric microorganisms for noncommercial purposes and (2) intend to release these constructs to the environment, are not required to submit PMNs or SNURs to TSCA. The statutory exclusion of noncommercial products from regulation under TSCA means that there is a gap in regulatory oversight (Background Paper 1989).

Risk Assessment Organization Units. The purpose of the risk assessments from the viewpoint of TSCA is to protect the public and the environment from any new chemical or any new microorganism. Proposed commercial use of GEMs in closed systems and in the open environment comes under the jurisdiction of TSCA if not regulated by some other authority.

In anticipation of future PMNs for recombinant DNA, the OTS consulted experts in fields such as biotechnology and ecology, both outside and at EPA laboratories on possible risks. The staff also performed mock reviews. When the actual PMNs were submitted, the PMNs for closed-system use were reviewed by expert consultants and ad hoc committees. For releases to the environment, reviews were conducted by formal committees and a Biotechnology Science Advisory Committee (BSAC) subcommittee.

The Chemical Control Division coordinates the scientific review. Personnel are drawn from the Exposure Evaluation Division (including field test and manufacturing facilities specialists), Health and Environmental Review Division, Information Evaluation Division, Chemical Control Division, and the Economics and Technology Division. Two separate assessments, for hazard and for exposure, are conducted. In general, the hazard and exposure elements are independently estimated and then brought together for a risk assessment.

Although EPA employees are primarily drawn from the divisions listed above, other agency experts are called upon as needed. The experts in the agency are mainly microbiologists, ecologists, molecular biologists, chemists, toxicologists, and environmental specialists.

Whenever needed, coordination is established with other agencies, such as USDA, FDA, and the Consumer Product Safety Commission, thus resulting in appropriate input from the various agencies with jurisdiction.

Experts are obtained from local universities, and formal requests for scientific reviews are made of the BSAC when needed. Consultants are carefully chosen to avoid conflicts of interest.

The Agency conducts a risk assessment as part of the review process for each application received. Each applicant must conduct an environmental assessment, which is included into the review and subsequent risk assessment. To date, no significant impacts have been associated with any recombinant DNA or RNA product.

Products subject to PMN can include either live or killed organisms, which may either be used in contained facilities and(or) released to the environment. There is some question about the risk of microorganisms that will be used only in closed systems. Some EPA personnel think that live microorganisms will inevitably escape to the environment; others think that containment criteria should be so strict that the escape of microbes will be improbable, if not impossible. Differences among the various submissions required different perspectives and expertise. For example, reviews for dead recombinant DNA bacterial constructs were different from those for live constructs. The hazard component of the assessment changes from infectious-disease hazards associated with live microorganisms to considerations of toxicity or elicitation of allergic responses associated with dead microorganisms. Review specialists are selected to meet the need for appropriate expertise. In this example, assurance that the product was nonviable would be required, calling for expertise in verification of sterilization procedures.

All the OTS risk assessments have been conducted on a case-by-case basis. Eventually, generic risk assessments and predictive models may be tools that will be sufficiently developed for use in regulation.

12.4 REGULATORY CONSIDERATIONS FOR THE FUTURE

12.4.1 Proposed TSCA Experimental Release Regulations

On February 15, 1989, EPA announced that it had prepared a set of regulations amplifying and expanding the provisions of the June 26, 1986 *Federal Register* notice. The Agency asked for public comment on the proposed regulations in the *Federal Register* (1989b) announcement. The proposed regulations included:

1. The establishment of the TSCA Experimental Release Application (TERA),
2. The use of Environmental Biosafety Committees (EBCs).

In keeping with the information gathering, gap-filling, and regulatory powers of TSCA, it was proposed that experimental release applications and Significant New Use Notices (SNUNs) would be adjuncts to premanufacture notices and Significant New Use Rules (SNURs). This would implement exemptions from formal agency PMN and SNUR requirements. Certain experimental releases would then only be reviewed by Environmental Biosafety Committees (EBCs), or an EBC and EPA, or immediately reviewed by EPA.

There is some question as to whether this is a modified permit process or whether a permit process should be instituted. In any case the "permit" would be specific to each submitting company, and other companies would have to develop their own submissions for an experimental testing review.

In addition, there is the unresolved issue of relevancy and binding of decisions among the various EBCs. For example, would decisions made by one EBC be relevant or binding on another EBC located in a very different geographical area, with different ecological and climatic conditions?

Opponents to the addition of new types of permits and a more complex regulatory structure believe that this could result in greater regulatory impediments.

Proponents contend that the proposed procedures will allow for more nongovernmental involvement, and less state and local regulatory action, and that this process would be no more cumbersome than chemical pesticide regulation at the state and local levels.

12.4.2 Future Use of Other Statutes and Organizational Units

Until now, only the "gateway" premarketing and licensing statutes have been used for the review of recombinant DNA constructs and products. This seems to be a reasonable approach because there are no pollution problems of air, water, or soil media associated with biotechnology in the United States.

However, there is the possibility that recombinant DNA microbes eventually will be used in sewage treatment and hazardous waste detoxification. In some cases, biotechnology may be called for as being the best or the most appropriate technology.

In a recent symposium on the potential use of biotechnology to reduce and detoxify chemical wastes, William Ruckelshaus (1987) described the early years of the EPA, as follows:

> In the late 1960s and early 1970s we were dealing with gross kinds of pollution problems—the problems you could smell, touch, and feel. The Cuyahoga River that goes through Cleveland burst into flames when somebody threw a match in it. Lake Erie was declared "dead" back in the 1960s. Air pollution problems were quite visible, rendering other things invisible. . . .

Discoveries of obvious conditions of pollution were soon followed by disclosures of more subtle and complex problems. The development of more sensitive and specific monitoring methods made it apparent that there were widespread low levels of environmental pollutants that had not previously been recognized. These seemingly inert chemicals in low environmental concentration resulted in continuous exposure with uncertain chronic health effects, including mutagenicity or carcinogenicity that were hard to assess.

For many chemicals, the risk assessment procedure is simple:

I. Identify the chemical hazard and its source
   A. Its dose/response toxicity to humans, animals, and plants

    B. Its chemical and physical characteristics
    C. Exposure assessment
II. Apply the remedial process
III. Monitor the subsequent breakdown products
    A. The transport and fate of the hazard and its products
    B. Their dose/response toxicity to humans, animals, and plants
    C. Their chemical and physical characteristics
    D. Exposure assessments
IV. Assess the risk and, for balance, assess the economic benefits and positive environmental effects

Developing and applying a variation of this risk assessment procedure to microbial degradation will require changes in skill mix and organizational structure for reviewing purposes, similar to the reorganizations and changes in the OTS and OPP.

12.4.3 Microbial Bioremediation—Regulatory Issues

Into the 1970s, much emphasis was given to identification of pollutants and their physical and chemical remediation and storage. Manufacturing and sewage wastes were usually either (1) immobilized in stationary landfills or waste lagoons, (2) diluted by pouring into water (above or below ground), or (3) subjected to chemical neutralization and physical processes (such as incineration). Combustion products were disposed of as were the wastes, via air or water. Whether in air, water, or land, the strategy was to use dilution as the solution to pollution.

Microorganisms were once considered the "biological incinerators" of almost all natural and synthetic organic chemicals, although some insecticide, organic, and plastic products were known to not be degradable (Alexander 1967 and 1981). Today, it is more often the case that, instead of mineralization, the end product of microbial xenoorganic chemical degradation is usually another organic molecule whose toxicity may be unknown and has to be assayed. Sometimes breakdown intermediates or end products are also toxic. For example, microbial degradation can lead to the activation of the pesticide lindane, which produces trichlorophenols (water-soluble and carcinogenic); mercuric ions have been converted to methyl mercury derivatives; perchloroethylene to vinyl chloride; and pentachlorobenzyl alcohol to trichlorobenzoic acid. Thus, breakdown products can be acutely or chronically toxic, carcinogenic, mutagenic, or teratogenic, resulting in different transport, fate, and effects in the environment (Sayler et al. 1982; Rochkind et al. 1986). As late as 1980, there was still a need to call for more research into the toxicity and ecological effects of nonmineralized breakdown products of synthetic organic chemicals (Alexander 1980 and 1984). The same is true today.

Nonetheless, the use of microbes to detoxify in situ in land fills, water sites, and at the end of the pipe is an attractive proposition. The use of biotechnology for waste degradation was soon recognized as an energy-efficient and easily applied technology and, by the 1970s, was gaining in use in the treatment of waste water, oil refinery wastes, paper and pulp wastes, and contaminated soils (Garden 1980; Tracy and Zitrides 1979).

Genetically engineered organisms are attractive to manufacturers for two main reasons:

1. The precision of genetic manipulation in recombinant DNA, such as by plasmid mediation, transformation, and transduction, is usually more accurate, and more well-known and controlled than mutation with non-recombinant DNA.
2. The patenting of recombinant DNA organisms and GEMs is more easily defensible than for organisms that have been randomly mutated.

12.4.4 Possible Regulation of Microbial Waste Degradation

Prior to 1984, there was no direct federal regulation of microbial waste reduction processes, probably because these treatments were viewed as being "natural" and safe. Up to that time, the formal risk assessment process only took into consideration the risk and source characterization of the original chemical and environmental considerations in terms of transport, fate, and effects.

On December 31, 1984, EPA announced a PMN review requirement for commercial GEMs that was destined for use in the environment under TSCA. This requires a risk assessment of the microbe and the microbial processes. As a result, the review process and the selection of products for review was altered. The review procedure now included microbial products and the processes by which they are generated at the early, research, and development stages (Federal Register 1984b). This PMN process affects products that will be reviewed and regulated under the Resources Conservation and Recovery Act (RCRA) and thus requires that clearance from both offices within EPA be obtained.

In 1984, the RCRA reauthorization did the following: (1) banned land disposal of hazardous wastes without pretreatment (this program is designed for phased implementation from 1986 to 1990; this ban will be imposed unless the permit applicant proves that the waste will not migrate); (2) required waste minimization; and (3) required hazardous waste site operators to agree to clean up contamination at the site.

These are requirements that seem to lend themselves to bioremediation, but implementation will remain uncertain until standards and criteria are developed. Will biotechnologies satisfy the Best Demonstrated Available Technology requirements under RCRA? Will they have significant ecological effects? In 1986, the Superfund law (Public Law 99-499) was amended and reauthorized. It required: (1) treatments that permanently and significantly reduce waste toxicity, mobility and volume; and (2) a research, evaluation, and testing program (Superfund Innovative Technology Evaluation–the SITE Program). RCRA review criteria and procedures will have to be developed for the review of biotechnology products.

12.5 CONCLUSIONS

At this time, new guidelines of retesting and risk assessment are still in the formative stages at EPA. New testing guidelines for microbes were formulated in July 1989 and are available from the EPA Office of Pesticide Programs. The new biochemical pest-control-agent guidelines have not been printed yet.

There are still a number of TSCA issues for recombinant DNA microorganisms that need to be resolved. In February 1989, EPA announced the intention to propose regulations and requested public comment. Some of the major issues being discussed were:

- Should EPA prescreen research and development field releases under the auspices of TSCA?
- Is there a satisfactory or logical distinction between commercial research and development compared to academic research and development?
- When live microorganisms capable of multiplication are tested, what is the difference in oversight of small field releases compared to large field releases?
- Are there microorganisms that should be exempt from research and development field release oversight?
- What procedures should be used to develop criteria for decision making and protocols for data development?

It is clear that much effort is still needed in streamlining and developing a consistent regulatory process. In some cases, it is conceivable that more than one EPA office and more than one agency will be involved in regulating the same product.

REFERENCES

Alexander, M. (1967) *New Scientist*, 31 August, 35, 439–440.
Alexander, M. (1980) *Microbiology* 41, 328–332.
Alexander, M. (1981) *Science* 211, 132–138.
Alexander, M. (1984) in *Genetic Control of Environmental Pollutants* (Omenn, G.S., and Hollaender, A., eds.), pp. 151–168, Plenum Press, New York.

Background paper (1989) Regulation of Research and Development Field Releases of Microorganisms under TSCA (Also includes excerpted public comments on R&D issues received in response to the February 15, 1989 *Federal Register* notice) Office of Toxic Substances, September 20, 1989.
Clean Air Act (1982) *United States Code* 42, §§ 7401–7642.
Clean Water Act (1982) *United States Code*, 33, § 466–466g.
Comprehensive Environmental Response, Compensation and Liability Act (1982) *United States Code* 42, §§ 9601–9657.
Federal Insecticide, Fungicide and Rodenticide Act (1982) *United States Code* 7, § 136–136y.
Federal Register (1979) *Fed. Regist.* 44, 28093–28095.
Federal Register (1982) *Fed. Regist.* 47, 53192–53221.
Federal Register (1984) *Fed. Regist.* 49, 50856–50907.
Federal Register (1986) *Fed. Regist.* 51, 23302–23393.
Federal Register (1989a) *Fed. Regist.* 54, 7026.
Federal Register (1989b) *Fed. Regist.* 54, 7027–7028.
Fenner, F., and Ratcliffe, F. (1965) *Myxomatosis*, Cambridge University Press, Cambridge.
Gardner, C. (1980) *Public Works* 3, 71–72.
Levin, M.A., Kidd, G.H., Schwartz, J.R., and Zaugg, R.H. (1983) *Applied Genetic Engineering: Future Trends and Problems*, Noyes Publications, Park Ridge, NJ.
Nester, E.W., Yanofsky, M.F., and Gordon, M.F. (1985) *Engineered Organisms in the Environment: Scientific Issues* (Halvorson, H.O., Pramer, D., and Rogul, M., eds.), pp. 191–196, ASM, Washington, D.C.
Resource Conservation and Recovery Act (1982) *United States Code* 24, §§ 6901–6987.
Rimington, A. (1989) *BioTechnology* 7, 783–788.
Rochkind, M.L., Blackburn, J.W., and Sayler, G.S. (1986) *Microbial Decomposition of Chlorinated Aromatic Compounds*, EPA/600/2-86/090.
Ruckelshaus, W.D. (1987) *Environmental Biotechnology: Reducing Risks from Environmental Chemicals through Biotechnology* (Omenn, G.S., ed.), pp. 333–340, Plenum Press, New York.
Sayler, G.S., Reid, M.C., Perkins, B.K., et al. (1982) *Arch. Env. Contam. and Toxicology* 11, 577–581.
Staff Report (1984) Subcommittee on Investigations and Oversight, Science and Technology Committee, U.S. House of Representatives, Serial V, February 1984, GPO, Washington, D.C.
Subdivision M of the Pesticide Testing Guidelines (1989) *Microbial and Biochemical Pest Control Agents: Part A, July 1989*, USEPA, Washington, D.C.
Toxic Substances Control Act (1982) *United States Code* 15, §§ 2601–2654.
Tracy, K.D., and Zitrides, T.G. (1979) *Hydrocarbon Processing* 59, 95–98

CHAPTER
13

Ecological Considerations in EPA's Review for Field Tests of Genetically Engineered Organisms

H. Reşit Akçakaya
Lev R. Ginzburg

The aim of this chapter is to evaluate the significance of ecological considerations in the review process of the U.S. Environmental Protection Agency (EPA) for field tests of genetically engineered organisms. Therefore, we will analyze the risk assessment statements issued by EPA for 12 cases involving field releases of such organisms. The EPA has so far received the majority of applications for field testing of biotechnology products. Two offices within EPA that have authority to regulate the release of genetically engineered organisms to the environment are the Office of Pesticide Programs (OPP) and the Office of Toxic Substances (OTS). They are both within the Office of Pesticides and Toxic Substances (OPTS), but they utilize different review processes for biotechnology applications. These processes will be briefly summarized in the next two sections below. For a more detailed discussion

---

We thank Marvin Rogul for his comments on the manuscript. This study was supported by the Electric Power Research Institute.

of these procedures, see Chapter 12. The third section presents detailed information about 12 of the cases that have been submitted to these two offices of EPA. The last two sections evaluate the regulatory process by identifying common ecological concerns expressed in EPA's review of these cases.

13.1 REGULATION BY THE OFFICE OF PESTICIDE PROGRAMS (OPP)

OPP has authority to regulate microbial pesticides under the mandate of the Federal Insecticide, Fungicide, and Rodenticide Act (FIFRA). This authority includes regulation of the testing, registration, and use of microbial pesticides, including those produced by using genetic engineering techniques. Most of the applications reviewed in this study were submitted to OPP.

Applications to OPP can take either of two forms: notification or request for an Experimental Use Permit (EUP). OPP must be informed by a "Notification of Intent" about any field tests involving genetically engineered microorganisms used as pesticides. If EPA has safety concerns about test conditions, the applicant is asked to apply for an Experimental Use Permit (EUP). Alternatively, an applicant may choose to apply directly for an EUP. The types of information that need to be submitted with an EUP application are listed in Table 13–1. This list is compiled from the standard form used for EUP applications (EPA form 8570-17) and from a number of applications.

Both types of applications are reviewed by the Hazard Evaluation Division (HED). The review is conducted by the staff of this division with input from the Scientific Advisory Panel (SAP), which is composed of outside experts. The end result of this review is a risk assessment in the form of a "Scientific Position Paper." This paper presents the conclusions of OPP about the proposed field test, together with justifications of this position and specific concerns and issues that have been addressed by SAP.

13.2 REGULATION BY THE OFFICE OF TOXIC SUBSTANCES (OTS)

OTS regulates the manufacture and use of chemical substances as mandated by the Toxic Substances Control Act (TSCA), originally enacted in 1976. This legislation covers chemicals that are not covered by other laws, including genetically engineered microorganisms (GEMs) used in the production of chemicals, even in closed systems. This excludes chemicals used *only* as food, pesticides, cosmetics, or medical substances, which are regulated by other laws. TSCA requires the manufacturers of chemical substances

TABLE 13–1 Information Needed for Obtaining an EUP

1. Name of product and total quantity for shipment/use
2. Donor and recipient organisms and vector or vector agent used
3. Date and location of importation, movement, and release
4. Expression of the altered genetic material and how that expression differs from the expression in nonmodified parental organism
5. The molecular biology of the system that will be used to produce the regulated article
6. Country and locality where the donor organism, recipient organism, and vector or vector agent were collected, developed, and produced
7. Purpose for the introduction of the regulated article, including the proposed experimental and/or production design
8. Processes, procedures, and safeguards that have been used to prevent contamination, release, and dissemination in the production of the donor and recipient organisms, vectors, or vector agents
9. Final and intermediate designations, uses, and distributions of the regulated article
10. Proposed procedures, processes, and safeguards that will be used to prevent escape and dissemination of the regulated article at each of the intended destinations
11. Proposed method of final disposition of the regulated article.

TABLE 13–2 Information Requested for a PMN

1. Chemical identity information
2. Production volume, intended use, and function
3. Degree of containment; copy of hazard warning labels
4. Human exposure during production operations and processes
5. Environmental release and disposal (amount and medium of release; control technology)
6. Optional information (comparative risk analysis, process chemistry, efficacy, industrial hygiene, engineered safeguards, and economic and noneconomic benefits)
7. Test data (such as physical and chemical properties, mutagenicity, carcinogenicity, and toxicity)

to notify OTS by filing a Premanufacture Notice (PMN) before manufacturing or distributing a new chemical. The types of information required in a PMN are listed in Table 13–2, which is compiled from the EPA PMN form (7710-25) and the related instruction manual.

OTS reviews PMNs and produces a risk assessment document. This document has the following sections: background information and the description of the engineered organism; assessment of worker exposure and environmental exposure (survival, transport, genetic transfer); hazard assessment (hazards of the engineered and nonengineered organisms to human, crop plants, and nontarget organisms); risk characterization; and additional data needed.

13.3 CASE STUDIES OF 12 APPLICATIONS

This section presents detailed information about some of the biotechnological applications to EPA. A "case" here is defined as application(s) by an organization involving a specific organism. Thus notifications and EUP applications by the same organization for different stages of field testing are summarized together as a single case. For a complete list of all separate applications, see Chapter 12. Only those cases that involve field testing of genetically engineered organisms and those for which sufficient information could be obtained in 1988 are included in this section. Thus, cases with insufficient information and new applications (or amendments to previous applications) after 1988 are not included. As such, the cases reviewed here are only a sample of all applications. (One of the cases involves a non-genetically engineered, nonindigenous organism and is included because of its similarity to other cases.) All the available information that is relevant to ecological risk assessment is presented for each case in a standard format. EPA position, justifications, and concerns are compiled from risk assessment and scientific position papers prepared by the relevant offices of EPA, and are presented in a form as close to the original wording as possible. The section on *Justifications* includes the reasons for the stated conclusion of EPA and the monitoring/contingency plans proposed by the applicant that were part of the reason for EPA's position. The *Concerns* section includes the specific issues and concerns raised by the members of review committees or scientific advisory panels and additional controls and precautions required by EPA for the specific case in question. A list of the 12 cases reviewed in this study is given in Table 13–3 and the cases are listed in the order of the date of their submission to EPA. Details of each case are given in the next several pages.

Case 1 *(Submitted November 1984)*

Applicant: Advanced Genetic Sciences, Inc.
Application type: Notification (November 1984) and EUP application (July 1985)

Organism: *Pseudomonas syringae* and *P. fluorescens*
Intended use: As a frost protection agent on strawberry plants
Progenitor strains: Ice nucleation active ice-plus strains of *P. syringae* and *P. fluorescens* (isolated from strawberry plants)
Genetic methods: The genetic material in *P. syringae* and *P. fluorescens* that codes for the production of ice nuclei has been deleted, pro-

TABLE 13–3 The 12 Cases Reviewed in This Chapter

Case	Applicant	Organism
1	Advanced Genetics Sciences, Inc.	*Pseudomonas syringae* (ice-minus) and *P. fluorescens*
2	Dr. Steven E. Lindow	*P. syringae* (ice-minus)
3	Monsanto Co.	*P. fluorescens* with *Bacillus thuringiensis* gene
4	McNeese State University	*Amblyospora aedis*
5	Ecogen, Inc.	*B. thuringiensis*
6	Rohm & Haas Co.	Tobacco with *B. thuringiensis* gene
7	BioTechnica International, Inc.	*Rhizobium meliloti*
8	University of Arkansas	*Colletotrichum gloeosporiodes*
9	Monsanto Co.	*P. fluorescens*
10	Crop Genetics International Corp.	*Clavibacter xyli* with *B. thuringiensis* gene
11	BioTechnica Agricultural, Inc.	*R. meliloti* and *Bradyrhizobium japonicum*
12	Monsanto Co.	*P. aureofaciens*

ducing non-ice nucleating (ice-minus) strains, as opposed to the naturally occurring ice-plus strains.

Field test procedure: Strawberry plants were sprayed with ice-plus bacteria. The purpose was to determine if the application of ice-minus bacteria to strawberry blossoms could effectively prevent or reduce growth of ice-plus bacteria such that the blossoms would be protected from frost injury when exposed to temperatures between 0 and −5°C.

Site: A single test site in the central coast area of California

Area: About 0.2 acres of planted area surrounded by a 15-m-wide bare-soil buffer zone

Plants used: Strawberry plants (a total of 2,400)

Amount of microorganism used: A total of about 8×10^{12} viable cells

EPA position: Hazard Evaluation Division (HED) concluded that this small-scale field test was unlikely to pose a significant risk to human health or the environment (November 1985).

Issues considered in this case (and consequently justifications and concerns) were similar to those in the EUP application of Dr. Steven E. Lindow (see Case 2).

Case 2 *(Submitted December 1984)*

Applicant: Dr. Steven E. Lindow

Application type: Notification (December 1984) and EUP application (December 1985)

Organism: *Pseudomonas syringae* strain CIT7DEL1B and strain TLP2-DEL1

Intended use: As a frost protection agent on potatoes

Progenitor strain: Two ice nucleation active (ice-plus) strains of *P. syringae* (CIT7 and TLP2)

Genetic methods: The genetic material in *P. syringae* that codes for the production of ice nuclei has been deleted, producing a non-ice nucleating (ice-minus) strain, as opposed to the naturally occurring ice-plus strains.

Field test procedure: Potato seed pieces were treated with ice-minus bacteria just prior to planting. After emergence of potato plant foliage, the plants were sprayed with ice-minus bacteria. The purpose was to evaluate the potential of ice-minus bacteria to control frost damage under field conditions.

Site: Tulelake Experimental Station of the University of California

Area: Two test plots comprising a total of about 2 acres, including experimental plots where ice-minus bacteria were applied (0.3 acres), control plots, and 20 feet of bare-soil buffer zone.

Plants used: Potato plants

Amount of microorganism used: A total of 7.2×10^{13} CFU for three years

EPA position: Hazard Evaluation Division (HED) could foresee no significant risk to human health or the environment resulting from this field study.

Justifications:

1. At the request of EPA, the application provided data showing that the ice-minus deletion mutants have no competitive advantage over their parental or other ice-plus strains when applied to potato leaves. Even in cases where the ice-plus strains were applied at much lower concentrations and 2 days later than ice-minus strains, they were not completely eliminated, although they stayed at a low concentration.
2. The ice-minus deletion mutant bacteria would escape from the test plot in low numbers only, and the number of ice-minus mutants on plants outside the test plot would be much lower than the indigenous ice-plus and other bacteria already present on these plants.
3. The ice-minus mutants exhibit great similarity to their ice-plus progenitors, except for the ability to form ice nuclei. In particular, they show identical susceptibility/resistance pattern to 23 different antibiotics.
4. At the request of EPA, the applicant submitted information on host plant specificity (pathogenicity) demonstrating that the parental strains and ice-minus deletion mutants would not be expected to pose a risk of plant pathogenicity.
5. Monitoring plans for detecting off-site dissemination (including aerial dissemination) via drift and insects were adequate.

Concerns:

1. A number of people have commented that ice-plus bacteria may have an important role in precipitation patterns, and this role may be altered as a result of the proposed test. However, HED concluded that because of the small scale of the field test, and because of the facts listed in 1 and 2 above, such an effect was unlikely.
2. HED recommended that the potato plants should be incinerated and any remaining plant debris should be disked into soil at the end of each experiment.
3. In addition to the soil and foliar monitoring proposed within the test plot at site A, foliage should also be monitored adjacent to each side of the test plot at a distance of 10 m from the edge of the treated area. Then, if either of the engineered strains are detected in these samples, a 10-m-wide concentric band around the buffer zone should be treated with Kocide 101.
4. In sampling vegetation outside the treated area, an effort should be made to sample weed or crop species known to support colonization by the ice-minus mutants.
5. If engineered strains are detected in any soil or vegetation sample outside the bare-soil buffer zone, then the applicant should consult with EPA prior to initiating the next experiment.

Case 3 *(Submitted January 1985)*

Applicant: Monsanto Co.
Application type: Notification (January 1985) and EUP application (October 1985)

Organism: *Pseudomonas fluorescens*
Intended use: Corn seed treatment against root-associated lepidopterans
Progenitor strain: Isolated from soil near the test site
Genetic methods: P. fluorescens was genetically engineered to contain the insecticidal delta endotoxin gene from *Bacillus thuringiensis* var. *kurstaki.*

Field test procedure: About 27,000 corn seeds treated with the engineered organism to be planted at the test plot
Site: On the applicant's farm in St. Charles, MO
Area: About 1 acre
Plant used: Corn
Amount of microorganism used: Each seed contained about 10^8 colony-forming units (CFU); a total of approximately 5.4×10^{12} CFU (9×10^{11} CFU of each strain) were to be applied in the experiment.

EPA position: Several unresolved problems remained; some of the studies might have to be repeated in order to adequately support a risk assessment.

Concerns:

1. *P. fluorescens* (a Gram-negative bacterium) would be unlikely to have acquired this gene naturally, since the gene originated in Gram-positive bacteria in nature. *P. fluorescens* can multiply in a wider range of environmental conditions than *B. thuringiensis* from which the gene was taken. However, the applicant adequately addressed the issue of the potential for gene transfer and amplification by inactivation of the transposition function.
2. Test results indicated that, at lethal doses, the engineered organism might colonize the susceptible insects (lepidopteran larvae) and outcompete the bacterial flora in the insects. In addition, bacteria isolated from the killed insects could infect and kill other insects. This insect colonization study should be repeated and expanded in order to better quantify an infective dose and a no-effect dose, and to understand the dynamics of the infective process.
3. Some of the nontarget species tests (mosquito larvae, honeybee larvae, hemiptera, aquatic insect/crustacean and fish studies) had major deficiencies and must be repeated.
4. A recent study showed adverse effects in mollusks from a strain of *B. thuringiensis* that produces the same gene as the engineered organism. In view of the presence of an endangered mollusk species in the Mississippi River and in tributaries near the test site, it would be prudent to test for mollusk effects.
5. The study that addressed the issue of above-ground exposure did not sufficiently simulate environmental conditions to be able to draw conclusions from the negative results. The potential for above-ground plant colonization needed to be quantified.

Case 4 *(Submitted May 1985)*

Applicant: McNeese State University (Dr. Roger S. Nasci)
Application type: Notification of small-scale field test

Organism: *Amblyospora aedis* (a protozoan)
Intended use: As a pesticide against mosquito, *Aedes aegypti*
Progenitor strain: *Amblyospora aedis* strain 1316, imported from Thailand in 1979
Genetic methods: No genetic methods used

Field test procedure: 400 female mosquitoes infected with *Amblyospora aedis* were to be released.
Site: A single test plot, 5 km from the next-nearest native mosquito *(Aedes aegypti)* population as located in a 1984 survey
Area: About 4000 square meters (diameter = 75 m.)
Animal used: Mosquito, *Aedes aegypti*

Amount of microorganism used: Not applicable

EPA position: Additional data requested from the applicant
Justification:
It was not possible to evaluate the proposed study as thoroughly as is done for data submitted in support of EUPs or registrations.
Concerns:
1. The strain was not native to this country. Even though it was claimed to be similar to a Puerto Rican species, there were not enough data to document this similarity.
2. There might be populations of mosquito species other than *Aedes aegypti* close to the test site.
3. Nontarget species, about which very little data were submitted, might also be affected.
4. Data submitted on the life cycle of the protozoan were not complete.

Case 5 *(Submitted May 1986)*

Applicant: Ecogen, Inc.
Application type: Six notifications and four EUP applications

Organism: *Bacillus thuringiensis* var. *kurstaki* and *azawai*
Intended use: Applied to crop plants as an insecticide against Lepidoptera
Progenitor strain: Naturally occurring strains from the United States, Europe, and Japan
Genetic methods: (1) Partially plasmid cured strains (heat-induced plasmid loss); (2) Transconjugants (receiving plasmids from donor strains)

Field test procedure: Small-scale field tests in which eight genetically altered *B. thuringiensis* products were to be applied to various crop plants in several sites
Sites: 24 sites in 13 states
Area: Less than 10 acres per strain; total of less than 20 acres in each of the two notifications
Plants used: Cotton, corn, tomatoes, vegetables, and forest trees
Amount of microorganism used: 5–10 gallons per acre

EPA position: No foreseeable risks; no EUP required prior to testing (April and May 1987) for the first six notifications; EPA review ongoing for the four EUP submissions of January/February 1988.
Justifications:
1. Progenitor strains were currently registered for commercial use; none of them produced beta exotoxin; there are no reports of significant adverse effects on humans or other nontarget species.

2. Progenitor strains were ecologically similar to naturally occurring strains that were well-characterized and would be subject to the same ecological constraints.
3. Plasmid-cured strains (plasmid loss) could occur spontaneously; transconjugants could arise in nature under certain circumstances.
4. Total area used in field tests was low, and all crops would either be used for research or destroyed (none would be used for food or feed).
5. There were no endangered species at risk from the field tests. California and Florida have endangered Lepidoptera, but closest plots were 20 miles from their habitat.

Concerns: Some of the parental strains were isolated originally from outside the United States (but they have been distributed and used internationally).

Case 6 *(Submitted August 1986)*

Applicant: Rohm & Haas Co.
Application type: Notification for small-scale field testing

Organism: Tobacco plant producing *Bacillus thuringiensis* toxin
Intended use: Crop plant resistant to two insect pests, tobacco hornworm (*Manduca sexta*) and tobacco budworm (*Heliothis virescens*)
Progenitor strain: A noncommercial variety of tobacco (Petite Havana SR-1)
Genetic methods: Insecticidal gene from the bacterium *Bacillus thuringiensis* var. *berliner* inserted into the tobacco chromosome by means of a "disarmed" *Agrobacterium* bacterial plasmid vector system

Field test procedure: Small-scale field test in which genetically engineered tobacco plants were infested with tobacco hornworm *(Manduca sexta*) and tobacco budworm (*Heliothis virescens*)
Site: One test plot in either Bolivar county, MS, or Dade county, FL, depending on the time of year the test took place
Area: 0.3 acres (72 × 161 ft)

EPA position: The test posed negligible risk to the environment and further data were not necessary. The proposed containment procedures were mostly adequate (but see *Concerns*).
Justifications:

1. The only insects (five species) known to feed on tobacco are considered to be pests.
2. The regions of the *Agrobacterium* plasmid that was inserted into the tobacco DNA were inactivated, hence there was little potential for adverse effects on plants.

3. There was virtually no possibility for the gene to be transferred sexually by means of pollen or seed since the plants will be prevented from flowering.
4. There was no evidence that bacteria or viruses can serve as vectors in the nonsexual transmission of the *B. thuringiensis* toxin gene to other organisms.
5. The exposure of humans to the engineered organism was negligible.

Concerns:

1. Nontarget organisms might be effected by exposure to engineered tobacco plants which now produce toxin. At the test termination, all plants in the test plot should be destroyed. There should be a strip of bare ground (no plant growth) around the test plot sufficiently wide to prevent the roots of the test plants from contacting roots of plants outside the test plot.
2. The *B. thuringiensis* toxin can affect a wide range of insects, and the host range of the noncrystalline form of the toxin has not been established.
3. In Dade county, FL there is an endangered butterfly (the Schaus Swallowtail Butterfly). Even though it does not feed on tobacco, testing at the Florida site should not be performed during the months of May and June when its larval stage occurs.
4. It was not absolutely clear that the disinfectant procedure for treating the abscised plant parts would be effective. This procedure should be tested experimentally.
5. The bacterial vector *Agrobacterium* that was used to insert the *B. thuringiensis* toxin gene might help mobilize the gene from the plant chromosome. The test should be terminated and the crop destroyed if any evidence of *Agrobacterium* infection is observed on the test tobacco plants.

Case 7 *(Submitted February 1987)*

Applicant: BioTechnica International, Inc.

Application type: Three Premanufacturing Notifications for field testing

Organism: *Rhizobium meliloti*

Intended use: To promote alfalfa yield increases by increasing nitrogen fixation by symbiotic microorganisms

Progenitor strain: Wild-type *R. meliloti* strain Rm1021

Genetic methods: Endogenous genetic material from *R. meliloti nif* genes, which control nitrogen fixation, were cloned and reinserted back into *R. meliloti* on a commonly used broad-host-range vector.

Field test procedures: A cell suspension of engineered organisms were to be applied to the soil using a low-pressure sprinkler immediately after seeds are planted. The plants would be monitored for 3 years and harvested (cut) 3 times in each growing season. The aim was to measure increases in the yield due to treatment with engineered *R. meliloti.*

Site: An agricultural station near Arkansaw, WI

Area: Less than 5 acres

Plants used: Alfalfa

Amount of microorganism used: 2×10^{13} cells per year (for 3 years)

EPA position: The risks from the release of the engineered *R. meliloti* during the course of the limited field tests proposed by the applicant were low, similar to those expected from the release of a nongenetically manipulated strain of the same species.

Justifications:

1. The application by a sprinkler might cause the indigenous strains to have more opportunity to compete for attachment sites compared to the traditional method of seed coating (but see *Concerns*).
2. Although the outcome of competition between engineered and indigenous microorganisms is not always predictable, introduced rhizobia often require multiple yearly reinoculations to successfully become established in legumes. The engineered organisms should behave no differently than nonengineered rhizobia in this respect.
3. None of the genetic manipulations should cause a difference in the host range of the engineered organism compared to that of the progenitor strain.
4. No adverse human health effects were expected as a result of this test. The risks associated with the production stage were less than those associated with the production of nonengineered strains of the same species.
5. The utility of antibiotic resistance in this case was seen to outweigh the more general concern for dispersal of antibiotic-resistance genes. Lack of gene transfer capabilities further lessens concerns about dispersal.
6. There were few conceivable undesirable effects, and they could also occur with nonengineered commercial products. No such effects associated with commercial inocula were known to the EPA.

Concerns:

1. The low-pressure sprinkler application might provide a slightly greater tendency to distribute the introduced strains away from the test site at the time of application since larger numbers of bacteria could become airborne than would during an application using seed coating.
2. Although enhanced nitrogen fixation is beneficial, effects such as weed-growth enhancement could be deleterious.

Case 8 *(Submitted April 1987)*

Applicant: University of Arkansas (Dr. TeBeest)

Application type: Notification for small-scale field testing

Organism: *Colletotrichum gloeosporiodes* f. sp. *aeschynomene* (a fungus)
Intended use: As a mycoherbicide to control Northern Jointvetch *(Aeschynomene virginica)*, a weed in rice and soybean fields
Progenitor strain: Registered (EPA Reg. No. 45639–134) for commercial use under the trade name Collego
Genetic methods: Chemically (ethyl methanesulfonate) induced mutation and selection for resistance to benomyl (a fungicide)

Field test procedure: Small-scale field test in which chemically mutated and experimentally selected benomyl-resistant strain of the fungus were to be applied together with benomyl (to enhance its competitive ability against native fungus strains)
Sites: A single test plot near Stuttgart, AR
Area: 0.36 acres, including control plots
Plants used: Rice (test plot also contained weed species *Aeschynomene virginica)*
Amount of microorganism used: 10 gal per acre on 0.14 acres

EPA position: No foreseeable risks; no further EPA oversight or additional data were needed; no EUP was required prior to testing
Justifications:
1. Test strain was chemically mutagenized from an already-registered microbial pesticide that could occur or easily arise in nature
2. Small, well-controlled study
3. No adverse effects on endangered species expected

Concerns:
1. This type of mutation could increase the host range of the fungus, which might then attack nonpest species (crop plants). If the benomyl-resistant strain was detected in lesions of such plants, the infected plant should be removed, and the area should be sprayed immediately with a fungicide other than benomyl.
2. The test area should be plowed at conclusion of the test and a fungicide that actually kills the *Colletotrichum* (if one can be found) should be used to control any foci of fungal contamination outside the test plot.

Case 9 *(Submitted June 1987)*

Applicant: Monsanto Co.
Application type: Premanufacturing Notification for a small-scale field test

Organism: *Pseudomonas fluorescens*
Intended use: As a monitoring organism; in the future as a potential pesticide
Progenitor strain: *P. fluorescens* strain Ps.3732RN

Genetic methods: Genes from *Escherichia coli* K-12 that code for the production of lactose permease and β-galactosidase (the *lac*ZY genes) were inserted into the genome of the parental strain by means of a delivery plasmid

Field test procedure: The seeds of the first planting (winter wheat) would be inoculated with the engineered organism. The next two crops would be planted at the same plot, but without prior inoculation. The aim is to test the efficiency of the inserted *lac*ZY gene as a marker to monitor the survival and location of the engineered organisms under field conditions, and to evaluate the performance and survival of the engineered organism, which might be used in the future for delivery of plant-beneficial substances.
Site: A single site at a research and education farm in Blacksville, SC
Area: Approximately 1.5 acres
Plants used: Wheat and soybean
Amount of microorganism used: A total of 4.8×10^{13} cells

EPA position: The proposed field test was not expected to pose an unreasonable risk to human health or the environment. Any risk associated with the proposed field test was expected to be small.

Justifications:

1. The available data supported the conclusion that the engineered organism would survive and become established at low levels in soil environments and it would have no competitive advantage over the nonengineered parental strain.
2. The ability to utilize lactose (given to the engineered strain by the inserted *lac*ZY gene) would not give any significant competitive advantage to the engineered organism, since lactose would occur very rarely in natural settings and even if it did become available, many other common soil microorganisms that can utilize lactose would compete with the engineered strain for this resource.
3. Laboratory experiments conducted in aquatic systems showed that within 15 days the nonengineered progenitor strain populations declined to low levels. Survival was enhanced substantially when the water was filter-sterilized first. This was consistent with other literature showing enhanced growth and survival of other Gram-negative microorganisms after competing microorganisms were removed.
4. The capacity of the inserted gene to be transferred to other organisms was limited because the gene was inserted into the chromosome and there was no residual plasmid and because the transposition genes of the delivery transposon have been deleted. This chromosomal insertion appeared to be very stable. Its transport to other organisms would be primarily under the control of naturally occurring transfer mechanisms.

5. The engineered organism was not expected to be a human health hazard because the nonengineered (progenitor) microorganism was nonpathogenic and nontoxic, and the genetic manipulations were so well characterized as to reasonably preclude the possibility of introducing unexpected traits. Furthermore, human exposure to the microorganism would be limited and controlled in this field test.

Concerns:

1. There was some uncertainty in predicting the fate, survival, and establishment of the engineered organism in the environment after its release, because it has never been used in the environment.
2. There was a possibility of horizontal gene transfer without expression. The genomes of suspected microorganisms should be probed.
3. Natural water samples inoculated with the engineered organism, as well as mixed-culture testing, should be evaluated to determine the significance of the microorganism's potential impact on water-quality testing programs.
4. The fish test protocols used were inadequate. They have low sensitivity for pathogenicity because of low concentrations used and the short duration of the test.
5. A recent study showed that chemically purified siderophore from the progenitor organism inhibited, but did not prevent, the uptake of ferric iron by the roots of peas and corn. The significance of lower iron uptake was not clear, but it might lead to slower growth.

Case 10 *(Submitted December 1987)*

Applicant: Crop Genetics International

Application type: EUP application

Organism: *Clavibacter xyli cynodontis*

Intended use: As a plant-associated insecticidal agent against the European corn borer

Progenitor strain: Crop Genetics isolate no. MD69a, isolated from bermudagrass in Westover, MD

Genetic methods: Clavibacter xyli cynodontis (C.x.c.) was modified by recombinant DNA techniques to contain the *Bacillus thuringiensis* var. *kurstaki* (*B.t.k.*) delta endotoxin gene and to express its protein.

Field test procedure: Corn plants inoculated with genetically engineered *C. xyli cynodontis* were to be planted at two sites. The purpose was (1) to determine *C.x.c./B.t.k.* recombinant levels in inoculated plants; (2) to monitor mechanical and natural spread of the recombinant organism to corn and other plant species; (3) to monitor the recombinant organism in plant residues; (4) to monitor the presence of the recombinant in runoff water and soil; and (5) to compare yields of colonized and control corn plants. All seeds were to be destroyed to prevent the transfer to subsequent crops.

Sites: Two sites in Ingleside and Beltsville, MD

Area: About 2 acres at each site, consisting of (1) a central test plant (corn) population; (2) a 20-ft-wide plant-free barren zone surrounding test plants; (3) a "trap" plant zone bordering the barren zone; (4) a security fence; and (5) a 30-ft-wide low-maintenance fallow zone outside the security fence.

Plants used: Corn (Dekalb T-1100) as test plant; corn and bermudagrass as "trap" plants; a total of 9,800 plants to be inoculated.

Amount of microorganism used: A total of 1.4×10^{13} cells

EPA position: The small-scale tests of *C.x.c./B.t.k* at the two Maryland sites posed no significant risk to humans and the environment.

Justifications:

1. *C..x.c.* is a natural resident of plants, and it was extremely unlikely that it could infect mammals. It has not been associated with any disease symptoms in humans. Therefore, human health risks were negligible.
2. Data submitted to EPA showed that the probability of gene transfer to other organisms should be extremely remote.
3. In a 70-generation growth experiment, it was shown that the engineered genes were lost from the parental *C.x.c.* at a rate that was orders of magnitude greater than the natural mutation rate.
4. The presence of the engineered gene decreased the growth rate of *C.x.c.*; i.e., in the absence of selective pressure, the toxin-producing *C.x.c.* would eventually be outcompeted by faster growing *C.x.c.* bacteria that have lost the gene.
5. The group of bacteria containing *C.x.c.* were not expected to grow well or to survive well in the environment except in association with plants.
6. The engineered strain *C.x.c./B.t.k.* had a relatively low order of toxicity to the target insect as compared to the parental strain of *B.t.k.* In addition the engineered gene would eventually be eliminated from the population (see 3 and 4 above). Therefore, it was anticipated that the toxic hazard to nontarget insects would be minimal.
7. There were no endangered lepidopteran species susceptible to *B.t.k.* delta endotoxin in Maryland.

Concerns:

1. Trap plants, used to detect infective *C.x.c./B.t.k.*, should be planted directly in soil rather than grown in containers, as proposed by the applicant. In this way, the plants would also be exposed to soil and runoff water that might contain the recombinant organism.
2. In all analyses involving potential dissemination of the recombinant (trap plants, plants in the fallow zone, corn residues in the experi-

mental plot), the applicant should use a method with higher sensitivity than the one they had proposed.

3. During monitoring, the majority of samples from the fallow zone should be taken from plants that were found to be susceptible to infection in tests prior to the field study.
4. The efficacy of all proposed disinfection procedures should be well documented prior to the initiation of the field trials.
5. Methyl bromide fumigation should be used at the end of the field test to kill all the bacteria. This would also kill any nematodes that could serve as vectors for *C.x.c.* Afterwards, an inoculum of naturally occurring bacteria should be introduced to the soil to ensure that the soil was not recolonized by *C.x.c./B.t.k.*

Case 11 *(Submitted March 1988)*

Applicant: BioTechnica Agricultural, Inc.

Application type: 11 Premanufacturing Notifications for field testing

Organisms: *Rhizobium meliloti* (7 PMNs) and *Bradyrhizobium japonicum* (4 PMNs)

Intended use: To promote alfalfa yield increases by increasing nitrogen fixation by symbiotic microorganisms

Progenitor strain: Wild-type *Rhizobium meliloti* and *Bradyrhizobium japonicum*

Genetic methods: Antibiotic-resistance traits were introduced using recombinant plasmids that contained either: (1) a transposon with traits for kanamycin and neomycin resistance, or (2) a DNA fragment (omega fragment) carrying traits for streptomycin and spectinomycin resistance. A piece of DNA added to the omega fragment from *Escherichia coli* would provide a site for integration of additional genes in the future.

Field test procedure: Detailed information about the field test was not available.

Sites: Field stations near Arkansaw, WI, and Mount Pleasant, IA

Plants used: Alfalfa and soybeans

EPA position: The proposed tests did not pose an unreasonable risk to humans or the environment.

The issues addressed in risk assessment for this case were similar to those for the application by BioTechnica International (case 7).

Case 12 *(Submitted July 1988)*

Applicant: Monsanto Co.

Application type: Notification of intent to conduct small-scale field test

Organism: *Pseudomonas aureofaciens*

Intended use: As a monitoring organism and as a pesticide against take-all disease

Progenitor strain: P. aureofaciens strain Ps.2-79, isolated from the rhizosphere of wheat plants in Washington state; it was selected for its pesticidal activity against take-all *(G.g.t.)* disease which infects wheat plants.

Genetic methods: Lactose fermentation (*lac*ZY) gene from *Escherichia coli* K-12 was inserted into the genome of the parental strain by means of a delivery plasmid.

Field test procedure: Wheat seeds coated with the engineered strain were to be planted at two sites. The monitoring would continue until the density of the engineered organism fell below 10^3 CFU/g of root.

Sites: Two sites: one at a research and education farm in Blacksville, SC; the other at an agronomy farm of Washington State University near Pullman, WA

Area: Approximately 1.5 acres at each site

Plants used: Wheat

Amount of microorganism used: A total of 1 to 5×10^{14} CFU on about 8 pounds of seeds (about 4×10^8 CFU/seed)

EPA position: No adverse environmental effects or human health risks could be foreseen from this small-scale field test.

Justifications:

1. The Tn*7* transposon carrying the *lac*ZY gene was assisted by a plasmid that was subsequently lost. Thus, Tn*7* was integrated into the chromosome and lacked the transposition function itself. This made it unlikely that the transposon would be mobilized from the chromosome and the *lac*ZY gene transferred to another host cell.
2. *P. aureofaciens* is not known to be pathogenic for mammals, including humans. There would be minimal human exposure, and researchers would wear protective clothing.
3. The parental strain has not been reported in association with plant diseases.
4. The parental strain was not known to be an insect pathogen, and it was not likely to disseminate from roots to the phyllosphere. Thus, the risk to beneficial insects was minimal.
5. Preliminary data suggest that there would be only a small number of engineered organisms in the upper centimeter of soil available for runoff. Thus it was unlikely that this particular field test would increase the level of *P. aureofaciens* that occur naturally in the environment.
6. There were no federally listed endangered or threatened species occurring in the counties where these field tests were planned.

Concerns:

1. The engineered organism was very similar to a nonpesticide strain for which the applicant got permission from OTS (case 9), but it was not certain they would have identical environmental survival and persistence characteristics.
2. Since some unspecified strains of *Pseudomonas* might infect fish, it was not possible to rule out the possibility that the organism could cause fish disease without testing the strain in question (but see 5 above).

13.4 EVALUATION OF THE CASE STUDIES

An analysis of the justifications given to support the position of EPA and the specific concerns expressed by panel members reveals that there are five major issues that are repeatedly addressed. These are explained below in order of their importance, as measured by the number of times they are stated in case studies:

1. *Competitive advantage:* This includes the similarity of the engineered strain to the progenitor organism in terms of survival, niche breadth ("host range"), and growth characteristics. This important ecological property is addressed in all of the 12 cases in one form or another. In some cases, this is in the form of expression of a concern because there is not enough information to evaluate competitive relationships between strains (for example: case 9, concern 1; and case 12, concern 1).
2. *Pathogenicity and toxicity:* This includes any adverse effects to humans and other nontarget organisms. Obviously, a prerequisite for allowing a small-scale field test is its safety for humans. As a result, this issue is addressed in almost all of the cases. Pathogenicity for nontarget organisms, especially crop plants, is another important characteristic. One special class of nontarget organisms, endangered species, will be discussed below.
3. *Genetic transfer:* Unintended transfer of genetic material (such as toxin or antibiotic genes) to naturally occurring microorganisms is another important issue that is addressed in most of the cases reviewed. Since the prerequisite for such an undesirable effect is the establishment, survival, or dispersal of the engineered organism, this effect is dependent on other issues discussed here.
4. *Dispersal:* Dissemination and transport of the engineered organisms to areas outside the test plot, especially to aquatic environments, is an issue addressed in most of the cases. This effect becomes a more important concern if the engineered organism can potentially survive and multiply in nontarget environments.

5. *Endangered species:* Occurrence of rare, endemic, or endangered species in regions close to the test plot is another issue that is addressed in most of the cases, even if the engineered organism is believed to be nontoxic and nonpathogenic for the class of organisms to which the endangered species belongs.

Three of these five issues (competitive advantage, dispersal, and endangered species) can be classified as ecological, and another (pathogenicity) has an important ecological component. Competitive advantage is the issue that is common to all cases and is also usually expressed in more than one form for several cases. For example, in case 7, there are three issues related to competitive advantage: competition between engineered and indigenous strains for attachment sites, conditions for successful establishment of engineered organism, and the effect of genetic manipulations on the host range (that is, the niche breadth) of the engineered organism.

13.5 EMPHASIS ON ECOLOGICAL CONSIDERATIONS BY THE EPA

An interesting conclusion of the analysis of case studies is the relation between the list of important issues or factors that are considered in risk assessment studies and the types of information requested from applicants by the EPA. A comparison of Tables 13-1 and 13-2 with these factors shows a discrepancy between data requests and the priorities set by the scientific panels: panels give a very high priority to ecological considerations while data requirements relate mostly to toxicity and pathogenicity. We believe there are two major reasons for this discrepancy:

1. The first reason is the difference between the rate of technological and regulatory developments. The data requirements are based on regulatory practices that were in effect before the development of biotechnology. In the case of OPP, the requirements for notifications and EUPs were developed for chemical pesticides and for microbial pesticides that were not genetically engineered. In the case of OTS, the requirements for PMNs were developed for chemical substances. Although there were revisions of these requirements, they do not fully reflect the new scientific and technological developments in the field. Now that several studies have passed the stage of small-scale field testing, current regulatory practices are expected to be even more inadequate. Large-scale field tests and, especially, commercial production and distribution of biotechnological products for agricultural and environmental use will require a completely new set of guidelines for ecological risk assessment. Since some companies are already at the stage of large-scale tests, EPA is expected to take

some steps soon to define new directions for the development of the regulatory process.

2. The second reason for the discrepancy between data requirements and issues important for biotechnology risk assessment is the fact that the ecological theory related to the factors discussed above (most importantly, competitive interactions) has not yet been applied to the prediction of their effects in cases of genetically engineered organisms. The application of ecological theory to biotechnology risk assessment is crucial for the development of new regulatory practices, and it will be stressed by the regulatory agencies. Since the major ecological issue in the cases discussed in this chapter appears to be one of competitive interactions, the development of a link between theory and practice in this area will be one of these major directions.

Several chapters in this volume address these ecological considerations: chapters in Part I discuss effects of introductions on nontarget species. Issues related to genetic transfer are addressed in Part II and III. The importance of dispersal is reviewed in Part II. Thus, recent studies seem to focus on issues that are the most relevant in assessing the ecological risks of the release of genetically engineered organisms.

CHAPTER 14

Regulation and Oversight of Biotechnological Applications for Agriculture and Forestry

Maryln K. Cordle
John H. Payne
Alvin L. Young

There are a multitude of potential agricultural applications for the advanced tools of biotechnology, such as improving yields, quality, and consumer acceptance of traditional agricultural products; producing new products; and reducing adverse environmental impacts of current technology. However, the implementation of any new technology, particularly one that has diverse and widespread applications, may create new risks to the environment and to public health.

The United States Department of Agriculture (USDA) has responsibility for providing both financial support of basic research and oversight of the safety and efficacy of the products of research. Important challenges for the scientific community include implementing the process of biotechnology research in a timely fashion and assuring the safety of environmental releases

The views expressed in this chapter are those of the authors and do not necessarily represent those of the United States government.

as research progresses from the laboratory to field trials and commercialization.

USDA is committed to providing leadership and funding for a science-based program aimed at minimizing the risks of field testing genetically modified organisms and encouraging the development of products that will sustain our agricultural productivity and competitiveness in world markets into the 21st century. Currently, the USDA is funding more than $100 million for biotechnology research. This funding is expected to increase substantially over the next few years.

USDA has broad regulatory authority to protect against the adulteration of food products made from livestock and poultry, to protect United States agriculture against threats to animal health, and to prevent the introduction and dissemination of plant pests. This authority is as applicable to genetically engineered animals, plants, and microorganisms as it is those produced through traditional processes.

The application of new techniques of genetic modification in developing products of agricultural importance is proceeding at an unprecedented rate. For example, during the 1989 growing season in the United States, 12 companies and 3 universities received permits and conducted small-scale field tests on a variety of plant species and genetic characteristics. Those field tests, located in 18 different states, involved transformed plants of alfalfa, cucumbers, potatoes, tobacco, tomatoes, cotton, and soybeans. For the most part, the plants were genetically modified to induce either herbicide, insect, or disease resistance. The results of field tests are just now being reported and will be invaluable in guiding the development of outdoor testing to be conducted in the future. In response to this progress, the USDA continues to review its existing policies and regulations so that they reflect current needs and scientific knowledge.

A prime component of the USDA program is public involvement. The public takes pride in the country's technological capability and looks to the scientific community to provide the means of solving current-day problems and improving the quality of life. Yet, at the same time, the public does not blindly accept new technology. Having come to understand some of the negative impacts resulting from the chemical revolution in agriculture, for example, the public has become more circumspect in accepting new technology and demands a voice in how biotechnology will be used in the agricultural systems of the future. Public confidence and acceptance is key if the promises of biotechnology are to become a reality. This means that the scientific community, government officials, and the communication media must reach out to the public by providing factual information, addressing the benefits and risks, and actively involving the public in the formulation of policies for oversight and regulation of biotechnology in order to build public confidence and trust.

Biotechnology, because of its diversity of applications, has raised concern about the adequacy of the federal government's infrastructure and regulatory statutes to appropriately deal with issues raised by the new technology in a timely manner, as well as the government's ability to formulate coherent, effective policies. For example, the application of biotechnology to agriculture involves not only activities directly managed by USDA but also involves the regulation of pesticides and toxic substances by the U.S. Environmental Protection Agency (EPA), the U.S. Food and Drug Administration's (FDA) responsibilities for food safety and the regulation of new animal drugs and animal feed, and activities of the National Institutes of Health (NIH) associated with their "Guidelines for Research Involving Recombinant DNA Molecules" (NIH 1986), which by Executive Order are mandatory standards for all federally funded research involving recombinant DNA molecules.

The development of effective policies for biotechnology and agriculture involves both coordination among diverse federal jurisdictions and support from a diverse community of interests. A coalition of understanding must be reached by considering the legitimate needs and concerns of, for example, the biotechnology, food, and forestry product industries; farm community; scientific and research community; environmental and other special public interest groups; and the Congress of the United States. The challenge is a significant one that will require addressing many important issues with an unprecedented level of coordination and public outreach.

Significant progress has already been achieved. The development of biotechnology is moving forward safely; many new products providing benefits to human and animal health have already reached the market place. Field trials of many promising new crop varieties are underway. The development of criteria for food safety evaluation of genetically engineered foods is now at the forefront of discussion, in anticipation that some of the commodities under development may soon be ready for marketing.

14.1 PRESENT ORGANIZATIONAL STRUCTURES AND COORDINATION

Because so many different parts of the federal government are involved in biotechnology research and regulation, interdepartmental coordination is vital in developing consistent policies and resolving jurisdictional matters. This has been achieved largely through the Biotechnology Science Coordinating Committee (BSCC), a committee of the Federal Coordinating Council for Science, Engineering and Technology (FCCSET) (90 Stat. 472; 42 U.S.C. 6651), which reports to the Office of Science and Technology Policy (OSTP), Executive Office of the President (OSTP 1985). The BSCC is composed of a senior policy officials from USDA, EPA, FDA, NIH, and the National Science Foundation (NSF). BSCC provides an important forum for interagency science policy coordination and promoting consistency in the development of agency review procedures.

BSCC has been effective in improving interagency communication and resolving jurisdictional issues. For example, in the area of animal health care products, questions about whether a product is a new animal drug within the jurisdiction of FDA or a veterinary biologic within the jurisdiction of the Animal and Plant Health Inspection Service (APHIS) of USDA are readily resolved through liaison between the agencies for those products falling within a "grey" zone in the respective statutes. Some products fall within the jurisdiction of both EPA and USDA; however, procedures are in place to determine which agency will act as the lead agency and to ensure coordination of the review with minimal duplication and inconvenience to the applicant. BSCC leadership and a spirit of cooperation at the operational levels between agencies have now effectively overcome many of the earlier difficulties with jurisdiction.

BSCC also provides a useful forum for identifying issues and for fact finding, such as interagency funding of studies like the report issued in 1989 by the National Research Council of the National Academy of Science (NAS) on "Field Testing of Genetically Modified Organisms: A Framework for Decision Making" (NAS 1989).

Given the various missions of the member agencies, it should not be surprising that BSCC has had a difficult time in reaching a consensus on many significant issues—such as those involving policies for risk assessment and risk management, how to define what organisms should be subject to certain levels of oversight, defining environmental release, and establishing coherent standards. The rapid pace of development in biotechnology and its diverse applications add to the complexity of these issues. Nevertheless, in 1990, BSCC will be examining these and related issues as it undertakes a review of the "Coordinated Framework for the Regulation of Biotechnology," which was published by the White House Office of Science and Technology Policy in the *Federal Register* of June 26, 1986 (OSTP 1986). The Coordinated Framework describes a comprehensive Federal policy to ensure the safety of biotechnology research and products. USDA's proposed rules concerning plant pests were included in the 1986 publication of the Coordinated Framework. Final rules covering the introduction of organisms and products genetically modified by recombinant DNA techniques that contain or use plant pests in their construction, were issued on June 16, 1987. These regulations are discussed in Section 14.4.1.

Within USDA, there are nine agencies involved with biotechnology. These include: the Animal and Plant Health Inspection Service (APHIS), the Food Safety Inspection Service (FSIS), and the Agricultural Marketing Service (AMS)—all under the Assistant Secretary for Marketing and Inspection; the Agricultural Research Service (ARS), the Cooperative State Research Service (CSRS), Extension Service (ES), and the National Agricultural Library (NAL)—all under the Assistant Secretary for Science and Education; the Forest Service (FS)—under the Assistant Secretary of Natural Resources and Environment; and the Economic Research Service (ERS)—

under the Assistant Secretary for Economics. International activities involving biotechnology may also involve the offices reporting to the Undersecretary for International Affairs and Commodity Programs, in particular, the Office of International Cooperation and Development (OICD) and the Foreign Agricultural Service (FAS). It should be noted that regulatory activities are separated administratively from research activities.

The Committee on Biotechnology in Agriculture (CBA), which includes the administrators of the USDA agencies with major activities involving biotechnology, provides coordinated interagency review and recommendations on departmental policies involving research and regulatory matters. This committee is cochaired by the Assistant Secretary for Science and Education and the Assistant Secretary for Marketing and Inspection. The committee is ideally constituted to seek policies that neither overregulate nor underregulate the development and use of products of biotechnology.

In December 1987, the Office of Agricultural Biotechnology (OAB) was established under the Deputy Secretary of Agriculture. Its functions are: to facilitate and insure the coordination of all USDA biotechnology activities; to provide staff support for the Agricultural Biotechnology Research Advisory Committee (ABRAC) and the CBA; to promote coordination through various public outreach activities; and to provide leadership in activities such as the development of research guidelines (USDA 1986), guidelines for international scientific exchange (USDA 1989), and a handbook for field testing (Purchase and MacKenzie 1990). In August 1989, the Office of Agricultural Biotechnology was transferred under the Assistant Secretary for Science and Education. Its function will continue as before.

ABRAC was established in 1988, under the Federal Advisory Committee Act (Pub. L. 92-462; 85 Stat. 770, 5 U.S.C. App. I), to provide advice to the Secretary of Agriculture, through the Assistant Secretary for Science and Education, on matters of biosafety in the development and use of biotechnology in agriculture. ABRAC meetings are open to the public and announced in advance in the *Federal Register.*

14.2 HISTORICAL PERSPECTIVE ON RESEARCH GUIDELINES

When recombinant DNA technology began to be more widely used during the 1970s, concerns were immediately raised about its safety. At an historic international meeting held in California in 1975, scientists involved in the new technology made a number of recommendations and agreed to stringently control their own research (Berg et al. 1975). Within the federal government, it was agreed that research guidelines were urgently needed. NIH was given the lead to develop guidelines for containment of research involving recombinant DNA molecules. These guidelines were first published in 1976 (NIH 1976). The recombinant DNA Advisory Committee

(RAC) of NIH at first reviewed all recombinant DNA experiments, and the Director of NIH issued approvals. Later on, as experience was gained, NIH delegated certain review responsibilities to local Institutional Biosafety Committees (IBCs), which currently review the adequacy of physical and biological containment, in accordance with the NIH "Guidelines for Research Involving Recombinant DNA Molecules" (NIH 1978; NIH 1986). Experiments involving deliberate release into the environment or human gene therapy are reviewed by the RAC and require approval by NIH. By Executive Order, adherence to the NIH guidelines is mandatory for all federally sponsored research. The NIH guidelines have been widely accepted by both the academic and the private sectors, have reassured the public, and have allowed research to proceed in a timely and safe manner.

The evolution of the current NIH guidelines has progressed from a situation where initially nearly all federally funded research was reviewed on a case-by-case basis by the NIH Recombinant DNA Advisory Committee and approved by NIH, to the current situation where only a small fraction of the recombinant DNA laboratory research requires NIH approval. As scientists have learned more about the safety of genetically engineered organisms, initial fears have proven excessive, the NIH guidelines have been repeatedly revised, controls on recombinant DNA research in the laboratory have relaxed considerably, and more decision has been placed at the level of the local IBCs.

As agricultural biotechnology research progresses, investigators will need to further evaluate potential new products outside the laboratory, and thus, the need for additional guidance for investigators has become evident.

14.3 DEVELOPMENT OF USDA RESEARCH GUIDELINES

In response to that need and a Congressional mandate[1], USDA, with advice from its federal advisory committee, ABRAC, is developing guidelines for conducting USDA-funded research outside a contained facility, e.g., outside a laboratory or greenhouse. It is the intent of the USDA that the NIH guidelines will continue to apply to agricultural research conducted within a contained facility, and that the USDA research guidelines will apply when the research moves outdoors. The USDA guidelines envision that the IBCs will make a decision on outdoor testing of low-risk genetically modified organisms.

USDA expects to soon publish a "Notice of Intent" to prepare an Environmental Impact Statement for the guidelines with extensive public in-

---

[1] The Secretary of Agriculture has the responsibility to assure the safety of agricultural research funded by USDA. In the specific area of biotechnology, the Secretary also has the responsibility to establish appropriate controls with respect to the development and the application of biotechnology to agriculture [Section 1405 (12) of the National Agricultural Research, Extension and Teaching Policy Act of 1977, as amended, 7 U.S.C. 3121 (12)].

volvement. This action will be taken to comply with the National Environmental Policy Act (42 U.S.C. 4321), since the intent of the guidelines is to establish fundamental principles that can be used to evaluate experiments with genetically modified organisms and to select practices for confinement of organisms, so that outdoor experimentation will have a negligible or otherwise acceptable impact on the quality of the human environment. Like the NIH guidelines, it is anticipated that the USDA research guidelines will use a review and oversight system involving local IBCs, ABRAC, and USDA, with the level of review and approval commensurate with risk.

The principles of the USDA guidelines for evaluating safety and environmental risk and for designing confinement practices to minimize risk should be useful for investigators and reviewing authorities whether or not the research is subject to regulatory approval. The oversight provisions of the guidelines will ensure that federally funded research with products or organisms not regulated under current laws or exempted under the guidelines, will be subjected to oversight at either the local IBC or federal level as is commensurate with risk. Voluntary compliance with the guidelines by the private sector will be encouraged and is expected to occur, as has been the experience with the NIH guidelines.

One of the challenges to USDA is how oversight activity is to be managed as the number of new products ready for field testing expands. Dramatic increases may be anticipated as the technology of cell culture progresses, and our understanding of genes controlling functions of agricultural significance increases. As previously noted, one alternative for reducing the burden of oversight is to make greater use of a local IBC system for more routine review of field trials for which precedents have been set and specific guidelines are available. However, under current statutes administered by federal regulatory agencies, pursing this strategy for deregulation of research may be difficult and will certainly raise questions involving delegation of authority.

Defining the scope of organisms to which the USDA research guidelines should apply is problematic for several reasons. The public's attention is focused on potential risks of research involving genetic modification through recombinant DNA technology and similar molecular methods. But those risks are not inherently greater than risks for research with organisms in general or experiments involving organisms genetically modified by traditional breeding methodologies (NAS 1987). The agricultural research community has always been aware of the potential risks of hybrids, and the need to contain that research until a full evaluation of the hybrid justifies deliberate release into the environment. Basic principles of safety assessment are relevant to all agricultural research, whether or not they involve organisms in which the genome has been deliberately modified. Unlike traditional methods of genetic modification, recombinant DNA technology and similar advanced methods can provide greater precision and control at the molecular level over the genetic modification. However, these newer technologies also can be used to produce a wider range of genotypes or phenotypes, and their behavior may be distinctly different from those seen in nature. Use of these techniques, thus, may result in a certain "newness" of phenotype and a degree of "unfamiliarity" concerning the behavior of the organism in the environment. Therefore, an uncertainty about whether its behavior may be associated with risk arises. In contrast, there is an extensive history of management and safety for the development of hybrids through classical breeding methodologies.

A system of oversight ideally should depend on the risk of the organism based on its phenotypic characteristics and not the process by which the genetic modification was made. In defining the scope of organisms subject to the USDA research guidelines, those organisms must be clearly defined, public concerns must be adequately addressed, the need for oversight must be justified, and the impact on research and development and the quality of the environment must be carefully assessed before adoption of guidelines.

ABRAC has suggested that the guidelines apply to research with organisms that result from deliberate genetic modification (i.e., insertion, deletion, rearrangement, or other manipulation of DNA or RNA), and that they not apply to the following organisms:

1. Plants resulting solely from: selection or natural regeneration; hand pollination or other managed controlled pollination; chemical or physical mutagenesis; or that are regenerated from organ, tissue, or cell culture, including those produced through selection and propagation of somaclonal variants, embryo rescue, protoplast fusion, or treatment that causes changes in chromosome number;
2. Animals that result solely from selection, artificial insemination, superovulation, embryo transfer, or embryo splitting;
3. Microorganisms modified solely by chemical or physical mutagenesis; the movement of nucleic acids using physiological processes, including, but not limited to transduction, transformation, or conjugation; and plasmid loss or spontaneous deletion[2]; and
4. Other specific exclusions as approved by USDA.

Additional exclusions are expected to result from recommendations of ABRAC, IBCs, or the research community. Exclusions are likely to involve organisms, independent of the method of genetic modification, for which there is sufficient familiarity to foresee environmental effects equivalent to those associated with past safe introductions into the environment of similar organisms in similar environments.

In recommending the scope of organisms for which the USDA research guidelines would apply, ABRAC took into account the extent to which

---

[2] If nucleic acid molecules produced using in vitro manipulation are transferred using any of the techniques listed in this exclusion or the two previous exclusions, the resulting organism is not exempt.

existing federal regulations, guidelines, and accepted standards of practice have successfully been relied upon to prevent adverse safety and environmental effects. It has sought to impose new requirements only to the extent necessary to address legitimate safety concerns.

From a broader perspective, BSCC also is examining the scientific basis of the scope organisms to be included in oversight for planned introductions or organisms into the environment, with the goal of developing a definition that agencies can use for guidance, as appropriate, in carrying out their oversight responsibilities. The outcome of this initiative undoubtedly will influence decisions on the scope of organisms subject to the USDA research guidelines.

14.3.1 Principles for Safety Assessment for USDA Research Guidelines

Although still under development, certain concepts and principles have been recommended by ABRAC for the USDA research guidelines. One of these principles is that the conditions for the safe conduct of research with genetically modified organisms should be considered in relation to conditions that are generally accepted for safely conducting field research with the unmodified organism. Genetic modification may affect some, but not all, traits of the organism. Genetic modification may have no effect on the safety of research with that organism or it may increase or decrease the safety. The safety assessment should be based on the phenotypic characteristics of the modified organism rather than on the nature of the process involved in making the modification. However, knowledge of the precise modification, including the process by which it was made, may allow better predictions of the safety of the organism and its products, so that appropriate confinement and other safety practices for the research can be selected.

ABRAC has recommended a step-wise safety assessment process that begins with a determination of the level of safety concern for the unmodified organism. The evaluation of the unmodified organism would be carried out within the context of the environment accessible to the organism at the site where the research is to be performed assuming it were to be released at that site. The evaluation would consider such factors as the organism's pest/pathogen status; its ability to become established in the accessible environment at the research site; its ecological interrelationship, function, and importance in the community; its ability to transfer genetic information; and the potential for monitoring and control.

Assessment of pest/pathogen status would consider possible adverse effects, such as lowered productivity of economically important organisms, damage or destruction of natural habits, and adverse effects on plant, animal, or human health. Because an unmodified organism that is not a pest/pathogen may become one through the exchange of genetic information with other closely related organisms, the assessment would consider the potential for exchange of genetic information between the unmodified organism and pest/pathogens in the accessible environment. Ecological characteristics that allow or inhibit the organism from becoming a pest or increasing in pest/pathogen status should be defined in the assessment.

Evaluation of the potential of the unmodified organism to become established in the accessible environment would consider factors such as known mechanisms of survival or persistence, including natural biotic and abiotic control; mechanisms of dissemination; effects of population size on establishment; and the competitiveness and aggressiveness of the unmodified organism.

The ability of the unmodified organism to cause adverse effects is related to its ecological relationships with other organisms. The assessment recommended by ABRAC, therefore, considers the importance of the unmodified organism to the structure of the community. For example, it is important to identify whether the organism is involved in any critical ecosystem function, whether that involvement is direct or indirect, and whether other organisms in the ecosystem can also fulfill this function. Ecological specificity, range of interactions, and geographic range need to be considered, along with characteristics that might broaden or narrow its niche. The habit of the unmodified organism, e.g., whether it is free-living, mutualistic, pathogenic, parasitic, or symbiotic, is informative regarding its ecological niche and possible effects on the environment should it escape from the research site. Habit also may affect the feasibility and methods for control of the organism if it were inadvertently released from the research site.

The assessment would consider the potential for introduction of genetic change in natural or managed populations, taking into account the intrinsic genetic stability of the genome, such as its ability to incorporate exogenous DNA, the presence of transposable elements, or of active viral elements that interact with the normal genome, or of mutations that have resulted in an unusual genotype. This assessment also would consider the size of the interbreeding population, its genetic diversity, and the probability for genetic exchange resulting from release at the research site.

The ability to control a release of the organism is a factor in the level of safety concern. This would be evaluated by considering experience from prior experiments; the availability of practical and reliable monitoring methods; and methods of control in the event of an inadvertent release. One must also consider the effectiveness of available controls, whether they have any adverse effects on the environment, and how costly or difficult it will be to institute those controls, if necessary.

Among the numerous factors described above in the assessment, some will be more important than others for particular organisms; relative importance would be considered in the assignment of the level of safety concern.

Based on the evaluation as briefly described above, ABRAC has proposed that the unmodified organism be assigned to one of five levels of safety concern. The safest organisms, those organisms found to have virtually no potential for adverse effects on human health or the environment, would be assigned to concern level 1. The highest risk organisms, i.e., ones for which adequate controls are not available to prevent serious effects on health or the environment outside a contained facility, would be assigned to concern level 5. Organisms with potential for harm graded as low, moderate, or high, would be assigned to levels 2, 3, or 4, respectively. ABRAC has developed guidance for the assignment of levels of safety concern and has developed a number of examples for which the assessment and rationale for the level of safety concern is presented (USDA 1990).

The second step in the assessment would consider how the genetic modification affects safety, i.e., whether it increases, decreases, or has no effect on safety. The evaluation would consider the process of genetic modification data on gene construction and expression, and the degree to which knowledge of the molecular biology and other information is available for predicting the safety of the modified organism relative to the unmodified organism.

The third step would combine the evaluation in steps 1 and 2, essentially using properties of the organism determined empirically, as well as by prediction from examination of the molecular biology of the modification, and would assign a level of safety concern for the genetically modified organism.

The fourth step would select the confinement level appropriate for the level of safety concern of the genetically modified organism, and provide guidance for the principal investigator in developing a biosafety protocol. That protocol must take into account not only the level of safety concern, but also the specific characteristics of the organism, the accessible environment, and the experimental design. Five classes of confinement practices are addressed: physical, biological, chemical, environmental, and scale. The agricultural research community has traditionally used confinement measures selected from these classes in designing safe research experiments, and has a great deal of experience and knowledge about their effectiveness in controlling the spread of organisms beyond the research environment. ABRAC has recommended that detailed confinement criteria by type of organism be developed as an appendix to the guidelines. Confinement level 1, for example, applies to organisms with virtually no potential for adverse effects, would not impose requirements beyond that generally accepted as good agricultural practice (GAP) for the particular organism in the experimental area. Other confinement levels are achieved by utilizing one or more classes of confinement and by varying the degree of stringency of confinement with the appropriate classes. Research with organisms classified as concern level 5, for example, would be subject to containment conditions defined by the NIH guidelines. Generally, before research can be conducted outside the laboratory or greenhouse with moderate- or high-risk organisms, review and clearance by a federal regulatory agency is required.

The last two steps recommended by ABRAC include instructions for preparing a submission for review (step 5), and for determining the appropriate, required level of oversight (step 6).

The level of oversight is based on the level of safety concern for the genetically modified organism. ABRAC has recommended that notification of an IBC is sufficient for research with organisms of safety concern level 1 (i.e., organisms of virtually no concern). The purpose of the notification is to provide the IBC an opportunity to inform the investigator, if it disagrees with the designation of the organism as safety concern level 1. IBC review and approval has been recommended for research with organisms of safety concern level 2 (i.e., low-risk organisms) with a notification to be sent to USDA so that the information can be entered into the National Biological Impact Assessment Program (NBIAP) data base. Generally, before research can be conducted outside the laboratory or greenhouse with moderate- or high-risk organisms (i.e., safety concern levels 3 and 4), review and clearance by a federal regulatory agency is required. Where that is not the case, ABRAC recommends that research with organisms of safety concern level 3 and 4 first be submitted to the IBC for review and approval, and then be submitted to USDA for review and approval, using the expertise of ABRAC when appropriate.

Finally, the guidelines would define an oversight process, commensurate with the concern about safety and effects on the environment. The oversight for various risk categories may include: categorical exclusions; notification requirements only; review and approval by an IBC; or review and approval from USDA, using the expertise of ABRAC when appropriate. It is expected that using the existing system of local IBCs, which currently handle the vast majority of review under the NIH guidelines, can be used for the USDA guidelines as well. However, some IBCs may have to extend membership to obtain the necessary expertise for review of agricultural research and its potential for environmental impacts. The NIH guidelines require a five-member IBC; ABRAC has recommended a six-member IBC. For research subject to review and clearance by a federal regulatory agency, additional oversight under the provisions of the USDA guidelines would not be necessary. Until USDA completes its development and adoption of the research guidelines, ABRAC may be used to review, on a case-by-case basis, the biosafety protocols of proposals submitted to USDA for funding of outdoor experiments and will provide recommendations to the funding agency.

14.3.2 Complementary Activities

To further assist the research community, USDA has supported preparation of a publication, *Agricultural Biotechnology: Introduction to Field Testing,* that presents a wide range of information that investigators need to know in planning outdoor experiments with genetically modified organisms (Purchase and MacKenzie 1990). The publication discusses not only confinement and safety issues, but also provides information on federal regulatory jurisdiction and basic regulatory requirements, guidance on public relations, and a discussion of some of the sociological and economic concerns about research with recombinant DNA technology.

Also, USDA's National Biological Impact Assessment Program has developed and will continue to expand a computerized source of information on agricultural/environmental biotechnology that can be reached toll free using a personal computer equipped with a modem. The extensive information in the data bases is designed, among other purposes, to assist an investigator in preparing a research proposal for biosafety review by a local IBC or federal agency.

14.4 USDA REGULATIONS FOR BIOTECHNOLOGY

The USDA policy on the regulation of biotechnology, consistent with the overall federal policy, does not view genetically engineered organisms and products as fundamentally different from those produced by conventional methods. The products of the new techniques of biotechnology are regulated under existing laws that apply to products of traditional technologies. To address the need for specific information necessary for the assessment of the products of the new technologies, a few new regulations have been promulgated, and some old ones have been updated.

Regulations at USDA are focused on the product and any risks posed by a specific use of that product, rather than on the process used in production. Thus, USDA regulates products of biotechnology through established statutory mandates, not because the products and organisms are the result of biotechnology per se, but as a consequence of being regulated products or organisms irrespective of the technology involved.

14.4.1 Plant and Plant Pest Regulation

Through regulations promulgated under the Plant Quarantine Act (7 U.S.C. 151–164, 166, 167) and the Federal Plant Pest Act (7 U.S.C. 150aa–150jj), APHIS regulates the movement into and through the United States of plants, plant products, plant pests, and any product or article that may contain a plant pest (i.e., regulated article) at the time of movement. The purpose of the regulations is to prevent the introduction, spread, or establishment of plant pests new to or not widely prevalent in the United States. A "plant pest" is defined as any living stage of any insects, mites, nematodes, slugs, snails, protozoa, or other invertebrate animals, bacteria, fungi, other parasitic plants or reproductive parts thereof, viruses, or any organisms similar to or allied with any of the foregoing, or any infectious substances that can directly injure or cause disease or damage in any plants or parts thereof, or any processed, manufactured, or other products of plants. Permits are required for the introduction, importation, or movement of organisms and products that may be or may contain a plant pest.

Regulations governing the introduction of conventional plant pests have been in place and permits have been issued under these regulations since 1959. Specific regulations were promulgated in 1987 (7 CFR Part 340) to cover the introduction of organisms and products produced through genetic engineering, i.e., genetic modification of organisms by recombinant DNA techniques that contain or use plant pests in their construction.

Permits are required for any introduction of a regulated article. "Introduction" means to move a regulated article into or through the United States, to release it into the environment, to move it interstate, or any attempt thereat. "Release into the environment" means use of the regulated article outside the constraints of physical confinement of a laboratory, contained greenhouse, or a fermenter or other contained structure. A "regulated article" is any organism that has been altered or produced through genetic engineering, if the donor organism, recipient organism, or vector agent belongs to any genus or taxon designated on a list of organisms in 7 CFR 340.2 and which meets the definition of plant pest, or is an organism whose classification is unknown, or any product that contains an unclassified organism and may be a plant pest.

The regulations have been amended to include a specific exemption from the need for a permit for some organisms that meet the definition of a regulated article. A limited permit is not required for interstate movement for genetic material from any plant pest contained in *Escherichia coli* genotype K-12 (strain K-12 and its derivatives), sterile strains of *Saccharomyces cerevisiae*, and asporogenic strains of *Bacillus subtilis*, provided all of the specific conditions listed in the regulation at 7 CFR 340.2(b)(1) are met. This exemption is consistent with exemptions contained in the NIH guidelines.

As scientific knowledge with the products of biotechnology increases and as we achieve familiarity with the interaction between modified organisms and their environment, additional specific exemptions are expected to be added to the regulations. These exemptions will be added either by APHIS at the agency's initiative, or through the public petition process described in detail later in this section. The amendment will be subject to the usual rules of the regulatory process with a proposal published in the *Federal Register* for public comment before a final amendment is adopted.

The information APHIS requires for the evaluation of an organism for movement or release to the environment must be provided in the application for a permit on APHIS Form 2000. The form requests information about the responsible person and the intended action. Specific information about the organism includes all scientific, common, and trade names, and all designations necessary to identify the donor organism(s), recipient organism, and the vector or vector agent used to move the genetic sequences. Also, a description is requested of the anticipated or actual expression of the altered genetic material in the regulated article and how that expression differs from the expression in the nonmodified parental organism. A detailed description of the molecular biology of the system that was used to produce the regulated article is required. It is necessary to indicate the country and locality where the donor organism, recipient organism, and vector or vector agent were collected, developed, and produced.

In addition to the information about the organism, information is requested about the use of the organism and, for release to the environment, questions are asked about the location. A detailed description of the purpose for the introduction of the regulated article is required, including a detailed description of the proposed experimental or production design. The information on the protocol includes the quantity of the regulated article to be introduced and the proposed schedule, scale, and number of introductions; and a detailed description of the processes, procedures, and safeguards that have been used or will be used at all intermediate and final destinations in order to prevent contamination, release, and dissemination of the regulated article. That discussion should include any confinement practices to be used in the field—including biological, chemical, physical, spatial, environmental, and temporal isolation. It is essential that the application include a description of the proposed method of final disposition of the regulated article.

The detailed descriptions that make up the answers to the questions on the APHIS application form may be included as attachments. APHIS will give advice to potential applicants on format and on the amount of detail appropriate for the individual questions for a specific organism. Consultation with the agency prior to an application is recommended, especially for those who have never applied before, because it saves time and effort for both the applicant and the agency by focusing the information provided by the applicant to that necessary for the analysis—avoiding unnecessary reporting of superfluous or redundant material.

Two copies of the application for interstate movement or importation must be submitted at least 60 days in advance of the proposed movement, and applications for release to the environment should be submitted at least 120 days in advance of the release. The first copy of the application should contain all information, whether confidential or not, with the confidential information clearly marked. The second copy of the application has confidential business information deleted. Confidential business information is protected from disclosure under the Freedom of Information Act [5 U.S.C. 552(b)(4)].

During the 120-day review period for applications for release to the environment, several specific APHIS actions are accomplished. A notice of receipt of the application is published in the *Federal Register* soon after the application is received. A copy of the application and preliminary review are sent to the state in which the test will take place within 30 days after the application is received. APHIS coordinates all reviews with agencies in the states in which the field test will occur. The state agencies have 30 days to comment after receiving the application.

An environmental assessment is prepared by the reviewing scientist in APHIS, and the environmental assessment is subject to peer review by the agency's scientific staff. When the environmental assessment results in a Finding of No Significant Impact (FONSI) on the environment, a notice of availability of the environmental assessment and FONSI is published in the *Federal Register*, and the permit is issued. The permit may contain special conditions that must be met in addition to those specified in the protocol submitted as part of the application. It is a condition of every permit for release to the environment that test data be collected and submitted to APHIS.

The environmental assessment is a key component in the permit review process. It contains a thorough accounting of the agency's analysis leading to a decision to issue a permit. Upon notice of its availability in the *Federal Register*, it is a public document, made available to anyone who requests it, free of charge.

Each environmental assessment is made up of a summary describing the purpose of the document, APHIS regulations, the conditions under which the permit is issued or denied, precautions against environmental risk, the background biology of the organisms, and the possible environmental consequences of the field test. The environment that could be affected by the field test is described and the precautions developed for protecting that environment, including field plot design, field inspection and monitoring, test plot security, and disposal plans are analyzed. The environmental consequences of the test are examined from all possible perspectives. Consideration is given to the biology of the recipient, donor, and vector, and to the potential for biological containment based on knowledge of this biology. Any possibility of risk to native flora or fauna is evaluated, with special consideration of organisms that are threatened or endangered. Any potential impact on human health is examined. It is through the environmental assessment that the public can be assured that APHIS has fully considered the possible consequences of releasing the regulated article into the human environment.

An innovative feature of the regulations under 7 CFR 340 is the provision within the regulations for a petition to amend the regulations. The petitioning process is expected to be used by applicants and other members of the interested public to seek changes to the list of organisms in 7 CFR 340.2 or to obtain specific exclusions of genetically engineered organisms from regulation. The petition to amend must contain a statement of grounds and supporting literature, data, or unpublished studies, as well as opposing views or contradictory data. A petition that meets these requirements will be published in the *Federal Register* for comment. When a petition is approved, the changes to the regulation are published with a complete response to public comments.

APHIS works closely with the FDA and the EPA in all regulatory efforts involving the products of biotechnology. In certain instances, APHIS and EPA coordinate reviews of proposed field releases when jurisdiction is shared. For example, APHIS sends copies of permit applications to EPA for transgenic plants that EPA considers to have pesticidal properties, such as those containing the *Bacillus thuringiensis* endotoxin. Representatives from these agencies meet regularly to discuss specific cases arising under their authorities. We expect this coordinated approach will enable federal agencies to anticipate any potential safety concerns with the range of bioengineered food and agricultural products currently under development.

14.4.2 Animal Biologics

APHIS has authority under the Virus-Serum-Toxin Act (21 U.S.C. 151–158) to regulate all veterinary biological products produced, used, or imported into the United States, including shipment and delivery for shipment, both interstate and intrastate. The term "biological products" is broadly defined and means "all viruses, serums, toxins, and analogous products of a natural or synthetic origin, such as diagnostics, antitoxins, vaccines, live microorganisms, killed microorganisms and antigenic or immunizing components of microorganisms intended for use in the diagnosis, treatment or prevention of diseases of animals."

APHIS issues licenses and permits for production, importation, sale, and experimental use of various types of biological products. Permits are required for the importation of products for research and evaluation, for their distribution and sale, and for transit through the United States to some other destination (9 CFR Part 104). A person may be authorized to ship unlicensed biological products for the purpose of evaluating experimental products by treating a limited number of domestic animals if APHIS determines that the conditions under which the experiment is to be conducted are adequate to prevent spread of disease and approves the procedures set forth in the request for such authorization [9 CFR 103.3 (a)–(g)]. U.S. Form 14–5 is used to apply for a permit.

Requests to ship unlicensed biological products for experimental field studies must include information to demonstrate that the experimental conditions will be adequate to prevent the spread of disease. General information that should be provided in a request for authorization includes:

1. Evidence that appropriate authorities from each state involved have granted permission.
2. Names of the proposed investigators to whom the products will be shipped and the quantities to be sent to each. Any changes from the tentative list must be subsequently provided.

3. A description of the product, including all active and inert ingredients, recommendations for use, and results of any preliminary research work.
4. A proposed plan covering methods and procedures for evaluating the product, maintaining records concerning quantities of the product shipped and used, and at the conclusion of the field studies, obtaining and summarizing results and submitting those to Veterinary Services as required.
5. Copies of the proposed label that must bear a statement, e.g., "Notice! For Experimental Use Only—Not For Sale."
6. Information on how the animals will be disposed of at the end of the experiment, including data showing that use will not result in the presence of any unwholesome condition in the edible parts of the animals if they are to be subsequently slaughtered for human food. Research investigators must provide a statement agreeing to furnish additional information, as required, prior to movement of all experimental animals from the premises.

Specific information that must be submitted for genetically engineered products or organisms includes detailed information on stability, genetic constructs and vectors, and the effects of any insertions and deletions on the organism. Products and organisms requiring submission of specific data are recombinant DNA-derived products, chemically synthesized antigens, monoclonal antibody products, and master seeds for recombinant DNA-derived products.

Biological products produced using genetic engineering are classified into the following three broad groups based on the characteristics of the products and safety concerns:

1. Inactivated products, such as recombinant DNA-derived vaccines, bacterin-toxoids, viral and bacterial subunits, and monoclonal antibodies.

These nonviable or killed products pose no risk to the environment and present no new or unusual safety concerns.

2. Live products, such as products containing live or infective organisms modified by insertion or deletion of one or more genes.

Precautions must be taken to assure that any addition or deletion of specific genetic information does not impart increased virulence, pathogenicity, or survival advantages in these organisms that are greater than those found in natural or wild-type forms.

Modifications also must not impart undesirable new or increased adherence or invasion factors, colonization properties, or intrahost survival factors, or compromise the safety characteristics of the organisms.

The genetic information to be added or deleted must be well-characterized DNA segments. Required licensing data may include base pair analysis, sequence information, restriction endonuclease sites, as well as phenotypic characterization of the altered organism. A comparison must be made between the genetically engineered organism and the wild-type form with respect to biochemical pathways, virulence traits, and other factors affecting pathogenicity.

3. Live vectored products, such as products using live vectors to carry recombinant-derived foreign genes that code for immunizing antigens and(or) other immune stimulants.

Characteristics of safety and transmission must be examined before questions and concerns dealing with safety to humans, animals, and release into the environment can be answered. To provide guidance to current or prospective manufacturers, APHIS has indicated certain points to consider, which are given below, for the use of recombinant DNA-derived products, chemically synthesized antigens, monoclonal antibody products, and master seeds. If available, investigators should submit the information described below with their applications for permits for environmental release experiments, depending on the nature of the product.

1. For recombinant DNA-derived products, the specific cloned nucleotide segment coding for the desired product or other foreign DNA segments must be defined in data supporting each license application. These data must also include a description of the source of the DNA and the nucleotide sequence. With regard to the vector or cloning vehicle, the mechanisms of transfer, the copy number, and the physical state (integrated or extrachromosomal) of the constructed vector inside the host cell must be described.
2. If the product consists of chemically synthesized antigens, the procedures used to increase or prolong an immune response, such as coupling to carrier proteins or addition of adjuvants, must be described. Immunological data derived from chemically synthesized peptides must be as definitive as those from natural antigens.
3. For monoclonal antibody products, a description of cell cloning procedures, preparation, and characterization of cell passages must be provided. The specificity and potency of monoclonal antibody will be compared to those of similar polyclonal antibody products where appropriate. Monoclonal antibody must be derived from Master Cell Stocks that meet the applicable requirements of 9 CFR 113.52.
4. The use of Master Seeds consisting of constructed plasmids or transfected cells requires submission of background information concerning the recombinant DNA procedures used to isolate, purify, and identify genetic material from one source and the modification used for inserting of this

material into a new host. A nucleotide sequence analysis should be provided to characterize adequately the foreign DNA used to code for a particular antigen. Tissue culture-propagated cells from vertebrate animals used for vector propagation and antigen production must meet the requirements of 9 CFR 113.51 or 113.52.

Before authorizing field testing of genetically engineered live products, APHIS prepares an environmental assessment to determine the potential effects of the test or tests proposed. Evaluation of data submitted is additionally used to determine conditions necessary for safe field testing and use. The environmental assessment and the Finding of No Significant Impact (FONSI) are published in the *Federal Register* and are available for public comment for 30 days prior to initiation of the field test. A second environmental assessment is published in the *Federal Register* prior to licensure.

No livestock or poultry used in any research investigation involving an experimental biological product will be accepted for slaughter at establishments subjected to the Federal Meat Inspection Act or the Poultry Products Inspection Act unless written approval from FSIS is provided. The requirements are discussed further under Section 14.4.1.

14.4.3 Animal Organisms and Vectors

The Animal Quarantine Statutes, i.e., The Act of February 3, 1903 (21 U.S.C. 111) in combination with the Virus-Serum-Toxin Act (21 U.S.C. 151–158), give APHIS the authority to regulate importation and interstate movement of organisms and vectors that may cause introduction or dissemination of contagious or infectious diseases of animals, including materials of animal origin that may carry agents of disease. The importation and shipment of organisms and vectors are regulated under 9 CRF 122. Permits are also required for importation and interstate movement (including subsequent usage) of organism or vectors, including cell cultures and hybridomas. Permit applications must describe the substances, their intended use, the location of the permittee, and safeguards. VS Form 16-3 is used to apply for a permit for importation or interstate movement. In certain cases, testing for adventitious disease agents is required. Testing is done at the USDA's Foreign Animal Disease Diagnostic Laboratory at Plum Island, New York. Usually 60 to 90 days is required for issuing a permit when testing at Plum Island is necessary. Applicants will be advised if safety testing is required and the estimated cost of the testing.

Permits also are required for introduction and field testing of genetically engineered organisms that, in their construction, use material from infectious, contagious, pathogenic, or oncogenic organisms. The type of information required in requesting a permit is similar to that described in Section 14.4.2.

14.4.4 Food Safety

The USDA's Food Safety and Inspection Service (FSIS) is responsible for assuring the safety, wholesomeness, and proper labeling of food products prepared from domestic livestock and poultry under the Federal Meat Inspection Act (FMIA) (21 U.S.C. 601 et seq.) and the Poultry Products Inspection Act (PPIA) (21 U.S.C. 451 et seq.). Under these acts, FSIS inspects cattle, sheep, goats, horses and other equines, domesticated birds, and food products prepared from these that are intended for use as human food. USDA's Agricultural Marketing Service is responsible for assuring the safety, wholesomeness, and proper labeling of egg products under the Egg Products Inspection Act (EPIA) (21 U.S.C. 1031 et. seq.). The regulation of all other food products falls within FDA's jurisdiction.

The food adulteration provisions of the FMIA, PPIA, and EPIA closely parallel those of the Federal Food, Drug and Cosmetic Act, so that, in effect, USDA enforces the standards for added substances (e.g., animal drugs, veterinary biologics, pesticide chemicals, and food additives) established by FDA or EPA.

Under the provision of the FMIA and PPIA concerning the slaughter of research animals, FSIS has developed regulations (9 CFR 309.17 and 381.75), which stipulate that no livestock or poultry used in any research investigation shall be eligible for slaughter at an official establishment until adequate data are provided to FSIS to show that edible products from the experimentally treated animal are safe for human consumption. These regulations cover the slaughter of experimental animals treated with recombinant DNA-derived products and transgenic animals.

In the review of a request for approval to slaughter experimental animals, FSIS would coordinate its review with the agency that has jurisdiction for the experimental product (i.e., APHIS for biologics, FDA for drugs, and food and feed additives, or EPA for pesticide chemicals). This assures that all investigational use provisions of the agency's regulations have been met, that the agency concurs that the edible products will be safe for human consumption, and that all restrictions, e.g., a withdrawal period prior to slaughter, have been met. Evidence that edible products from an investigational animal at time of slaughter are safe for human consumption and not adulterated may require that certain phenotypic, biochemical, and microbiological parameters not be exceeded. Generally, the data provided to the agency with jurisdiction for the experimental product involved will be sufficient for FSIS's evaluation.

Investigational animals approved for slaughter will be subject to the same inspection standards as noninvestigational animals, except as may be provided as a condition of approval for slaughter. It is expected that most food animals genetically modified by modern biotechnological techniques will not differ substantially in appearance, behavior, or general health from currently inspected food animal species, and therefore would be subjected to the same inspection procedures as traditionally bred animals. FSIS is aware, however, that some animals produced by new technology, such as mosaics, chimeras, and some hybrids, may differ substantially from animals that are inspected currently under FMIA and PPIA. If such animals are intended for slaughter and human food use, FSIS will have to determine on a case-by-case basis whether the animals are covered under FMIA or PPIA, and if not, whether the acts should be amended to require inspection.

FSIS also has jurisdiction over substances that may be used in processing meat and poultry products. No substance may be used in preparing meat and poultry products unless it is approved and listed in 9 CFR 318, 319, and 381, and its use conforms with the approved conditions. New substances or new uses of approved substances are approved by FSIS only if the substance has previously been approved by FDA for use as a food or color additive, or the substance is generally recognized as safe. Its use must be in compliance with applicable FDA regulations and cannot cause the product in which it is used to be adulterated or misbranded or otherwise not in compliance with FMIA or PPIA and regulations issued under those acts. The use must be functional and suitable for the product and at the lowest level necessary to accomplish the stated technical effect.

Those engaged in agricultural research with food crops and animals need to know the food safety evaluation criteria that the respective agencies will use to grant clearance, so that they can obtain the necessary data. The International Food Biotechnology Council (1126 Sixteenth Street, N.W.; Washington, D.C. 20036), a consortium of private sector entities, with an interest in food biotechnology, is finalizing a report that is expected to include recommended criteria and data to be evaluated in determining the safety of genetically engineered foods and food additives of nonanimal origin. The council hopes that its report will be useful as the federal government develops its policies for this important area. Food safety is likely to be a high priority for discussion within the BSCC.

14.5 SUMMARY

USDA regulatory activities, in conjunction with those of other federal agencies, have allowed agricultural research involving the new tools of biotechnology to move forward safely. Coordination between the agencies within the BSCC or at the operating level has, for the most part, resolved jurisdictional questions. As experience is gained from the results of field testing, the necessary level of federal oversight should be continually reexamined, so that we neither overregulate nor underregulate applications of modern biotechnology. The development of USDA research guidelines will supplement the NIH guidelines and provide sound scientific principles for the design of safe experiments to be conducted outside a laboratory or greenhouse.

The review in 1990 of the Coordinated Framework for Biotechnology by the Executive Office of the President is expected to result in a more coherent policy among the federal agencies as the BSCC examines significant issues. Food safety is one of those issues that is just now moving to the forefront. Public outreach and involvement in the development of policies affecting agriculture, food, and the environment are essential for consumer acceptance of the products of biotechnology.

REFERENCES

Berg, P., Baltimore, D., Brenner, S., Robin, R.O. III, and Singer, M.F. (1975) *Science* 188, 991.

National Academy of Sciences (NAS) (1987) *Introduction of Recombinant DNA-engineered Organisms into the Environment: Key Issues*, National Academy Press, Washington, D.C.

National Academy of Sciences (NAS) (1989) *Field Testing Genetically Modified Organisms: Framework for Decisions*, National Academy Press, Washington, D.C.

National Institutes of Health (NIH) (1976) "Recombinant DNA Research Guidelines," *Federal Register* of July 7, (41 FR 27902).

National Institutes of Health (NIH) (1978) "Guidelines for Research Involving Recombinant DNA Molecules," *Federal Register* of December 21, (43 FR 60108).

National Institutes of Health (NIH) (1986) "Guidelines for Research Involving Recombinant DNA Molecules," *Federal Register* of May 7, (51 FR 16958).

Office of Science and Technology Policy (OSTP) (1985) "Coordinated Framework for Regulation of Biotechnology: Establishment of the Biotechnology Science Coordinating Committee," *Federal Register* of November 14, (50 FR 47174).

Office of Science and Technology Policy (OSTP) (1986) "Coordinated Framework for Regulation of Biotechnology," *Federal Register* of June 26, (51 FR 23303).

Purchase, H.G., and MacKenzie, D.R., eds. (1990) *Agricultural Biotechnology: Introduction to Field Testing*, Mississippi State University Press, Starksville, MS.

U.S. Department of Agriculture (1986) "Advanced Notice of Proposed USDA Guidelines for Biotechnology Research," *Federal Register* of June 26, (51 FR 22352).

U.S. Department of Agriculture (1989) *Guidance for U.S. Researchers Involved In International Exchange on Agricultural Biotechnology*, U.S. Government Printing Office, Washington, D.C.

U.S. Department of Agriculture, Office of Agricultural Biotechnology (1990) *Minutes: Agricultural Biotechnology Research Advisory Committee, Classification of Unmodified Organisms Working Group*, U.S. Government Printing Office, No. 89–WG–02, Washington, D.C.

CHAPTER 15

Ecological Risk Assessment and European Community Biotechnology Regulation

Rogier A.H.G. Holla

In Europe, as in the major part of the industrialized world, the new developments in biotechnology have given rise to intense debate about the risks involved for public health and the environment. As a result, most European countries have been reviewing their regulations and discussing regulatory measures to guarantee appropriate risk assessment prior to the use of genetic engineering techniques in contained systems and to deliberate release of genetically engineered organisms.

Ecological effects and the geographic ranges of organisms transcend national boundaries. There is, therefore, a need to coordinate risk management and regulation. At the level of the European Community (the EC), directives have been proposed to harmonize the regulation of biotechnology. "European Community" and "European Communities" (EC) are now the common phrases that embrace the European Economic Community (EEC), the European Coal and Steel Community (ECSC), and the European Atomic

The author wishes to express his gratitude to Mr. Paul Kearns for his editorial suggestions, and to Mr. Mark Cantley, Dr. Goffredo Del Bino, Professor Ulli Jessurun d'Oliveira, Professor Christian Joerges, and Professor Ernst von Weizsaecker for their comments and suggestions concerning the author's research into biotechnology regulation.

Energy Community (Euratom). Present members of the EC are: Belgium, Denmark, the Federal Republic of Germany, France, Greece, Ireland, Italy, Luxembourg, the Netherlands, Portugal, Spain, and the United Kingdom. Most of the organs of the EC are integrated. These joint organs include the Commission, the Council, the European Court of Justice, and the General Assembly (European Parliament).

This chapter gives an overview of the involvement of the EC in biotechnology regulation. In particular, the focus will be on the regulation of the deliberate release of genetically engineered organisms.

15.1 APPROACHES TO REGULATION IN EUROPEAN COUNTRIES

At present, environmental regulation of biotechnology differs rather drastically among the various countries that form the EC. The situation ranks from complete absence of specific regulations, via guidelines to be applied on a voluntary basis, to comprehensive legislation (Holla 1988; Gibbs et al. 1987; Mantegazzini 1986). Of course, in almost all countries a variety of health and environmental protection laws are applicable to biotechnological processes and products. For instance, legislation usually covers risks relating to foodstuffs and food additives, pharmaceuticals, chemicals, pesticides, and so on. But these laws are normally limited to ordinary commercial products, and they were not designed for monitoring releases into the environment of genetically engineered organisms.

Since regulatory harmonization is intended for the EC, it would be of limited value to give a detailed description of existing and planned regulation for single member states. A brief outline will suffice.

In Belgium, Greece, Italy, Luxembourg, Portugal, and Spain, there are no specific regulations with respect to environmental applications of genetic engineering. In view of the forthcoming EC harmonization, these countries will probably not develop their own licensing systems, but will implement the proposed EC Directives once they have been adopted. A Directive gives certain rules without prescribing their legal form. Member states are obliged to implement Directives in their own legal system. Directives are adopted by the Council of the EC upon proposal of the Commission. Other European countries adopted guidelines and(or) regulations in a variety of forms. Most countries have guidelines for genetic engineering research, based upon the Guidelines of the U.S. National Institutes of Health (NIH). These guidelines are, in the majority of cases, to be applied voluntarily; the United Kingdom, however, has adopted mandatory guidelines; in the Federal Republic of Germany (FRG), the guidelines applied may be either mandatory or voluntary depending on the funding of the research.

Denmark was the world's first country to proclaim a comprehensive statute regulating the environmental use of biotechnology. This "Environment and Gene Technology Act" came into force in June 1986 (Act no. 288 of June 4, 1986). The act establishes a licensing system for the development of biotechnology-derived products, which includes a procedure for risk assessment and inspection. Deliberate releases are in principle prohibited. However, the Danish Minister of the Environment may approve releases in special cases. Before such an exemption is given, an assessment is made of the possible harmful effects on the environment (on a case-by-case basis). Detailed prescriptions may be laid down for individual releases (Balslev et al. 1987).

In the FRG, a Parliamentary Commission issued a report about the adequacies of existing laws that pertain to biotechnology (Enquete Kommission des 10. Deutschen Bundestages 1987). The report recommends making the existing guidelines mandatory for all research, and establishing a five-year moratorium on the deliberate release of genetically engineered microorganisms (GEMs), to which exemptions would be possible on a case-by-case basis.

In France, Ireland, the Netherlands, and the United Kingdom, review procedures exist for proposed experiments on a case-by-case basis.

15.2 TOWARD COMMUNITY-WIDE REGULATION

Against the background of the different national positions in Europe, the Commission of the EC became increasingly interested in the option of Community-wide biotechnology regulation. Several arguments underlie the Commission's commitment to providing adequate regulation of biotechnology:

1. The health and environmental impacts of biotechnology might easily cross national frontiers–national regulations cannot protect against risks from genetically engineered organisms.
2. EC regulation would offer the scale of the common market for biotechnology products, and thereby provide a more economically attractive environment for European innovation in biotechnology.
3. Pooling the data relevant to the regulatory assessment of risks from biotechnological innovations at the European level would mean a more rapid development and adaptation of regulations in the light of experience.
4. European Community regulation would mean a more efficient use of resources, both in the underlying research, and in the development and application of regulations.

It is not surprising that economic incentives play a major role, because the objectives of the EC are primarily economic; the treaties establishing the EC originally did not even expressly permit Community institutions to act in the field of environmental protection. However, it soon became clear that a harmonization of national environmental policies was necessary to abolish obstacles to the objectives of the treaties. At first, an environmental policy developed on an ad hoc basis. From the beginning of the 1970s, a Community environmental policy gradually became more expressly formulated.

In 1987, the Single Europe Act (SEA) amended the treaties that established the EC. New provisions (articles 100A, 130R, 130S, and 130T) in the EEC Treaty, introduced by the SEA, provide for a legal basis for environmental policy.

The proposals of the EC commission for harmonization of biotechnology regulation are based, in particular, on new article 100A of the EEC Treaty, which allows for the adoption of measures for the approximation of national provisions with the aim of establishing the internal market by 1992. Under this procedure, the Council acts, by qualified majority, on any proposal from the Commission, in cooperation with the European Parliament and the Economic and Social Committee of the EC.

Although the SEA introduced the possibility of qualified majority voting procedures, deliberation structures still rely to a great extent on consensus, since, at present, the EC has almost no means of direct enforcement. Rather the EC seeks to harmonize legislation of its member states and to coordinate policies and budgets. Implementation of EC regulation is generally left to the member states. (For a general comparison of EC and U.S. policy, with specific reference to environmental policy, see Rehbinder and Stewart 1985.)

15.3 INVOLVEMENT OF THE EC IN THE REGULATION OF BIOTECHNOLOGY

Involvement of the European Communities in the regulation of biotechnology dates from the 1970s. In 1978, the Commission proposed a Council Directive establishing safety measures for recombinant DNA work (Commission of the EC 1978). But memories of Galileo were still too strong; restricting scientific inquiry proved to be politically unacceptable, and the proposal was rejected. In 1982, the Commission raised the subject again with a Proposal for a Recommendation. This time, the proposal was accepted and resulted in a Council Recommendation concerning "the registration of work involving recombinant deoxyribosenucleic acid" (Commission of the EC 1982). A notification procedure was proposed in order to register laboratories wishing to undertake work involving recombinant DNA techniques, research projects envisaged, safety evaluations, and control measures.

In 1983, in a Communication from the Commission to the Council, the necessity for a coherent European response to the challenges of biotechnology was recognized (Commission to the EC 1983). In February 1984, a decision of the Commission established the Biotechnology Steering Committee (BSC) to coordinate the implementation of the Commission's initiatives in biotechnology. The BSC is comprised of the following Directors-General of the Commission:

- Internal Market and Industrial Affairs;
- Competition;
- Agriculture;
- Environment, Consumer Protection, and Nuclear Safety;
- Science, Research and Development; and
- Information, Market, and Innovation.

The BSC is supported by the Concertation Unit for Biotechnology in Europe (CUBE), which acts as the secretariat to the BSC, monitors worldwide developments in biotechnology, and provides information and assessments.

Some interservice working groups cover specialized areas for the BSC. The working group dealing with regulatory aspects is called the Biotechnology Regulations Interservice Committee (BRIC).

15.4 A FRAMEWORK FOR THE REGULATION OF BIOTECHNOLOGY

The Commission announced proposals for Community regulation of biotechnology in 1986, to be drafted by BRIC (Commission of the EC 1986a). Such a framework would provide, as the Commission stated, a high and common level of human and environmental protection throughout the Community, and so prevent market fragmentation by separate unilateral actions of member states.

In April 1986, a "high-level" meeting of Community and member state officials took place, organized by BRIC, to discuss the regulation of biotechnology in the Community (Commission of the EC 1986b). The Commission services involved have also consulted with the industries most involved with biotechnology; the chemical, agrochemical, pharmaceutical, and food industries have in fact submitted a joint report to the Commission, setting out their views on the need for Community-wide regulation of biotechnology (European Committee on Regulatory Aspects of Biotechnology (ECRAB) 1986; ECRAB was formed by representatives of five industrial organizations).

The work of BRIC resulted in the drafting of a "Regulatory framework for the use of genetically modified* organisms" (Commission of the EC

* In the EC, the term "genetic modification" is used to indicate what in the United States is known as "genetic engineering." Certain European countries use the term "genetic manipulation," e.g., the United Kingdom.

1988). This regulatory framework consists of two Commission proposals for Council Directives: one on the contained use of genetically engineered microorganisms, and the other on the deliberate release into the environment of genetically engineered organisms. They were proposed to the Council by the Commission in March 1988.

The proposed directive on the contained use of genetically engineered microorganisms includes issues such as levels of physical and biological containment, accident control, and waste management in industrial applications. Risk assessment associated with these issues requires the adoption of good working practices, containment measures, and safety precautions for routine release in normal operating conditions, e.g., as wastes or in airborne emissions, as well as the prevention of accidental releases of genetically engineered microorganisms and the limitation of consequences of any such accidents.

For this book, the scope of the second part of the regulatory framework, the deliberate release of genetically engineered organisms, is by far the most interesting. It is the risk assessment that is required for this application that will eventually give us knowledge about the behavior of genetically engineered organisms in the environment.

15.5 REGULATION OF DELIBERATE RELEASES

Part 2 of the regulatory framework contains a proposal for a "Council Directive on the deliberate release to the environment of genetically modified organisms" (Commission of the EC 1988). The proposed Council Directive (PCD) seeks to establish a case-by-case notification and endorsement procedure. Every member state is required to designate a competent authority (CA) or authorities, that will receive and evaluate notifications of planned releases and be responsible for carrying out the requirements of the Directive and its annexes. The composition of this authority as well as its position in the national administration fall under the discretion of the member states. Before carrying out a release, the person responsible for it has to submit a notification to the CA of the member state within whose territory the release is to take place. This notification must include a technical dossier supplying information necessary for risk evaluation and a statement evaluating the impacts and risks for people and the environment. The CA must review and evaluate the notifications and decide upon endorsement. No release may be carried out before a CA has endorsed it.

The PCD proposes to establish two different procedures: one for experimental releases, where each CA is fully responsible for the releases carried out in its member state, and a second one for commercial production of genetically engineered organisms for a given use, where agreement with the other member states is needed before the product may be endorsed. As justification for establishing two different procedures, the Explanatory Memorandum to the PCD gives "the clear quantitative difference in the level of risk between experimental and commercial releases." Experimental releases are carried out under very controlled conditions, strictly limited in space and time, and closely monitored, whereas commercial ones are limited only in area and conditions of use (Commission of the EC 1988).

Genetically engineered organisms are defined in the PCD as "organisms of which the genetic material is altered in a way that passes the natural barriers of mating and recombination." Annex I to the PCD lists the techniques of genetic engineering covered by this definition: "Genetically modified organisms are organisms obtained by such techniques as recombinant DNA, micro-injection, macro-injection, micro-encapsulation, nuclear and organel transplantation, genetic manipulation of viruses."

15.5.1 Releases for Experimental Purposes

Upon receipt of a notification of a proposed release for experimental purposes, the CA examines its conformity with the PCD and the adequacy of risk assessment and safety measures, and writes its own risk assessment report. One of the ideas behind the notification procedures is that it should promote a dialogue between the notifier and the CA. This dialogue should serve to clarify all those cases where the application of the PCD is in doubt (Commission of the EC 1988).

Within 15 days of receipt, the CA must transmit a summary of the notification to the Commission, who will forward it to the other member states. The CAs of these states may ask for the full dossier, and they can give suggestions to the CA who received the notification.

The CA of the member state in which the release is to take place remains competent to decide on the endorsement. For purposes of research and development, the PCD only seeks to harmonize the review procedure and the information requirements. Member states are allowed to maintain their current attitude towards release experiments. Countries where no endorsement takes place, however, will be obliged to introduce a review procedure.

Within 90 days of receipt, the CA has to respond, either by deciding on endorsement, or by asking the notifier to provide further information or verification tests.

15.5.2 Releases for Commercial Purposes

A product to be placed on the market in one member state enters the common market of the EC and must be able to circulate without trade barriers. Consequently, the PCD requires mutual agreement for this type of release; a common policy must be developed. No product may be endorsed without all member states having had the right to object. However, taking into account the high specificity expected for most of these products, and the diversity of environments within the Community, the endorsement is made valid only for the use of the product under very specific conditions and, where relevant, in specific geographical areas.

Manufacturers or importers have to submit a notification to the CA of the member state in which they want to place a certain product on the market. This notification has to contain a technical dossier and a risk impact statement (as for experimental releases), taking into account the more diverse sites of release and uses of the product. In addition, supplementary information is required, in particular specific conditions of use and handling and a proposal for labelling and packaging; Annex III to the Directive clarifies these requirements.

As in the case of experimental releases, the CA evaluates the information provided and writes its own risk assessment report. Again, there is a period of 90 days to ask for additional information or suggest further testing or changes in the conditions of use. When the CA is satisfied with the adequacy of the risk assessment and the safety measures, and with compliance with the PCD, it sends a copy of the notification dossier, and a summary of the dossier to the Commission. The Commission forwards the summary of the dossier, together with other relevant information, to all other member states. Then, for a period of three months, CAs of other member states may consult the CA that received the original notification, ask for information, and suggest further testing, data required, or other measures.

The PCD proposes to give the Commission the unprecedented executive competence to make decisions about endorsements itself, in the event of a failure by the CAs of the member states to reach an agreement, if one or more of the CAs feel ("on the basis of scientific evidence") that placing the product on the market may pose risks to people or the environment. A Consultative Committee, composed of representatives of the member states, would in such a case present an opinion, of which the Commission "shall take utmost account." Competence of this nature would go further than what was common up to now, and it is doubtful whether the Council will adopt this proposal in its current form.

Another proposed competence that might be considered too far-reaching by the Council is the "adaptation of the annexes of the PCD to technical progress." This is done by the Commission after hearing the consultative Committee. Annex I has major implications for the scope of the PCD, and competence to change the annexes would confer upon the Commission the power to extend the scope of this EC law itself, without explicit permission of the member states.

15.5.3 Exemptions to the Scope of the PCD

All releases for experimental purposes fall within the scope of the PCD. The member states are allowed to provide for derogations from the procedures prescribed in the PCD, but they may not alter the obligation to assess the risk of individual releases, nor the obligation to submit information to the

Commission. For commercial sale of products containing genetically engineered organisms, however, the situation is rather different. This procedure does not apply to:

1. Medical products;
2. Veterinary products;
3. Food stuffs and animal feed and their additives;
4. Plants and animals produced or used in agriculture, horticulture, forestry husbandry, and fisheries, the reproductive material thereof and the products containing these organisms; or
5. Any products covered by Community legislation that requires a specific risk assessment.

This means that many categories of products to be placed on the market will be exempted from the PCD. The main groups of products falling under the PCD are those of microorganisms to be used for nonveterinary agricultural or environmental purposes, including microbial pesticides and microorganisms for degradation of wastes. For the exempted product groups nos. 1, 2, and 3, there are traditionally rather strict review procedures (see e.g., Commission of the EC 1986c). Group no. 4, however, is not fully covered by well-established review procedures that have to be passed before a product can be marketed. Furthermore, because agriculture is a politically sensitive area, it is particularly difficult to regulate adequately.

15.5.4 Risk Assessment Information Requirements

Along with a notification proposing a release of genetically engineered organisms, a technical dossier must be provided containing the information necessary for risk evaluation. A five-page second annex to the PCD sets out the requirements of this dossier (Commission of the EC 1988). These include:

1. Information relating to the parental organisms, the recipient organism, and the genetic modification;
2. Characteristics of the genetically engineered organism affecting survival, multiplication, and dispersal, and predictions with respect to its interactions with the biological systems;
3. Information on the geographical location and local characteristics of the site, and a comparison of the natural habitat of the recipient organism with the proposed site(s) of release;
4. Conditions of the release; and
5. Monitoring and control techniques and emergency response plans in case of an unexpected spread of the organism.

The annex should not be viewed as a checklist, for not all items are relevant in every case. It is intended as a comprehensive list of different aspects relevant for risk evaluation, from which a notifier will pick those items that are relevant to the individual case. The level of detail of the information to be provided and its quality will be a function of the type of release: a field test is likely to rely on bibliographic data and assumptions, whereas a notification for a product should be based on experimental evidence.

Although Annex II mentions many types of information that could be relevant to risk assessment, the emphasis is on short-term effects at the site of release, such as pathogenicity, disruption of ecosystems, and transfer of novel genetic traits to other species with undesired side-effects. Factors like the possibility of dispersal to other ecosystems and the genetic characteristics of the receiving ecosystem do not receive much attention in the information requirements, whereas some recent review recommendations urge that these factors be taken into consideration (Tiedje et al. 1989).

15.5.5 Criteria for Adequate Risk Assessment

The PCD gives "points to consider" but no suggestions for criteria for approval or rejection of the endorsement. If it is left completely to the member states to develop criteria, considerable variations may be expected. The information requirements focus on ecological risks, but the directive by no means prohibits the additional evaluation of social or ethical considerations. Furthermore, there are no requirements or suggestions as to the composition of the CAs in the member states. This might result in biased risk assessment and in divergence in different member states, and is a potential source of harmonization and coordination problems. Scientists with a microbiological background usually have quite different basic ideas about the risks involved in releases from scientists with an ecological background.

The CAs are to a great extent dependent on the notifier for the provision of data. The notifier is expected to draft a statement evaluating impacts and risks to people and the environment. It is assumed that he or she will include information that might influence decisions about endorsement or further information requirements in a negative way. The interest of notifiers in endorsement, however, does not make them very reliable reviewers of their own proposals.

15.6 CONCLUSIONS AND RECOMMENDATIONS

The PCD is an important step towards assuring adequate regulation of environmental applications of genetically engineered organisms for the whole territory of the EC. In a single document, companies and the public can locate information on data requirements and the review procedure.

Jurisdiction is not based on a complex division into product areas or research funding areas, but on a territorial principle: in every member state, a CA is responsible for receiving notifications and deciding upon endorsements. The Directive promotes a flexible approach and constructive dialogue between notifier and CAs. A harmonized regulatory framework removes barriers to trade and provides the biotechnology industry with a more attractive environment for innovation. It stimulates a transborder environmental policy and faster and more efficient risk assessment and data accumulation.

However, there is still much work to be done. The PCD only includes establishment of a common procedure for review of proposed releases of genetically engineered organisms. A consistent policy still needs to be established throughout the EC. Coherent criteria for decisions about endorsements need to be developed. The task of the Commission's services should not be limited thereby to mere exchange of information, but should include active stimulation of development of a common policy and of consistent treatment and evaluation of release proposals.

The procedure for marketing products consisting of or containing engineered organisms contains exemptions for many categories of products. Assessment of adequate coverage with risk review in these product areas should be undertaken. Networks for cooperation and exchange of information should be established between review committees in these areas and the CAs so that they can mutually profit from review expertise and data accumulation.

We should not be lulled by the thought that the biotechnology sector is sufficiently regulated because there is a review procedure for individual experiments (i.e., we must not forget the forest when looking at the trees). Long-term changes of a social nature, enhanced by the development of biotechnology, may have major ecological impact. Increased possibilities for patenting biological inventions, along with industrialization, increase of scale, and intensification of production in biotechnology, are likely to change patterns of land utilization (von Weizsaecker 1987). Control over land use and agricultural resources is shifting from farmers to industrial managers and bioengineers. As the biotechnology industry develops, continuing regulatory oversight, as well as long-term research and monitoring, will be necessary for responsible risk management.

REFERENCES

Balslev, M., Kornum, N., Moth, M., Holm Olsen, A., and Meulengracht Olsen, A. (1987) in *Biotechnology and the Environment: International Regulation* (Gibbs, J., Cooper, I., and Mackler, B., eds.), pp. 221–228, Stockton Press, New York.

Commission of the EC (1978) in *Official Journal of the EC* No. C 301, 15 December 1987, pp. 5 ff.

Commission of the EC (1983) *Biotechnology in the Community,* COM(83)672, Brussels.
Commission of the EC (1986a) *A Community Framework for the Regulation of Biotechnology,* COM(86)573, Brussels.
Commission of the EC (1986b) *Biotechnology Regulation; Meeting of Commission Staff with Member States Officials,* Biotechnology Regulations Interservice Committee, BRIC/2/86, Brussels.
Commission of the EC (1986c) *The European Community, and the Regulation of Biotechnology: an Inventory,* Biotechnology Regulations Interservice Committee, BRIC/1/86, Brussels.
Commission of the EC (1988) *A Regulatory Framework for the Use of Genetically Modified Organisms,* COM(88)160, Brussels.
Council of the EC (1982) in *Official Journal of the EC* No. L 213, 21 July 1982, pp. 15 ff.
Enquete Kommission des 10. Deutschen Bundestages (1987) *Chancen und Risiken der Gentechnologie,* Deutscher Bundestag, Bonn.
Environment and Gene Technology Act, Denmark (1986) Act no. 288 of June 4, 1986.
European Committee on Regulatory Aspects of Biotechnology (1986) *Safety and Regulation in Biotechnology,* Brussels.
Gibbs, J., Cooper, I., and Mackler, B. (1987) *Biotechnology and the Environment: International Regulation,* Stockton Press, New York.
Holla, R. (1988) *The Regulation of Deliberate Release of Genetically Engineered Organisms for Environmental and Agricultural Applications in the United States of America and the European Communities,* European University Institute, Florence.
Mantegazzini, M. (1986) *The Environmental Risks from Biotechnology,* Frances Pinter, London.
Rehbinder, E., and Stewart, R. (1985) *Environmental Protection Policy,* vol. 2 of the series *Integration through Law; Europe and the American Federal Experience* (Cappelletti, M., Seccombe, M., and Weiler, J., general eds.), Walter de Gruyter, Berlin.
Tiedje, J., Colwell, R., Grossman, Y., et al. (1989) *Ecology* 70:298–315.
von Weizsaecker, E. (1987) *Environmental Aspects of the Deliberate Release of Genetically Engineered Organisms,* Paper presented at the Workshop "EC Regulatory Harmonization of Genetically Engineered Organisms" of the Institute for European Environmental Policy, Cervia, Italy.

CHAPTER

16

Ecological Risk Analysis of Biotechnological Waste Decontamination

Anthony D. Thrall
Robert A. Goldstein

Recent advances in molecular biology are enabling established and start-up firms to create genetically engineered microorganisms (GEMs) for a variety of new applications, including the on-site treatment of environmental wastes at costs well below that of current methods. These companies expect to have products ready for field testing in the next few years.

An alternative approach called genetic ecology is being investigated by the Electric Power Research Institute (EPRI). In this approach, environmental conditions are altered to amplify the expression of desired genes among indigenous microorganisms.

The new technology will only be applied, however, if the environmental risks—e.g., from the partial or unintended degradation of some substance, or the undesired amplification or increased invasive potential of some gene system—are shown to be acceptably low.

The authors thank Lev Ginzburg for many stimulating discussions, Applied Biomathematics for supporting work, and Drs. Richard Lenski, Morris Levin, and Simon Silver for thoughtful review comments, all of which facilitated the research planning described in this chapter.

In this chapter we describe EPRI's plans to identify possible risks of biotechnological waste decontamination, and to determine whether and how these risks can be reliably assessed.

16.1 APPLICATIONS OF BIOTECHNOLOGY TO WASTE TREATMENT

It is useful to distinguish two approaches used in modern biotechnology. The first approach, which has already had stunning successes in medicine, is to create genetically engineered microorganisms (GEMs). A more recent approach is a modern adaptation of artificial selection in which environmental conditions are altered to amplify the expression of desired genes among indigenous microorganisms. This approach is called genetic ecology (Olson and Goldstein 1988; Goldstein et al. 1988).

16.1.1 Coal Tar Deposits

In the next few years, researchers are likely to make substantial progress toward biotechnological solutions to a wide range of utility problems, such as the desulfurization and liquefaction of coal, production of new forms of fuel, prevention of microbially induced corrosion, and on-site biodegradation of wastes. One of the most promising areas of application is the on-site treatment of coal tar deposits.

During the first half of this century, gas manufactured from coal and oil was widely used in the United States for street lighting and space heating. The coal tar residue (which at that time was considered harmless) that could not be sold or reprocessed was usually deposited in shallow ponds near the processing site. Today, some of the chemicals contained in the coal tar are designated as toxic in various state and federal regulations. This has prompted utilities and other site owners to investigate the public health risks of these sites and, in some cases, to take action to reduce risks.

If treatment is needed, the cost for a single site can be as much as several million dollars, or even tens of millions of dollars. In many instances, more than half the cost is to excavate and incinerate or store the contaminated soil. Biotechnology offers the possibility of decontaminating the soil on site, thus eliminating the cost of soil excavation, transport, incineration, or storage.

16.1.2 Other Uses

Heavy metal decontamination and recovery are two additional, near-term, potential applications of genetic engineering and genetic ecology; that is, metal-tolerant microorganisms can be used to absorb and sequester metals, and to transport and transform metallic ions to reduce potential availability and toxicity to macrobiota (Goldstein et al. 1988). Many metal-tolerance mechanisms are encoded on plasmid-borne genes and could also be located on transposable elements, making them easy to manipulate using current genetic engineering techniques (Duxbury 1984).

Another potential area of development is desulfurization of coal and oil. Chemoautotrophic bacteria, such as species of *Thiobacillus* and *Sulfolobus,* are known to convert insoluble metal sulfides to soluble ions of sulfur. *Thiobacillus* appears to be especially suited to genetic engineering since a majority of strains contain plasmids capable of being altered, and some *Thiobacillus* genes have been expressed in *Escherichia coli,* a bacterium used as a vector in genetic engineering (Monticello and Finnerty 1985).

One of the most promising applications of genetic engineering and genetic ecology is the biodegradation of complex organic compounds such as those found in coal tar deposits and crude oil. Several naturally occurring bacteria, such as species of *Pseudomonas* and *Flavobacterium,* are known to attack various components of crude oil. Although no individual species can degrade all oil constituents, a strain of *Pseudomonas* has been altered to metabolize several crude oil components. This is an example of how one organism can be endowed with the ability to carry out several different carbon-degradative pathways (Office of Technology Assessment 1988).

16.1.3 Prospects for Genetic Engineering and Genetic Ecology

At large firms such as Monsanto and at smaller firms such as Ecova, teams of scientists are developing GEMs for these various applications, including the biodegradation of polychlorinated biphenyls (PCBs) and other organic pollutants (OTA 1988). These companies expect to have products ready for field testing in the next few years.

Using genetic ecology techniques, the Electric Power Research Institute (EPRI) is conducting a multi-year project to develop the scientific foundation for on-site enhancement of indigenous microbial genetic systems that govern the degradation of organic wastes. Naphthalene has been selected as the coal tar constituent on which the investigation will focus, because the genes controlling degradation are well characterized and the biochemical degradation pathways are well understood for the bacteria involved. It is believed that study findings will be applicable to other polycyclic aromatic hydrocarbons (PAHs) contained in coal tar. Using DNA probe technology, the team will measure genetic potential, i.e., the relative abundance of the naphthalene-degradation operon, at representative waste sites. Concurrently, project researchers will conduct laboratory microcosm experiments to search for the environmental conditions that most strongly influence the proliferation and expression (function) of the desired genes. Based on this information, investigators will formulate and field-test possible management strategies.

16.2 THE NEED TO IDENTIFY AND ASSESS ENVIRONMENTAL RISKS

There are two reasons for assessing potential risks of biotechnological waste decontamination now, while the technology is being developed. The first is to ensure that the risks are investigated thoroughly and presented clearly to all concerned during this period in which regulatory policy is being formed. The second reason is that, once the technology and regulations are in place, all the users will want to know as much as possible about the benefits and liabilities of the technologies that will be available to them.

For an individual electric power utility, for example, an adverse event might well entail substantial costs. For the electric power industry, the failure to anticipate the hazard of some technique applied under particular circumstances could result in an incident analogous to the one that occurred at Three Mile Island that would effectively bring a very promising technology to a halt. To make informed decisions and to promote safe technologies, we must determine what can go wrong and estimate the chance of these undesirable events. Moreover, these findings must be communicated to regulatory agencies and the public as effectively as possible.

16.2.1 Concerns of the Public and the Scientific Community

In the first phase of modern commercial biotechnology, which began in the 1970s, scientists have created various GEMs that operate as biochemical factories to produce insulin, growth hormones, and other antibodies and drugs. Initially, there was some scientific and public debate as to whether this new technology would endanger either the health of laboratory workers or the local ecology, should the GEMs escape the confines of the laboratories. The debate prompted the U.S. Food and Drug Administration (FDA) and the National Institutes of Health (NIH) to develop the "Guidelines for Research Involving Recombinant DNA Molecules" (Milewski 1987). Under these guidelines, GEMs are strictly contained in the laboratories in which they are produced and are also designed so as to be unable to survive in a natural environment. This physical and biological containment has essentially resolved the debate.

In the second phase of modern commercial biotechnology, the area of application has extended from medicine to agriculture and industry. The new applications require the engineered organisms themselves, rather than their metabolic products, to be released into the environment in large numbers, and to be able to survive in nature in order to perform their prescribed function. The necessity of releasing the engineered organisms to the environment has rekindled concerns, expressed by both Congress and environmental groups, about the possible dangers of biotechnology. It has been argued that the new organisms, having no natural controls, might spread beyond the intended area of application and cause the unwanted extinction of indigenous organisms, or that they might proliferate and produce unforeseen effects on the physical, chemical, and biological characteristics of pre-existing ecosystems.

16.2.2 Federal Regulatory Policy

Responding to these concerns, the U.S. government developed in June 1986 a "Coordinated Framework for the Regulation of Biotechnology" (Federal Register 1986). The document explains how existing statutes should be applied to the regulation of biotechnology and outlines the jurisdiction of several governmental agencies over various types of biotechnological products (Kingsbury 1988; see also Chapter 12). To foster the establishment of common standards for approving biotechnological products, the framework promotes the sharing of viewpoints among the different agencies. At present, each agency reviews requests for such approval on a case-by-case basis, typically with the assistance of an expert panel.

There seems to be general agreement both inside and outside the government that a judicious balance must be maintained in the regulation of the new technology. On the one hand, potential dangers must be carefully investigated and new products must be monitored both before and after their commercial acceptance if we are to avoid a catastrophe. On the other hand, if our regulatory policy is overly cautious, we may lose this technology to other countries whose policies are more conducive to its development. On the whole, EPA's Office of Solid Waste has promoted the use of biotechnology to control pollution and decontaminate waste sites (Williams 1987).

To encourage careful scientific assessments of potential biotechnological dangers, the EPA's Office of Research and Development (ORD) conducted a workshop in January 1988 entitled "The Integration of Research and Predictive Model Development in Biotechnological Risk Assessment." Acting on the recommendations of the workshop, the ORD has requested proposals to investigate the potential risks of biotechnology. This research will complement a project being conducted by the Environmental Criteria and Assessment Office (ECAO) entitled "Risk Assessment of Applications of Biodegradative Microorganisms for Hazardous Waste Destruction."

Other organizations participating in the discussion of regulatory policy in the United States and abroad are environmental groups, industrial associations such as the European Biotechnology Coordinating Group, and scientific advisory groups such as the Recombinant DNA Scientific Advisory Committee in the United States and the Genetic Manipulations Advisory Group in the United Kingdom (Ager 1988; Poole et al. 1988).

16.2.3 Needed Research

Environmental biotechnology is only now being developed, and reliable methods for evaluating its risks are needed. Fortunately, there is yet some time to identify the risks and develop methods of evaluating them, and we can learn from the protocols being established for agricultural biotechnology, which is somewhat further along in its development.

In agricultural biotechnology, a GEM is applied to a plot of land, and the perimeter of the plot is monitored to determine whether the GEM is capable of spreading beyond the area of application. Such testing certainly seems prudent at this early stage of development, but other tests are needed for several reasons. First, there is a practical limit to the duration of any field test, and therefore the test may be incapable of detecting an event that occurs with low, yet non-negligible probability. Second, in some cases the adverse event to be detected, such as the loss of soil fertility, may be more complex than the mere spread of the GEM beyond the target area. In these cases, sole reliance on empirical methods would require complex and costly experiments. Third, we need a theoretical basis from which field results can be generalized, because the conditions of application are likely to vary considerably.

The search for microorganisms or genetic systems that perform desired functions will become much more efficient as we learn how these systems function. Similarly, as we learn how these systems interact with the surrounding microbial ecosystem, we will extend the basis of our risk estimates from circumstantial evidence to scientific knowledge. Finally, to enable utilities to apply this emerging technology with confidence, we need to carefully develop hypotheses whose experimental confirmation or refutation will lead us further along the path of understanding microbial ecosystems.

16.3 SCIENTIFIC ISSUES

16.3.1 Formulation of Risks

The possible unintended adverse effects of applying environmental biotechnology usually involve either a single species of microorganism, or some larger subset of the microbial ecosystem. As discussed in the next section, our first task will be to carefully formulate the risks of such applications and plan their investigation. Described here are some initial steps toward such a formulation.

16.3.1.1 Single-Species Effects. The prime examples of the first type of effect are the unintended invasion of an ecosystem by a new species or the unintended extinction of an existing species (National Academy of Sciences, 1987). For plant and animal ecosystems, this single-species type of criterion is used almost exclusively to evaluate the success or failure of conservation efforts or to assess the ecological impact of industry. Possible questions to be investigated are as follows:

1. What is the probability that a GEM or an amplified set of genes used for waste decontamination will invade a natural community away from the site of application?
2. When the organism completes its function (such as degrading a specific toxic compound at a coal tar site), will it die off or will it use other energy sources and continue to grow?
3. Will the organism interact with other species in the community in such a way as to inhibit the biodegradation of other compounds or to cause the accumulation of intermediate products that are nonbiodegradable and toxic?
4. What is the probability that an engineered or amplified gene sequence will be incorporated in the genome of a widespread microorganism capable of surviving in environments other than waste deposits?
5. What is the probability that alteration of the environment to amplify one gene or gene complex will inadvertently amplify others?

The invasion or extinction of any single microbial species at a site may be an overly conservative criterion for declaring a proposed biotechnology to be environmentally harmful. That is, the alteration of species within a microbial ecosystem may have no effect on animal or plant life.

16.3.1.2 Disruption of Ecosystem Functions. Alternatively, we might define adverse effect as the disruption of some function of the microbial ecosystem, i.e., some characteristic of the system that could affect animal or plant life (Vitousek 1985). For example, we might ask the following: If the events described above in nos. 1 to 5 do occur, what will be the effect of these events on overall ecosystem functions, such as primary productivity, nutrient cycling, and soil fertility?

Although such questions address practical concerns, they must be disaggregated to permit scientific investigation—for example, soil fertility could be measured in various ways. It should also be kept in mind that we might overlook some function whose importance is not yet recognized. In fact, both types of adverse effect, e.g., invasion and disruption of function, need to be investigated since the former may often be a prerequisite for the latter. For example, a gene system designed to biodegrade coal tar could possibly degrade an oil reserve, but only after it has invaded the oil reserve.

16.3.2 Characterizing the Stability of Microbial Ecosystems

Because environmental technology and the assessment of its risks are such new areas of investigation, it seems prudent to start with models of very simple microbial ecosystems, investigate these simple models both theoretically and experimentally, and gradually develop more realistic models as we improve our understanding. More specifically, we must address the question of whether complexity itself makes actual microbial ecosystems more impervious to invasion and disruption of function than the simple systems that can only be maintained artificially. We will consider three types of complexity: physical, chemical, and biological.

16.3.2.1 Physical Complexity. To estimate the probability that a species or gene system, such as the naphthalene-degradation operon, will disperse beyond the area where it was introduced or amplified, it is important to understand how the initial spatial configuration of the system is likely to change over time. Microbial competition, compared to that of plants and animals, is characterized by extremely fast growth and equilibration, as well as a high degree of spatial heterogeneity. In addition, microbes are dispersed primarily by moving water, and therefore their dispersal in the outside environment is largely determined by precipitation rate and soil hydrology. This causes the spatial distribution of microbial communities to be very patchy and to change episodically. Moreover, because microbial communities usually reside on the surface of soil particles, microbial interactions may be restricted to a two-dimensional space. If so, growth and dispersal would be more limited in the outside environment than in a laboratory microcosm that has been made homogeneous throughout its three spatial dimensions (see, e.g., Chao and Levin 1981).

16.3.2.2 Chemical Complexity. Coal tar deposits and the surrounding soil are chemically complex, and this complexity may offer a wide range of opportunities for the survival of an introduced microorganism or the dissemination of a gene system that has been encouraged by altering some aspect of the environment (see, e.g., Miller et al. 1988). By the same token, this complexity may well enhance the stability of the original microbial ecosystem. Thus the chemical complexity of the environment could increase the probability of invasion and decrease the probability of disrupting the functioning of the original microbial ecosystem.

16.3.2.3 Biological Complexity. The strong interdependence of two species, as in competitive interactions described by systems of differential equations, can cause wide variations in the abundance of the two species, i.e., such interdependence can cause two-species systems to be highly unstable. This has been shown theoretically and confirmed empirically (Gardner and Ashby 1970; May 1973). But such isolated two-species systems are rare in nature, and their convenience for theoretical analysis has overemphasized their relevance to actual ecosystems. Typically, ecosystems do not contain such unstable two-species subsystems. Usually, every species is weakly dependent on a large number of the other species in the system, and this arrangement is much more stable (Ginzburg et al. 1988). Even in the case of a system consisting of only a few species, Gottschal et al. (1979), for example, have shown that both the stability of the system and the outcome of the interaction between two species can depend on the presence of a third species.

If laboratory experiments were to show that a potential biotechnological treatment of coal tar engendered a strong interdependence between two microorganisms, one might conclude that the use of biotechnology would destabilize the microbial ecosystem. This conclusion would be premature, however, in the absence of information about the interactions of other species in the system and the sensitivity of these interactions to the proposed biotechnology.

16.3.3 The Reliability of Computer Simulation Models

There is general agreement about what constitutes a good scientific theory, e.g., it should be testable and sufficiently compelling that its confirmation or refutation would be regarded as useful information. Such criteria have evolved over several centuries of scientific investigation. On the other hand, our experience with computer simulations of complex phenomena is much more recent, and there is much less agreement about appropriate criteria or protocols for evaluating such models. The usual purpose of a scientific theory is to improve understanding, whereas the usual purpose of a computer simulation model is to make predictions, and thereby guide decisions.

16.3.3.1 Reliability of Low-Risk Predictions. In the present context, we are attempting to guide the development of biotechnological waste decontamination so that it will be not only cost-effective but also ecologically safe. The primary goal is to ensure that adverse events that are difficult to control—e.g., the spread of microorganisms that substantially reduce soil fertility—are extremely unlikely. It is not sufficient that some risk model assigns low probability to such events, rather we need to be highly confident that the model is correct in this respect. Thus our first priority in evaluating a microecological risk is to clarify the model's reliability as a guide to decisions and policies.

Conceivably, it might be possible to demonstrate the accuracy of the low probabilities calculated by the model. For example, if we knew that the adverse event could only occur under certain conditions whose low probability was known, we might artificially maintain those conditions in an experiment, and then observe the conditional frequency with which the adverse event occurred.

It seems more likely, however, that the model's reliability in this respect will not be capable of demonstration. We might then attempt to infer the model's reliability based on other demonstrations, or make decisions about the biotechnology that are less reliant on model calculations (perhaps at some additional cost that is nevertheless cheaper than continued model development), or extend the base of knowledge supporting the model by conducting further scientific investigations. Katz and Murphy (1987) provide a simple illustration of how information of poor quality may be of no value to a decision maker, and how the value of information can rise sharply as its quality improves.

At some point, we will need to assess the model's reliability by comparing calculations—of, for example, the invasive potential of an amplified or engineered gene system—to actual experimental observations. This task may seem routine, but in fact it has been a major challenge to the techniques of statistical inference.

16.3.3.2 Model Complexity. Models are usually judged according to two criteria: scientific plausibility and compatibility with data ("goodness-of-fit"). Both criteria encourage the development of complex models, but the predictive ability of such models can be easily overestimated. Intuitively, one would expect the most reliable model to be somewhere between the extremes of simplicity and complexity (Kemeny 1953). Rissanen (1978, 1982, 1985, and 1986) has made great strides toward establishing this principle quantitatively in his continuing work on modeling by shortest data description.

A related issue is that the model is often partially derived from the data in ways not accounted for in most model evaluation protocols. The result is a bias toward overconfidence in the model (Efron 1986; Hodges 1987). It would be fairer to compare the model to a data set unknown to the model developers, but this is usually not practical or even possible. An alternative is to "cross-validate" the model, i.e., to withhold a subset of the data from model-fitting, and then compare this subset to model predictions (Stone 1974 and 1977; Efron 1983). Dawid (1984 and 1986) describes another method of evaluating a modeling system that uses a sequence of experiments to update model parameters and predict the next experiment's outcome.

16.3.3.3 The Uncertain Reliability of Probabilistic Predictions. Models that predict detailed outcomes for prescribed conditions can be compared to data more easily than models that make probabilistic predictions. In the latter case, it may be appropriate to compare the frequency distributions of the model output and the data, respectively, and the Kolmogorov-Smirnov procedure is often used for this purpose. This is a "robust" procedure that avoids assumptions about the shapes of the two distributions. Further robustness is offered by resampling or "bootstrapping" the model output and the data (Romano 1988).

16.3.3.4 Effects of Observational Errors. If errors of measurement, observation, transcription, etc. are ignored, discrepancies between model output and data may be incorrectly ascribed entirely to the limitations of the model. Even when the possibility of imperfect data is acknowledged, it may not be clear how to infer the reliability of the model from the model's agreement or lack of agreement with flawed data.

What is needed is a formulation, or at least an idea, of the various sources of model-data discrepancies. It may then be possible to estimate what the magnitude of the discrepancies would be under ideal conditions, were all the data free of errors.

As an example, Sen (1987) modifies the usual Kolmogorov-Smirnov statistical test of the hypothesis that two sets of numbers represent the same probability distribution, by considering a broader hypothesis, namely that the two sets of numbers represent two probability distributions having the same shape, but centered at possibly different points. Sen's modified test procedure might be applicable, for example, if the set of measurements were off by an unknown constant amount due to possible miscalculation of the measurement device. In general, however, such accounting for the possibility of erroneous data is an open challenge to current statistical theory and practice.

16.4 EPRI's RISK ANALYSIS RESEARCH STRATEGY

The ultimate goal of EPRI's planned risk analysis research is to develop a sound scientific basis for regulatory policies and company decisions concerning the ecological risks of using emerging biotechnologies in the electric power industry. The challenge is to achieve the research goal as efficiently as possible from our present state of knowledge, even though we have yet to even specify which adverse events, out of the myriad possibilities, most deserve to be investigated.

The strategy proposed here is first to conduct a two-year exploratory investigation whose goals are to identify possible risks of biotechnological waste decontamination and, second, to determine whether and how these risks can be reliably assessed. Figure 16-1 shows the major components of the planned research. An initial review of scientific concerns has already been completed. In the proposed two-year exploratory phase, the research team would formulate risk hypotheses, plan a joint theoretical and experimental investigation, and conduct the investigation. If the results of the exploratory study are promising, a model of the risks would be developed, evaluated, and disseminated to utilities and regulatory agencies. The re-

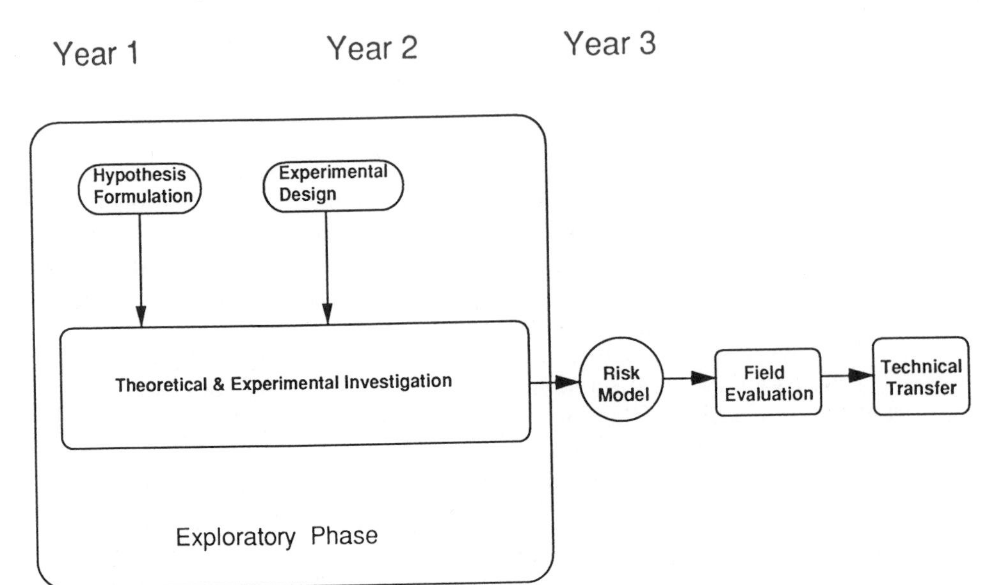

FIGURE 16–1 Diagram of proposed GPRI-sponsored risk analysis research.

search team will work in tandem with the developers of the biotechnology methods, especially in the formulation of risk hypotheses and the use of field experiments to evaluate risk models.

16.4.1 Review of Scientific Concerns

At present, the U.S. Environmental Protection Agency (EPA) has no general set of criteria for allowing or disallowing the use of genetically engineered microorganisms, but rather uses a Biotechnology Review Panel to consider each case individually. Under contract to EPRI, Applied Biomathematics, Inc. (AB) has reviewed the deliberations of the panel for those cases where the release of GEMs was allowed (Akcakaya and Ginzburg 1988). The purpose of the review was to characterize the panel's concerns for various types of cases.

16.4.2 The Exploratory Phase

In the first year of the exploratory phase, EPRI will conduct a scoping study to: (1) focus issues by surveying developers of the technologies; (2) address these issues in a workshop of experts from government, industry, and universities; and (3) formulate hypotheses, establish research priorities, and plan a course of investigation.

Also in the first year, we will initiate a joint theoretical and experimental investigation to help to refine the hypotheses and research plans to be developed in the scoping study. The research team will investigate the influence of chemical and biological complexity on the risks of species invasion and disruption of ecological function. This work will complement similar studies of spatial and temporal complexity that the EPA is conducting. Using the recommendations of the scoping study, this investigation will continue through the second year.

16.4.3 Development, Evaluation, and Dissemination of a Risk Model

In Year 3, following the completion of the exploratory phase, we will decide whether and how to proceed to develop a model of the ecological risks of biotechnological waste decontamination at utility sites. Model development will require additional theoretical work, along with experimental work to diagnose model components, and perhaps a "person-year" of computer programming. The model would then be evaluated in field experiments or case studies that will require careful design so as to artificially increase the probability of adverse events to permit researchers to compare observed frequencies to calculated probabilities. After the model has been brought to a sufficiently reliable, or at least realistic, state of development, the model would be introduced to utilities and other interested parties through workshops and publications. Finally, the computer program and supporting documentation would be distributed by the Electric Power Software Center (EPSC).

16.5 CONCLUSION

Emerging biotechnologies based on genetic manipulation can save the utility industry and other industries a substantial portion of the billions of dollars spent each year on waste treatment, but only after the environmental risks have been determined to be acceptably low.

The research outlined in this chapter—to identify possible risks of biotechnological waste decontamination, and to ascertain whether and how these risks can be reliably assessed—is the first step toward this objective.

REFERENCES

Ager, B.P. (1988) in *Planned Release of Genetically Engineered Organisms (Trends in Biotechnology/Trends in Ecology and Evolution, Special Publication)* (Hodgson, J., and Sugden, A.M., eds.), pp. 42–44, Elsevier Publ., Cambridge.

Akcakaya, H.R., and Ginzburg, L.R. (1988) *Risk Analysis for Biotechnology: A Review of EPA Practices,* EPRI Project Report No. RP2391-7, Palo Alto, CA.

Chao, L., and Levin, B.R. (1981) *Proc. Nat. Acad. Sci. USA* 78, 6324–6328.

Dawid, A.P. (1984) *J. Roy. Statist. Soc. Ser.* B 47, 278–292.
Dawid, A.P. (1986) *Prequential Data Analysis. Research Report No. 46,* Dept. of Statistical Science, University College, London.
Duxbury, T. (1984) *Adv. in Microb. Ecol.* 8, 185–235.
Efron, B. (1983) *J. Amer. Statist. Assn.* 78, 316-331.
Efron, B. (1986) *J. Amer. Statist. Assn.* 81, 461-470.
Federal Register (1986) *Fed. Regist.* 51, 23302–23393.
Gardner, M.R., and Ashby, W.R. (1970) *Nature* 228, 784.
Ginzburg, L.R., Akcakaya, H.R., and Kim, J. (1988) *J. Theor. Biol.* 133, 513–523.
Goldstein, R.A., Olson, B.H., and Porcella, D.B. (1988) *Environ. Technol. Letters* 9, 957–964.
Gottschal, J.C., deVries, S., and Kuenen, J.G. (1979) *Arch. Microbiol.* 121, 241–249.
Hodges, J.S. (1987) *Statist. Sci.* 2, 259–291.
Katz, R.W., and Murphy, A.H. (1987) *Am. Statistician* 41, 187–189.
Kemeny, J.G. (1953) *Phil. Rev.* 62, 391–415.
Kingsbury, D.T. (1988) in *Planned Release of Genetically Engineered Organisms (Trends in Biotechnology/Trends in Ecology and Evolution, Special Publication)* Hodgson, J., and Sugden, A.M., eds.), pp. 39–44, Elsevier Publ., Cambridge.
May, R.M. (1973) *Stability and Complexity of Model Ecosystems,* Princeton University Press, Princeton, NJ.
Milewski, E. (1987) in *Application of Biotechnology: Environmental and Policy Issues* (Fowle, J.R., III, ed.), pp. 55–90, AAAS Selected Symposium No. 106, Washington, D.C.
Miller, R.D., Dykhuizen, D.E., and Hartl, D.L. (1988) *Mol. Biol. Evol.* 5, 691–703.
Monticello, D.J., and Finnerty, W.R. (1985) Microbial desulfurization of fossil fuels. *Ann. Rev. Microb.* 39, 371–389.
National Academy of Sciences (1987) *Introduction of Recombinant DNA-Engineered Organisms into the Environment: Key Issues,* National Academy of Sciences Press, Washington, D.C.
Olson, B.H., and Goldstein, R.A. (1988) *Environ. Sci. Technol.* 22, 370–372.
Office of Technology Assessment (1988) *New Developments in Biotechnology. 3. Field-Testing Engineered Organisms: Genetic and Ecological Issues,* Congress of the United States, Washington, D.C.
Poole, N.J., Mahler, J.L., and Heusler, K. (1988) in *Planned Release of Genetically Engineered Organisms (Trends in Biotechnology/Trends in Ecology and Evolution, Special Publication)* (Hodgson, J., and Sugden, A.M., eds.), pp. 45–46, Elsevier Publ., Cambridge.
Rissanen, J. (1978) *Automata* 14, 465–471.
Rissanen, J. (1982) *Circuits, Systems, and Signal Processing* (special issue on *Rational Approximations*) 1, 395–406.
Rissanen, J. (1985) in *Encyclopedia of Statistical Sciences,* vol. 5 (Kotz, S., and Johnson, N.L., eds.), pp. 523–527, Wiley, New York.
Rissanen, J. (1986) *Ann. Statist.* 14, 1080–1100.
Romano, J.P. (1988) *J. Am. Statist. Assoc.* 83, 698–708.
Sen, P.K. (1987) in *Goodness-of-Fit* Revesz, P., Sarkadi, K., and Sen, P.K., eds.), pp. 493–510, North-Holland, Amsterdam.
Stone, M. (1974) *J. Roy. Statist. Soc. Ser.* B 36, 111–147.
Stone, M. (1977) *Biometrika* 64, 29–35.

Vitousek, P.M. (1985) in *Engineered Organisms in the Environment: Scientific Issues* (Halvorson, H.O., Pramer, D., and Rogul, M., eds.), pp. 169–175, American Society for Microbiology, Washington, D.C.

Williams, M.E. (1987) in *Environmental Biotechnology: Reducing Risks from Environmental Chemicals through Biotechnology* (Omenn, G.S., ed.), pp. 373–379, Plenum Press, New York.

CHAPTER 17

On Making Nature Safe for Biotechnology

Mark Sagoff

In the late 1940s, an ecological restorationist named Henry Greene reestablished, on 40 acres in the state of Wisconsin, plants that had not grown there for at least 200 years. The Greene Prairie was followed by many other restorations in the American Great Plains, some of which are much larger (Turner 1988). Some scholars recommend, indeed, that "[t]he federal government . . . convert vast stretches of the region to a use so old it predates the American presence–a 'Buffalo Commons' of native grass and livestock" (Popper and Popper 1989). In this way, the government could do something worthwhile with an area that covers roughly 20% of the contiguous United States, but contains less than 2% of the nation's people, and which is rapidly losing population.

Many demographers believe that the Great Plains will become almost totally depopulated in the next 30 years. In Oklahoma, the government is already restoring vast areas of unused farmland to prairie in order to form a new national park. The Director of the U.S. Park Service comments: "Fifty

While writing this chapter, the author received support from the National Science EVS Program, Grant No. BBS 8619104, and from the Maryland Sea Grant Program. The views expressed are those of the author only and not necessarily those of any funding agency. A few passages of this chapter are taken from Sagoff (1988b).

years from now, people could go there and sense what it was like for the pioneers, crossing the prairie to the West. That is what's exciting about it" (Peterson 1987).

To turn farmland back to prairie, ecologists must search for the original species, which they often find in recondite places, for example, in forgotten cemeteries and along abandoned railroad rights-of-way. In these nooks and crannies, native grasses and flowers have found refuge; the rest of the American heartland, as one reporter notes, "has been converted into the nation's breadbasket, growing corn, wheat, and soybeans" (Peterson 1987). Domesticated species have almost completely replaced the wild plants and animals that once covered a continent. There is a lesson to be learned in this about the likely effects of bioengineered species that we may now introduce into the environment.

17.1 THE EFFECTS OF DOMESTICATED SPECIES

Why is it that domesticated species, for example, hybrids of corn and wheat, dominate the prairies and the plains, while native grasses, flowers, and weeds now occupy only a minute fraction of their original range and habitat? Why do domesticated animals such as cattle abound, while the native buffalo would have become extinct, were it not for human intervention? The ecosystems that dominated the continent two centuries ago have all but disappeared and managed agro-ecosystems have replaced them. Now, because of chronic commodity surpluses, public officials want to turn the hands of time back two centuries. How did this happen?

One might answer that domesticated species—plants and animals produced by artificial selection—proved to be biologically more fit than native species, and that is why they forced wild flora and fauna into cemeteries and abandoned rights-of-way. This answer appeals to the theory of evolution: beef cattle and broiling chickens, on this account, were so well adapted to the wild that they outcompeted native animals, such as bison and passenger pigeons, for scarce resources. Likewise, the square tomato, the double-breasted turkey, and the seedless grape—which human beings have bred for the supermarket and dining table—must be better suited than native plants to the natural environment, for otherwise these domesticated species would not now abound in places native grasses and animals once dominated.

A century ago, society simply did not have the techniques of ecological risk assessment that biologists today can use to predict the environmental fate of domesticated organisms. If scientists had these methodologies 100 years ago, they could have tested domesticated animals in containment facilities and determined their likely ecological impact before introducing them into the environment. If scientists could have assessed the danger that the dairy cow and the feedlot pig posed to bison and other wild species, we might not have released them into the environment and thus we might not have brought native species to the verge of extinction.

Many ecologists fear that history may repeat itself—this time with respect to genetically engineered organisms. Animals and plants now being engineered genetically for particular economic purposes (boneless fish, for example) might (like seedless grapes) escape containment facilities or otherwise be released into the wild. There, they may outcompete wild species (those, for example, that have bones and seeds) and thus upset larger ecosystems, just as beef cattle outcompeted bison on the plains. How can we keep history from repeating itself in this way?

Fortunately, ecologists today stand ready to assess the environmental risks associated with genetically engineered organisms before they (the organisms, not the ecologists) are released into the wild. Ecologists, then, might test novel creatures in containment facilities and thus warn society in advance that a species may run amok in nature, reproducing endlessly and disrupting natural ecosystems. This assessment will be particularly important if there still is a "wild" or a "nature" of any sort remaining where engineered organisms can be released—if, in other words, nature has not been uprooted everywhere and replaced by high-technology, vertically integrated, intensely managed, centrally controlled, bioindustrial agro-ecosystems.

Ecologists can also serve society by devising management systems for the genetically engineered organisms that are unable to survive on their own in the environment. Ecologists can assist the agricultural industry to do what it has always done, which is to domesticate nature, that is, to control or to destroy wild and native species, in order to make the environment safe for economically valuable introduced organisms. Thus, ecologists may serve society in two ways: first, by testing novel organisms for risks they may pose to native species and, second, by controlling or eradicating native species because of risks they may pose to domesticated organisms.

A visitor from outer space might tell another story to explain why domesticated species dominate the prairies and the plains, while native grasses, flowers, and weeds occupy only a minute fraction of their original range and habitat. The alien might point out that a symbiotic relationship—the sort of mutualism one often finds in ecological communities—exists between human beings and domesticated species. Species that exploit this mutualism (so the alien might observe) easily colonize wilderness, semi-arid, and other areas not initially suitable for them. Indeed, a species able to succeed economically will almost certainly, by forming a mutualistic relationship with human beings, be able to succeed biologically as well.

The space visitor might buttress this explanation by showing students back home photographs of forests cut, dams constructed, groundwater mined, rivers polluted, ranches fenced, feedlots established, transportation networks built—in short, an entire ecology and resource system transformed—all to ensure the success of plants developed by the breeder's art and science. These photographs will also show that cows, pigs, chickens, and other artifacts of agricultural breeding dominate ranges that once belonged to wild and native animals. On this basis, the alien might convince his students that commercially fabricated chickens are biologically the most fit animals ever produced, since their populations increase every year, even though they are almost all harvested, at a young age, by human beings. Domesticated animals, within their symbiotic relation with human beings, appear to be marvels of natural selection—since they withstand every assault by nature or by human beings.

A visitor from outer space, then, might sympathize with the concern many ecologists and environmentalists have expressed "regarding the potential for genetically engineered organisms to displace resident species of the receiving community. . . ." (Tiedje et al. 1989). The space visitor might join ecologists on earth in stating that "[m]utualistic interactions, in particular, underlie many critical ecosystem functions" so that "[e]ngineered organisms that alter mutualistic associations, either by design or unintentionally, require careful evaluation" (Tiedje et al. 1989).

The mutualistic association between genetically engineered organisms and human beings requires careful evaluation. Plants and animals created by biotechnology feed and clothe us, and otherwise ensure our comfort and serve our needs. In return, human beings will utterly transform the environment to make it support these organisms. Ecologists play an important role in this mutualism. They can tell us whether a novel organism can survive in nature unaided or whether, as is likely to be the case, there must be a management system. And ecologists—as we shall see—may assist biological engineers in devising management systems to replace natural ones or, at any rate, in making natural ecosystems safe for biotechnology.

If relative reproductive success is the criterion for biological fitness, then nothing beats the products of companies such as Pioneer Hi-Bred and DeKalb. Hybrids of corn, wheat, and soybeans, and other products of conventional agricultural breeding, must be vastly more fit than native species, on that criterion, for, otherwise, "amber waves of grain" would not have replaced the bluestem and stipa, the compass plant and downy phlox, the puccoon, the coneflower, and the dropseed. Wherever one looks, domesticated plants and animals, aided by artificial insemination, in vitro fertilization, tissue culture, monoclonal variation, cloning, embryo transfer, micropropagation, pesticides, antibiotics, and price support programs, do very well, while native species display much lower rates of reproductive success.

Although there is a lot of controversy on the question of whether ecology can be a predictive science, ecologists can offer one general prediction with certainty about the biological fate of genetically engineered organisms. The universal law or principle is this: an engineered species that is successful economically will be successful biologically as well. This is because humans—often with the aid of their ecological expertise—will manage or eradicate any wild species that offers resistance to a profit-making domesticated one. The principle can be stated this way as well: nothing in nature can resist a species that is produced and sold at a profit, even in the short run.

One can justify this principle either as an inductive inference from history (e.g., the example of the hybrids that ate the heartland) or as a logical deduction from the structure of the relation between man and nature—exploitation. It is fair to predict on both *a posteriori* and on *a priori* grounds, then, that genetically engineered plants and animals, insofar as they are economically valuable, will crowd out natural and wild species, forcing them into zoos, botanical gardens, and banks for the storage of genetic material. We may choose to preserve examples of wild genotypes when it is inexpensive to do so, for, as we are often advised, you never know when you might need those genes. The upshot, however, is that economic value, real or conjectural, assures the creation and survival of species; random mutation and natural selection have become all but obsolete.

Oddly enough, the large literature that discusses the likely effect of bioengineered plants and animals on the environment focuses primarily on the possibility that these species will compete biologically with those in nature—but this is a comparatively minor issue. Did corn compete biologically with hoary puccoon or side oats? Did wheat succeed because it is better adapted to the prairie environment than the black-eyed Susan? Is biological fitness the reason soybeans replaced prairie clover; does competitive exclusion explain why the white-violet-chocolate prairie orchid cannot be found? The plants produced by Pioneer Hi-Bred and DeKalb rule the prairies and plains, and those developed by Monsanto, Biotechnica, Crop Genetics International, and a hundred other firms, will dominate the landscape of the future. Do we look to evolutionary ecology or to agricultural economics to explain this phenomenon?

Only a visitor from outer space, who classifies human actions as evolutionary rather than as cultural facts, could imagine that domesticated species compete or need to compete with wild or native species in a biological sense. Anyone born on this earth would have to agree that economic not biological competition accounts for—and will continue to account for—the environmental fate of nearly all domesticated plants and animals.

The goal of biotechnology is to improve upon nature, to replace natural organisms and processes with artificial ones, in order to increase overall social efficiency and profit. The point is to control nature to make it safe, ultimately, for investment. And that means *making nature safe for biotechnology.* The most efficient way to control the future is to invent it. That is why we spend more to produce economically valuable engineered species than to protect economically useless endangered ones. And that is also why we continually turn whatever natural and wild ecological systems we may have—from rain forests to savannas to estuaries—into carefully managed and engineered (and therefore predictable and profitable) bioindustrial productive systems.

Ecologists who write about the "safety" of engineered macroorganisms miss this point, in part, because they come at adaptation from the wrong direction. They think in terms of the adaptation of the organism to the environment. This is the Darwinian tradition. With genetic engineering, however, adaptation goes in the other direction: you adapt the environment to the organism. That is what agriculture, silviculture, and aquaculture are all about. That is what pesticides, tractors, chainsaws, and flamethrowers are for. One might worry, of course, that we might destroy a wild plant or animal for which we might find a use someday. But we can save examples in gene banks—and we may be able to make genomes just like them artificially when we understand how these organisms can be useful to us.

It is ironic that in the era of genetic engineering ecologists still think about environmental policy in evolutionary terms, that is, in terms of how a novel organism may adapt to given circumstances. This question—the question of biological fitness—has become moot. The only question is economic profitability. The question is how we can adapt given environmental circumstances to novel organisms.

Ecologists may look for economic opportunities themselves in testing the safety of engineered organisms—but the main chance lies elsewhere. The future for ecologists lies in environmental engineering, not environmental protection; it lies in making nature safe for biotechnology, not the other way around. The invisible hand of the market has superseded the invisible hand of evolution. The genetic code, for all intents and purposes, might as well be written on ticker tape.

17.2 THE EFFECTS OF BIOTECHNOLOGY ON WILD AND NATURAL ECOSYSTEMS

Genetic engineering and other forms of biotechnology will affect the environment primarily in two ways. First, quasi-industrial systems of production based on recombinant DNA (rDNA) technology, being economically efficient and profitable, will replace "wild" or "natural" ecosystems. Second, the commodity surpluses that result from these rDNA-based industrial systems of bioproduction will cause land that is intensively farmed to revert, ironically, to a "wild" or "natural" condition.

Consider the fate of tropical rain forests. The primary argument against cutting and clearing these wonderful ecosystems for agriculture has been economic or broadly utilitarian: few if any crops will grow in their place. Dr. Daniel Janzen, a well-respected conservation biologist, has recently noted that rDNA techniques are likely to provide varieties of economically valuable plants and animals that will do well on land rain forests now occupy. "When genetic engineering gives us crop plants and animals that thrive in the various tropical rain forest habitats, it is 'good-bye, rain forest' " (Allen 1987).

Consider estuaries, rivers, and other aquatic ecosystems of the United States. In the past, pollution has been a problem in aquatic ecosystems, in part because of its deleterious effect on recreationally and commercially desirable fish and other species. We may now think in terms of lowering the oxygen needs of these creatures, changing their behavior, or otherwise altering their genomes so that they will tolerate and, indeed, thrive in polluted waters (Greer 1984). In the past, we tried to control pollution to accommodate plants and animals. Now, new rDNA techniques give us the power to control plants and animals to accommodate pollution.

The new biotechnologies may also allow us to improve, transform, and perfect native species for the fishing rod or the dining table. Biotechnologists at the University of Washington, for example, have crossed rainbow trout with steelhead to produce "a deep-bodied fish with better flesh" that bites at a lure, making it excellent for both recreational and commercial purposes (Donaldson and Joiner 1983). "We like to think our results compare with the poultrymen's development of the broad-breasted turkey," these authors have written.

Because the genome of fish opens easily to genetic manipulation, permitting virtuoso performances, some commentators argue that native species should be replaced by varieties designed, for example, to grow faster and larger, to have a longer shelf life, or to reproduce and grow rapidly in artificial and polluted environments (Klausner 1985). When genetic engineering perfects the salmon, oyster, and bass, so that these and other species can be grown and harvested cheaply and efficiently, economic arguments for preserving "natural" species and even ecosystems will become even more tenuous and conjectural than they already are.

Many commentators observe that industrial aquaculture may soon replace hunting and gathering as the principal method of acquiring fish. Harold Webber, for example, argues that large-scale industrial mariculture based on biotechnology, which can push fish populations far beyond the carrying capacity of "natural" ecosystems, will soon render fisheries obsolete. Webber notes that we depend on traditional fisheries only because the "results of recent research and development in the biological sciences have not yet been integrated into the broader context of large-scale, vertically integrated, high technology, centrally controlled, aquabusiness food production systems" (Webber 1984).

Webber calls the substitution of industrial for "natural" methods of fish production in aquatic environments "Vertically Integrated Aquaculture" (VIA). He argues that genetic engineering means "good-bye" to the natural estuary and "hello" to VIA, at least if we are concerned about improving the efficiency of production to increase profits and supplies of commodities over the long run.

Economically, estuaries will remain valuable as sewers, liquid highways, and impoundments for aquaculture, to be sure, but these uses are consistent with, indeed, they often require, large-scale changes in the biological and ecological conditions. There is nothing about the biotechnological conversion or domestication of a bay or estuary that need conflict, moreover, with its recreational use. Vacationers, indeed, may expect developers not only to provide marinas, hotels, and amusement parks, but also to stock the water with the "latest" fish.

According to one knowledgeable observer, biotechnology will have its greatest effect on the way wood and other forest commodities are produced. "Forestry will probably realize the greatest improvement of any crop from biotechnology" (Powledge 1984). Genetic engineering, moreover, can compensate for the native species that are lost when forests fall to the chainsaw and axe. Native genetic material not only can be preserved in gene "banks" but also manipulated, varied, and improved in endless ways. "This requires genetic manipulation to evolve vigorous and fast-growing trees with a short reproductive cycle which can be mass propagated" (Bajaj 1986). It may be "good-bye" to forest, then, and "hello" to Vertically Integrated Silviculture.

Yet forest scientists moving into genetic engineering are concerned about regulation. The reason is not risk: trees stand still and their spread is easily controlled. The reason is not economic: biotechnologists will greatly improve forest productivity and lower costs, for example, just by increasing the photosynthetic efficiency of trees. The reason, of all things, seems to be ethical and aesthetic. People are unwilling to say "good-bye" to natural forest ecosystems. "The policy issues are serious," says one forest biotechnologist. "People are concerned about what you do to trees. People have affection for forests" (Powledge 1984).

Biotechnology, of course, is not new. For millenia, primitive forms of biotechnology, for example, plant selection and the use of pesticides and fertilizers, have allowed mankind to achieve abundance where it would have faced starvation, had it preserved the ecological or biological status quo. The hybrids that ate the heartland represent one of the great success stories of biotechnology: these crops feed the world. How do current rDNA techniques differ, in principle, from the methods by which mankind has always controlled and conquered nature?

The difference is this: in the past, people saw biotechnology as a pitchfork they used to drive nature out, while knowing that she would soon return. Today, biotechnology is like a wedge we can permanently thrust into nature. The difference, then, is that our conquest of nature is now unconditional; it is complete. In the past, we might let nature take her own way a little, thinking she understood her affairs better than we. In the future, we are more likely to trust our knowledge, and we will be able to force nature to go our way rather than her own.

And yet, and yet . . . people have affection for forests, for estuaries, and for other remnants of our evolutionary and ecological heritage. People want to live in harmony with nature and not just triumph over it. People respect nature and—what is more important—respect themselves for caring about its authenticity and integrity, and not just its efficiency and profitability.

One might even say that Americans identify themselves as a nation that cares about the environment—that considers nature as an object of moral attention rather than economic subjugation, or as a condition of our consciousness as a people, not merely as a vehicle for satisfying consumer wants or demands.

When we think about the risks associated with biotechnology, therefore, we must ask: risks to what? To the environment? To the ecological and evolutionary status quo? To wild and natural ecosystems? If rDNA organisms are more efficient or desirable or profitable than native species, why should we be concerned about native species? Which species threaten which? Which are better? Which should we protect?

17.3 NEW FISH FOR OLD

To understand the choice we must make between protecting nature from or managing it for novel organisms, consider fishing, which is, on some accounts, America's favorite pastime. For some of us, fishing essentially involves knowing and appreciating our surroundings, a love of nature for its own sake, and a respect for wildlife. Albert Miller begins his well-known book, called *Fishless Days, Angling Nights,* with this representative statement: "Fortunately, I learned long ago that although fish do make a difference—*the* difference—in angling, catching them does not." The secret of fishing "is to be content to not-catch fish in the most skillful and refined manner" (Miller 1971).

In another book about angling, *A River Never Sleeps,* Rodrick Haig-Brown agrees: "I want fish from fishing, but I want a great deal more than that, and getting it is not always dependent on catching fish." What is wanted is found in the attitude or philosophy of the angler, which is, to take joy even in the big fish that gets away. One can more easily and cheaply buy a trout at a store than catch it with a fly, of course, so it cannot be the acquisition of fish—or their number or weight—that makes people value fishing. What, then, is the value of this sport?

The point of recreational fishing is not to fill one's creel, but to free one's mind, to expand it by contemplating nature, and to refine it through the skills and virtues, primarily patience, one must learn to meet wild species on their own terms. Izaac Walton, in 1654, said it this way:

> He that hopes to be a good angler, must not only bring an inquiring, searching, observing wit, but he must bring a large measure of hope and patience, and a love and propensity to the art itself; but having once got and practiced it, then doubt not but angling will prove to be so pleasant, that it will prove to be like virtue, a reward to itself (Walton 1975).

The immense literature on fishing, from antiquity to the present, regards "[a]ngling and butchering fish . . . as two totally different occupations." Angling challenges knowledge and tests character; if the point were simply to amass fish, a net would be more convenient for the fisherman and cause less annoyance to the fish. Thomas Stoddart, writing more than a century ago, continues:

> The true angler I would describe to be one who follows the art as a science, who cultivates it, . . . by carefully studying the habits of the fish he wishes to capture—their food and feeding hour—their customary and occasional haunts—the effects of different states of weather or sizes or colors of water upon their tastes—together with a hundred other matters essential to be known, before he can venture to claim for himself the reputation of an adept in the craft (Stoddart 1870).

Many who fish, of course, know nothing about angling, and they often fish competitively, to show off to others. Plutarch tells the best-known fish story of this kind, and one of the most revealing, about Antony and Cleopatra:

> He [Antony] went out one day to angle with Cleopatra, and, being so unfortunate as to catch nothing in the presence of his mistress, he gave secret orders to the fishermen to dive under water, and put fishes that had already been taken upon his hooks; and these he drew so fast that the Egyptian perceived it. But, feigning great admiration, she told everybody how dexterous Antony was, and invited them next day to come to see him again. So, when a number of them had come on board the fishing-boats, as soon as he had let down his hook, one of her servants was beforehand with his divers, and fixed upon his hook a salted fish from Pontos. Antony, feeling his line give, drew up the prey, and when, as may be imagined, great laughter ensued, "Leave," said Cleopatra, "the fishing-rod, general, to us poor sovereigns of Pharos and Canopus; your game is cities, provinces, and kingdoms" (*Life of Antony* 29, 2).

Many anglers today are like Antony: they want to catch fish and are not particular about how they do it. This also was true a century ago, when Thoreau described people who came to Walden to fish, but who did not think they were well compensated for their time unless they made a good catch:

> . . . though they had the opportunity of seeing the pond all the while. They might go there a thousand times before the sediment of fishing would sink to the bottom and leave their purpose pure; but no doubt such a clarifying process would be going on all the time (Thoreau 1975).

Fishery managers know that the fishing public is less interested in purity of purpose or clarity of mind—which Nature provides without charge—than in catching fish. Accordingly, managers across the nation engage, like Antony's divers, in the practice of placing fish that are already caught onto the hooks of fishermen. *Audubon Magazine* describes one of the many ways this is done:

> Rather than make smarter fishermen they made dumber fish. They crossed a brook trout, actually a species of char, with a brown trout, a true trout, to get a garish, idiot fish that gobbles everything in sight. "Tiger trout," they call it. The public snatches them out as fast as the managers dump them in, which translates in the lingo to "a good return." "It has very little holdover potential because it's stupid," declares Charley Heartwell of West Virginia's Department of Natural Resources. "Fishermen," he says, "love it" (Williams, 1987).

Love it they do, if a report in *Field and Stream* is any indication. That magazine carried an article, "They're Making Tomorrow's Fish Today," praising dozens of "made-to-order" hybrids now common in recreational fish management (Dalrymple 1986). Consider a fish called the "wiper," popular in many states, which combines the genetic elements of a freshwater white bass and an anadromous striper. To breed this fish, managers stun and inject a human sex hormone into female stripers, the eggs of which they then squirt into a pan and mix with the milt of a few white bass, similarly obtained. High school students can do it. One biologist calls this species "the fish manager's perfect invention" (Dalrymple 1986).

Managers also recognize the importance of building resistance to pollution into hybrid fish. Twenty years ago, in an essay titled "Man-Made Fish," one official set out the program that others have since perfected:

> People are becoming more aware of the dangers of pollution. . . . Realistically, I think we can expect some thermal, chemical, and organic pollution to be with us for some time to come. It is, however, likely that we can breed desirable races of fish that will tolerate . . . alterations in the environment and still develop (Burrows 1971).

The "saugeye," a cross between the sauger and the walleyed pike, for example, does very well in Tennessee, Florida, Ohio, and other states where pollution depletes natural walleye populations. In New York, fisheries managers have developed a hybrid strain of brook trout that exhibits high acid tolerance, making it ideal for stocking in place of the species that acid rain has eliminated from the Adirondack lakes. These hybrids can reproduce. The October 1986 *Sport Fishing Institute Bulletin* advises: "Fishery personnel must produce fish strains that tolerate deteriorating habitat." Why try to control pollution to accommodate nature? It may be more efficient to control nature to accommodate pollution.

The U.S. National Forest Service relies on "economic analysis to determine fishery values, supplies, and demands for planning purposes" (Robertson 1988). The economic analysis in question has nothing to say about what Thoreau called the "clarifying process," or about virtues like hope and patience, or about love, or about the art of fishing itself. Willingness to pay is the only virtue it considers important. The Forest Service therefore promotes efficient game management, including "fish habitat improvement projects," by "constructing fishways, . . . fertilizing lakes, and neutralizing acid" (Robertson 1988).

The Chief of the Forest Service states that habitats are "managed to maintain viable populations of all native and desired non-native fish, and to produce species of recreational and commercial value to meet demands for use" (Robertson 1988). In other words, to sell the national forests to the public, the Forest Service guarantees that no one who comes to hunt or fish, however incompetent, will go away empty-handed. To the fisherman, it promises more catch per cast; to the hunter, more buck for the bang.

Anyone who wants to catch a school of hatchery-raised trout need only toss a handful of dog food on a stream and the fish will flock to it, since they are reared on those pellets (Dalrymple 1986). Fisheries managers have contrived to do nearly everything but instruct divers, as Antony did, to attach the fish to the hook. The director of a research station in Texas comments:

> So far we've just started. There are possibilities in these lab creations to build fish of endless diversity to fill the needs of the new millions of fishermen coming to us. The whole idea in modern management is to make available fish crops to be caught that can be replaced swiftly and economically. Hybrids are perhaps the most exciting aspects of today's fishery management programs (Dalrymple 1986).

What's wrong with this? Why should anyone be concerned about the release of genetically engineered fish into the nation's lakes, ponds, rivers, and streams? Will hatchery-raised "gainsburger" trout, once cut off from their daily regimen of doggie pellets, prove formidable competitors to native fish, who know how to feed themselves? Will cowalskis (one part cohoe, for taste, one part walleye, for size, and one part muskie, or muskellunge, for fight), once perfected in the laboratory, run amok in the wild?

These hybrid fish threaten fisheries and fishing more for logical than for ecological reasons. First, domesticated fish, being already in captivity, in principle cannot be caught. They can be caught, in other words, only in the absurd sense that Antony caught the salted fish from Pontos. One might as well try to climb a mountain from the top, or hunt in a slaughter house. We may worry, then, not because hybrid fish will outcompete native species and thus make fishing more difficult, but because they will make fishing too easy. We should worry that these fish, being in captivity, as it were, even before they are caught, make fishing logically, aesthetically, and culturally absurd.

Second, the public and its officials should be concerned about the hazard engineered fish pose to the refinement of taste, the nobility of purpose, and the clarity of mind of the fishing public. Put-and-take fisheries are to angling as prostitution is to love; they are, as one sportsman put it, "a cathouse proposition" (Gingrich 1965). Plainly, fishing in the wild (like courting a lover) requires appreciation, dedication, ability, knowledge, character, perseverance, and luck, which it amply rewards, while with prostitution, as with managed fisheries, success is simply a matter of willingness to pay. It appears that the kind of economic analysis on which the Forest Service bases its policies relies wholly on the willingness-to-pay criterion. This is why the fisheries policy may make economic but not moral, aesthetic, cultural, or logical sense.

Third, we may wonder what value or advantage attaches to plants that tolerate pollution and that withstand a deteriorating habitat. The principal value seems to be that these animals make pollution less costly and, therefore, more attractive. As long as we can engineer substitutes, why worry about the disappearance of wild and native species? Why make technology safe for nature? The profitable course, the efficient course, the rational course more often is to make nature safe for technology.

17.4 BIOTECHNOLOGY AND THE RETURN OF LAND TO NATURE

"My biggest fear is not that by accident we will set loose some genetically defective Andromeda strain," Senator Albert Gore has written; rather he is afraid that the nation will drown in a sea of surplus agricultural commodities. "The Green Revolution made America the world's breadbasket, but it has also brought on an age of intractable overproduction," Gore observes. "Unless we plan more carefully, the Gene Revolution could do the same—on an even grander scale" (Gore 1987).

Senator Gore backs up his assessment by referring to now-familiar problems, e.g., the effect of bovine growth hormone on the dairy industry. He describes the situation as follows:

> Scientists are working on Supercows, Superpigs, even supersized salmon. Other experiments have led to multiple births, more rapid growth, and higher resistance to disease. Unless we can somehow find a way to create very hungry Superhumans, each of these advances may produce nothing but *glut* (Gore 1987).

Politicians are concerned with this aspect of biotechnology, of course, because staggering surpluses of farm commodities already overhang world markets. In the United States, price support payments that keep farmers in business producing these surpluses cost the government about $26 billion a year (Galston 1985). Gore quotes Robert Kalter, an agricultural economist at Cornell, who predicts that "the unparalleled speed and magnitude of the expected productivity gains" due to biotechnology will overwhelm already saturated world markets. The government may not be able to maintain price supports given this rocketing agricultural production.

Genetic engineering may achieve some of these productivity gains at an environmental cost, for example, through an increased use of chemical pesticides. This is a risk to the environment worth considering. Both Calgene and Monsanto have used rDNA techniques to develop resistance in plants to glyphosphate herbicides (Roundup, as manufactured by Monsanto); accordingly, farmers and others now can use much more of the chemical to protect herbicide-resistant crops, lawns, or whatever from weeds. Likewise, biotech firms are developing crop resistance to atrazine, which is the largest-selling herbicide in the United States. According to one estimate, atrazine resistance in soybeans will increase sales of the chemical by about $200 million annually (*Genetic Engineering News* 1987). In the past, petrochemical companies engineered pesticides to make them compatible with crops. Now they can engineer crops to make them compatible with pesticides (Hayami and Ruttan 1971).

Agricultural biotechnology will also affect the environment indirectly by making production more independent of the functioning of natural ecosystems. Phosphate-solubilizing and nitrogen-fixing plants now in the experimental stage, for example, can compensate for poor soil conditions; why worry, then, about erosion (Batie 1983)? Likewise, drought-resistant crops, also being developed, substitute for water. Environmentalists, then, may soon lose important utilitarian or prudential arguments for protecting natural ecosystems (Crosson 1983). These arguments become harder and harder to defend as we find cheap technological substitutes for nature's gifts.

Another way that agricultural surpluses will affect the environment is by forcing land out of production. Farmers who adopt the new technologies early may profit by underselling their less-innovative competitors; eventually, however, increased production will drive prices lower, eliminating all but the most efficient producers. In the past, farmers who have survived each turn in the "technological treadmill" have bought out their less innovative neighbors; this is less likely to happen in the future because no one will buy the crops that would or could be produced on that land (Cochrane 1985).

Senator Gore suggests one reason for this: the inelasticity of demand. Unless we engineer wealthy superhumans with vast appetites (an interesting way to maximize consumption), we cannot hope to sell all the food we already produce. And biotechnology will plainly increase yields per cow, pig, and acre. The amount of farmland in production in the United States has decreased drastically since the 1940s, when hybridization and other innovations greatly increased yields. This trend is very likely to continue (Salquist 1987).

The principal reason that farms may go out of production, however, has to do with the possibility that the factory will replace farms and fields as the location where food and fiber are produced. The principal problem, in other words, may not be surpluses of farm commodities but rather of the industrial substitutes biotechnology will create for those commodities.

In the past, most agricultural products have been grown on farms and then processed in factories. In the future, factories may "grow" as well as process food and fiber—or the two functions will be one. As one scientist has said: "We have to stop thinking of these things as plant cells, and start thinking of them as new microorganisms, with all the potential that implies" (Curtin 1983; Chalef 1983).

The substitution of industrial for agricultural products is a familiar story, e.g., in the replacement of cane sugar by artificial sweeteners such as Nutrasweet and by high-fructose corn syrup (van den Doel and Junne 1986). Plant-produced milk proteins, known collectively as casein, may soon do to cheese what oleomargarine has done to butter (Jinenez-Flores and Richardson 1987). Likewise, synthetic fabrics like nylon and orlon have captured much of the textile market. Spices, fragrances, and flavoring agents are now being produced in vitro: vanilla is one example (Wheat 1986). Corporations competing to culture major agricultural crops in vitro—coffee, tea, rubber, cocoa, cotton, etc.—eventually may replace Third World growers as the principal suppliers of these commodities (van den Doel and Junne 1986).

For virtually any field-grown crop for which there is a market, an industrial process can fabricate a chemical equivalent through cell culture (Yanchinsky 1985). Cotton cells propagated in vitro grow fibers from both ends; citrus and tomato pulp vesicles have likewise multiplied in culture, suggesting the possibility of industrially produced puree, sauce, and juice (Busch and Lacy 1988). Feed additives (e.g., lysine) produced by enzymatic action, when added to cheap surplus grains, yield a nutritious fodder that might undersell soya exports from South America (Pond 1983). One need not multiply examples; many more can be found in the literature (Goodman et al. 1987; Zaitlin et al. 1985).

One might speculate, therefore, that some day most fields will be planted to produce perennial crops raised for biomass. This biomass, or sugars derived enzymatically from it, will be piped to metropolitan centers to sustain a massive tissue culture and fermentation industry. This industry will design, mass produce, and finish food and fiber products to satisfy the tastes and preferences of the consuming public. This will require a smaller farming sector than we support today, and it may make the economy less and less dependent for food on the natural environment.

From the point of view of private investment, the industrialization of agriculture presents exciting opportunities, for it allows corporations in the United States, Europe, and Japan to produce the raw materials—like coffee, rubber, spices, cotton, etc.—they now purchase and then finish and package for the retail market. This is an entrepreneur's dream. A firm producing rubber, let us say, by culturing cells in large vats, might at first price the product to undersell, e.g., Malaysian farmers; at the beginning, then, profits may be small. But the firm can raise prices after the Malaysian growers, undersold in the market, abandon their fields and then, because of starvation, political instability, or whatever, are unable to get their fields back into production. What an investment opportunity! To be sure, Third World economies will have to adjust while markets clear, but mass starvation apparently is the price of progress. Besides, theory tells us that a perfectly competitive market maximizes welfare in the long run, however much misery, dislocation, and death actually result.

One might adopt an ethical rather than an economic perspective, however, to evaluate the global industrialization of agriculture. We should then not endorse every use of genetic technology to lower the prices of commodities by substituting capital for labor. Farmland that comes out of production can be restored to an earlier "natural" condition or be employed some other way. But what will happen to farmers, especially peasants in the Third World? If production of cocoa, coffee, vanilla, and other such crops can be accomplished more efficiently in vitro than al fresco, how are people in the Third World to make a living? How will they pay even low prices for food and thus make a dent in the immense agricultural surpluses biotechnology may create?

Questions such as these have nothing to do with the unpredictable risks of biotechnology. They have nothing to do with the possibility, in other words, that some engineered creature will sneak out into the woods and run amok. Indeed, the ethical concerns this chapter has addressed have little if anything to do with inadvertent, unintentional, or accidental side effects of biotechnology. Rather, they concern the profitable, predictable, and intentional effects of biotechnology. Moral responsibility centers on consequences we intend rather than on those we cannot foresee and that happen by accident. The ethical issues—the ones that appeal most urgently for political and regulatory appraisal—do not arise from the possible failures but from the likely successes of genetic engineering. By concentrating our attention on risk assessment we simply abandon rather than address the major concerns of moral responsibility.

17.5 ECOLOGICAL SCIENCE AND THE FATE OF NATURE

The major effects of biotechnology on the environment, this chapter has argued, will be twofold. First, many "wild" or "natural" ecosystems may be converted to species and processes suitable to large-scale highly controlled aquacultural or silvicultural production. Second, as agricultural surpluses begin to be seen as infinite, and as the factory replaces the field as the location where food and fiber are fabricated, the agrarian economy must shrink, and many farms go out of production. Highways, commercial strips, and suburban "sprawl" might absorb some of this land in the United States, but a great deal of it may revert to "nature."

Novel genomes and genetic processes appear nearly unproblematic, indeed, miraculous, from a narrow economic point of view, since their development offers fine opportunities for profitable investment, and they will greatly increase the efficiency of agro-economic production. Nevertheless, there could be problems. A group of scientists writing in *Ecology* suggest that organisms engineered in 1990, for example, may themselves become obsolete, say, in 1995. These less-desirable recombinant organisms, however, may by then be entrenched in the environment, thus producing a "negative economic outcome" (Tiedje et al. 1989). This argument suggests that managers should enlist ecologists in converting wild to managed ecosystems. Ecologists can help to identify the most efficient ways to transform wild systems into carefully managed processes for industrial bioproduction.

At least one ecologist sees environmental engineering–the conversion of natural to engineered organisms and systems–as the future of his science. Why study natural history if the point is to change or replace it? He writes:

> Ecologists are the people most fit to develop the conceptual directions of biotechnology. We are the ones who should have the best idea as to what successful plants and animals should look like and how they should behave, both individually and collectively. Armed with such expertise, are we going to continue investing nearly all our talents in Natural History? . . . Or should we take the forefront in biotechnology, and provide the rationale for choosing species, traits, and processes to be engineered? I suspect this latter approach will be more profitable for the world at large as well as for ourselves (Forcella 1984).

Other prominent ecologists point out, correctly, that ecology may never become a "hard," "mature," "mathematical," and "predictive" science as long as it studies the haphazard and chance results of natural history. J.L. Harper (1982), for example, argues that in tightly controlled and very simplified systems, as may be achieved in agriculture, prediction is possible because complexity has been diminished. Harper notes that in agricultural ecosystems one may be able to discover underlying laws for ecological management, because "environments are simplified and variation . . . is reduced by activities such as drainage or irrigation, fertilizer application, cultivation, pest and disease control, and by the use of a limited number of crop varieties." Ecological models will work here–they can be precise, general, and realistic all at once–in part because the criterion of success against which these models may be tested, i.e., more efficient yields, is clearly understood. Harper (1982) comments: "In agricultural ecology precision, realism and generalization become more compatible in scientific enquiry. It may well be for this reason that the development of ecology as a science in the post-descriptive phase will come from the study of man-managed ecosystems."

"Biotechnology proclaims the transformation of biological theory into engineering," an anthropologist writes; biotechnology controls "biological processes for the purposes of the production of goods and services" (Fleising 1989). The *point* of biotechnology, then, is to replace the blind, random, historical processes of evolution with well-engineered and designed processes of bioindustrial production. The effect of biotechnology on the science of ecology, then, will be to create fewer opportunities for ecologists who are concerned with the products of natural history and more opportunities for ecologists who can do "hard" quantitative and predictive work on artificial biotechnologically based species and systems. Genetic engineering may give us–at least it aspires to give us–a safer, more predictable, and more efficient living world, just as civil, industrial, chemical, and electrical engineering have given us a safer, more predictable, and more efficient physical world. There is no reason to suppose that genetic engineering–like other forms of engineering–will not succeed rapidly in fulfilling this aspiration.

If the point of biotechnology is to replace natural with artificial systems, whenever it is economically efficient to do so, the effect of biotechnology on ecology is to direct that science away from the study of natural ecosystems and toward the study of engineered and artificial ones. Many ecologists, however, may have come to their discipline because of a love of nature; they may therefore resist the idea that they should give up natural history in order to become environmental engineers. They may wonder, then, how to serve society as scientists and, at the same time, to study and protect natural history. How can ecologists both create the future and preserve the past?

The most obvious way to answer this question is to assume that wild and natural biological communities are economically valuable–to suppose, in other words, that the rational or prudent thing to do is to preserve the products of evolution rather than to replace them wholesale with those of bioengineering. If one assumes, perhaps *a priori,* the rationality of protecting the habitat of wild and native species, for example, then one must be concerned about the risks novel organisms may pose to those creatures. If these "risks" appear to be the only obstacles that may prevent genetic engineering from transforming both natural history and the sciences that study it, then it is understandable that ecologists emphasize those risks. How else can they maintain ecology as an historical science and preserve the flora, fauna, and biological communities that they have traditionally studied?

The problem with this approach, however, may be that bioengineered organisms pose a *direct* risk to native species that is almost inconsiderable compared to the *indirect* threat that this paper has described. The indirect threat, which arises from the ability of engineered species to outcompete native ones economically if not biologically, occurs because environmental engineers will continue to destroy natural species and their habitats to build managed systems that are safe for domesticated organisms. These engineers will argue, with some plausibility, moreover, that prudence and economic rationality requires us to exchange old species for new. And ecologists who love the past and respect the historical nature of their science, presented with this argument, may then resort to invoking risks and dangers so conjectural and ideologically self-serving that they bring ridicule on themselves.

The principal problem with biotechnology, as Senator Gore observed, is surplus, not risk. The cycle, of innovation/increased efficiency/greater productivity/lower prices/the shake-out of less-innovative producers/and further innovation, never stops—and this cycle is the primary vehicle through which biotechnology will change the face of the environment. As both victors and vanquished in this "technological treadmill," aquacultural and silvicultural firms, after borrowing huge sums to transform "wild" ecosystems to vertically integrated cultivation, will find themselves in Chapter 11 bankruptcy, and their operations may revert to "nature," since only the most efficient and best-capitalized of these ventures can survive. Thus, even if biotechnology pushes nature out, bankruptcy will invite her back. Whether nature will endure to pick up the pieces—and whether the pieces will be worth picking up—is an imponderable question. It cannot be considered here.

17.6 THE POLICY ISSUES

This chapter has argued that genetic engineering is likely to affect the environment in two major ways: first, by converting wild ecosystems, such as rain forests and estuaries, into managed quasi-industrial processes for bioproduction and, second, by creating massive surpluses that will then cause investors to abandon many of these agricultural, silvicultural, and aquacultural ventures. What are the policy issues? How should regulatory agencies respond?

An extremely influential view among economists holds that a regulatory agency such as the Environmental Protection Agency (EPA) should be concerned primarily or, indeed, exclusively with improving the overall efficiency of the economy. (Other agencies might act to transfer wealth to the poor and otherwise enhance distributive justice.) According to this philosophical approach, regulatory actions are justified insofar as they bring market spillover effects, or "externalities," like pollution, into the pricing mechanism, so that the private and social costs of production will not diverge (Okun 1974). In this way, regulation will help markets to allocate materials and resources to their most valued employments, which is to say, to those producers and consumers who are willing to pay the most for their use.

This approach acknowledges that regulations are necessary to prevent "spillovers" or "externalities," in other words, hidden and involuntary harms that ecological changes may impose on individuals or on society. The idea, in other words, is to protect mankind from the environment. It is not necessarily to protect the environment from mankind.

The wide acceptance of this idea may explain why the regulation and assessment of *risk* has been the only policy issue that has received significant attention either in the press or in the professional and academic literature. This emphasis on risk pays no moral attention to nature or its protection *per se*; value attaches only to human health, wealth, and safety on a willingness-to-pay basis. As far as this approach is concerned, nature can go into the hopper of biotechnology as long as there is a market for everything that comes out. Torts are what we want to prevent, ultimately, and investments are what we want to protect.

Ecologists implicitly accept this market or economic model of public policy insofar as they take safety or risk to be the only, or at least the principal, reason for regulating the substitution of artificial for natural species and systems. This insistence on risk as the target of regulation accepts economic rationality as the touchstone of environmental policy. On this basis we should have to conclude, however, that if a domesticated animal—for example, a tame bear that children can pet in the parks—is safer or less risky than its natural counterpart, then we should systematically replace wild animals with substitutes whose personalities have been "improved" by genetic engineering. Similarly, as long as put-and-take fisheries are economically efficient and environmentally safe, there can be no reason to preserve wild species for the few—what are they willing to pay?—who might like to catch or to study the products of natural rather than of human origin.

When ecologists and environmentalists join the risk bandwagon, they accept, at least implicitly, the principle that artifice should replace nature whenever the benefits appear to outweigh the risks, which they generally will, as the greater efficiency of agriculture over hunting and gathering attests. Environmentalists then must emphasize uncertainties and conjectural dangers and needs—you never know if you'll need those genes—to find reasons to preserve wild and native species that instrumental and economic rationality would plainly send into extinction. Accordingly, the aesthetic, moral, cultural, and historical values that require the preservation of nature become lost in the intricacies of arcane arguments over conjectural and speculative risks.

The major environmental impacts of biotechnology will generally be intentional, profitable, and, most important, they will result from free competition in reasonably efficient markets. Why worry, then, about these impacts? Why worry about altering ecosystems—for example, converting wild to put-and-take fisheries—if these changes are economically rational and profitable over the long run? The emphasis on risk suggests we do not intend to protect the environment from changes that are economically beneficial. There seems to be little reason, then, to regulate—or do anything but celebrate—the "fast food" franchising of the natural environment.

The substitution of rDNA species for natural species, ecosystems, etc., *per se* would not be a problem; it is profitable; it is, indeed, a goal. The problem, if there is one, will arise because of unforeseen effects on aspects of nature that still have an economic value—those for which cheap technological substitutes have not yet been found. In other words, when we talk about risks to the environment, when we worry about "externalities" and side-effects, we refer to threats only to those aspects of nature we have not yet learned to reengineer or replace, which is to say, those aspects that biotechnology has not so far rendered obsolete. These aspects of nature are bound to decrease in number and importance—so why care about damage to the environment?

Presumably, there are economically valuable natural systems and species that are not easily replaced by engineered species and systems, but what are these? What are the aspects of nature upon which genetic engineering is unlikely to improve or for which it will not provide adequate substitutes? In the absence of any research pinning down which goods and services (if any) nature may provide more efficiently or economically than biotechnology, it is impossible to know how environmental impacts translate into impacts on human wealth and welfare. The destruction of wild species and ecosystems, even on a spectacular scale, does not count as a risk, at least not in this approach, unless those species and systems are economically more valuable than the improved bioengineered models that may take their place.

The task for regulation, on this account, is first to identify which, if any, communities or species produced by evolution can survive economic competition with bioengineered substitutes (as the hoary puccoon could not survive economic competition with wheat and corn); then regulators can act to protect those species and systems. A second task for regulation, in this approach, would recognize the likelihood that engineered organisms—fish that bite better, trees that grow faster, birds that sing more prettily, microbes that fix nitrogen more efficiently, or whatever—will be superior economically to the species they may be intended to replace. Regulators could then act rationally by seeing to the destruction of economically obsolete natural species, or they may leave this to private entrepreneurs, to make the environment safe for biotechnology.

Since our welfare—indeed, our entire agricultural economy—is likely to depend on genetically engineered organisms, it may be these organisms, rather than their "wild" cousins, we should try to protect. We might devise mesocosms, sentinel species, and other indices and models, then, to predict the probable effects of natural creatures on engineered organisms, rather than the other way around. Then we will eradicate natural species that threaten engineered ones. This is an approach that may make sense if our goal is to allocate materials and resources to their most profitable uses. It is the research agenda we should follow in order to maximize wealth or welfare, even if we must say "good-bye" to the natural environment.

17.7 ANOTHER VIEW OF RISK

So far, we have described the regulatory and research goals that EPA would advance were it to act solely with an eye to promoting human health, safety, and welfare. Welfare, as it is understood by economists, requires the allocation of resources to those who will pay the most for their use. Random mutation and natural selection are unlikely to allocate resources to their highest valued uses in that sense; natural history and economic theory hardly coincide. Accordingly, if EPA were to promote welfare as its primary goal, it would encourage rather than prevent the progress of biotechnology in replacing natural with managed processes and systems.

What would the regulatory and research agenda be like if EPA, instead, conceived of itself as an *environmental* protection agency? Its task would then include protecting the evolutionary and ecological status quo, even if that meant keeping economically beneficial rDNA organisms out of the natural environment. EPA might then conclude that the emphasis on risk is something of a red herring. To approve the anticipated and deliberate environmental impacts of biotechnology while reviewing the "spillover" or inadvertent effects is—to change the metaphorical animal—to swallow a camel while straining at a gnat.

The policy options would then be fairly obvious. EPA would strive to keep engineered organisms out of the natural environment, or it would permit only those introductions that will not intrude upon or excessively alter natural history. The reason, of course, would not be to promote the efficiency of the market. The reason would not be prudential. It would be moral. The goal would be to preserve the nation's ecological and evolutionary heritage.

This means making sure that genetically engineered organisms introduced into the environment will not survive there, or if they may survive, it will not be to the detriment of any naturally occurring species. It means keeping ecologically viable recombinant macroorganisms out of "wild" ecosystems even if they are preferable, economically speaking, to the natural species they could replace. Aquaculture, silviculture, and the like would go on in areas separated from surrounding ecosystems, so that "improved" species would not find their way into the larger environment. Likewise, novel microorganisms entering the environment in sewage or agricultural runoff would be screened as well as possible, on a case-by-case basis, to assure they will not affect the microbial communities they will enter in ways that might disrupt natural ecosystem processes.

We would then approach the protection of today's comparatively wild or natural ecosystems with the history of the prairie in mind. Why not preserve these environments now, rather than wait until they have been run through the "technological treadmill"? The efficiency of markets—the allocation of resources to those who, at some time, are willing to pay the most for them—is not everything. There is another approach, one that treats nature as an object of moral attention, which makes room for aesthetic, cultural, and political arguments. This approach may not be the recommendation of individual self-interest or even prudence, but it is consistent with national pride and self-respect.

Ecologists may find in this approach a surer and more edifying ground than economic efficiency on which to base the usefulness of their science. The medical sciences are not ashamed to take a normative or ethical goal—

health—as their purpose; indeed, it would be odd to say that medicine is not "scientific" because it advances a morally desirable goal. Likewise, the environmental sciences could see the health or integrity of natural communities and ecosystems as the end purpose of ecology; whatever knowledge advances that end would be important ecological knowledge, even if it served no particular economic advantage. The question of whether ecology will evolve toward the medical model (dealing with natural history and the integrity of its products) or toward an engineering model (maximizing the long-run benefits nature offers humans) has become a major controversy. We cannot consider it further here.

If the ecological sciences concern themselves ultimately with the health and integrity of nature (rather than with the efficiency of production over the long run), the agenda for ecological modeling and risk assessment falls into place. We need models—including meso- and microcosms, sentinel species, and other ecological indices—that will provide some assurance that we can protect the integrity of natural ecosystems from changes owing to biotechnology. This research agenda would help bring EPA in line with its statutory authorities, which require it to protect the nation's ecological and evolutionary heritage. The Clean Water Act, for example, requires EPA to set standards that assure "the protection and propagation of a balanced, indigenous population of shellfish, fish, and wildlife." The objective of the act "is to restore and maintain the chemical, physical, and biological integrity of the Nation's waters."

The notion of "integrity," of course, is a cultural, moral, and aesthetic concept—it has no direct biological significance. Related terms such as "equilibrium," "balance," and "stability" are probably likewise religious or ethical at bottom, even though scientists have tried to find technical meanings for them. This effort may never perfectly succeed. But the task of ecological modelers, indeed, the task of ecological science, must be to give concepts such as these that will enable regulators to integrate environmental science with our culture, our aspirations, our history, and our law (Sagoff 1988a).

The day may come a hundred years from now when ecological restorationists will return a bit of estuary, forest, river, or whatever to its original condition as a museum or a monument of the past. They will look in nooks and crannies for creatures that evolution rather than engineering produced—the world of born rather than made. Any creature whose phylogeny could be traced back more than a hundred years may seem a rare find, a kind of antique. Wealthy connoisseurs may keep a few of these natural species in private zoos and aquariums.

This is the future we may expect if we think only in economically rational and instrumental terms—in terms of the goods and services for which individuals are willing to pay. To do anything else—to make biotechnology safe for nature rather than the other way around—we should have to take ethical, cultural, aesthetic, and historical values seriously. But this means worrying less about the risks of biotechnology and more about its successes–less about our welfare as consumers and more about the security and decency of our souls.

REFERENCES

Allen, W. (1987) *Genetic Engineering News* 7(10), 10.

Bajaj, Y.P.S. (1986) in *Biotechnology in Agriculture and Forestry: Trees I* (Bajaj, Y.P.S., ed.), pp. 1–23, Springer-Verlag, New York.

Batie, S. (1983) *Soil Erosion: Crisis in America's Cropland?* The Conservation Foundation, Washington, D.C.

Burrows, R.E. (1971) in *Sport Fishing USA* (Daults, D., and Walker, M., eds.), pp. 344–349, U.S. Department of the Interior, Washington, D.C.

Busch, L., and Lacy, W.B. (1988) in *Biotechnology and the Food Supply* (National Research Council, Food and Nutrition Board, eds.) National Academy Press, Washington, D.C.

Chaleff, R.S. (1983) *Science* 219, 676–682.

Cochrane, W. (1985) *American Journal of Agricultural Economics* 67, 1002–1009.

Crosson, P. (1983) in *The Cropland Crisis: Myth or Reality?* (Crosson, P., ed.), pp. 4–16, Johns Hopkins University Press, Baltimore.

Curtin, M.E. (1983) *Bio/Technology* 1, 657.

Dalrymple, B.W. (1986) *Field and Stream* 54, 88–92.

Donaldson, L.R., and Joiner, T. (1983) *Scientific American* 249(1), 51–58.

Federal Water Pollution Prevention and Control Act of 1972. (1972) Clean Water Act. 33 *United States Code* Sec. 1251.

Fleising, V. (1989) *TIBTECH* 7, 52–57.

Forcella, J. (1984) *Bulletin of the Ecological Society of America* 65, 434–436.

Galston, W. (1985) *A Tough Row to Hoe: The 1985 Farm Bill and Beyond,* Hamilton Press, Lanham, MD.

Genetic Engineering News (1987) Technologies and Market Forces Shape the Form of Agribiotech Products, February, 16–19.

Gingrich, A. (1965) *The Well-Tempered Angler,* Knopf, New York.

Goodman, D. Sorj, B., and Wilkinson, J. (1987). *From Farming to Biotechnology: A Theory of Agro-Industrial Development,* Blackwell, Oxford.

Gore, A., Jr. (1987) *Genetic Engineering News* 7(5), 4.

Greer, J. (1984) *Maryland Sea Grant Newsletter* 6(2), 10–13.

Harper, J. (1982) in *The Plant Community as a Working Mechanism* (Newman, E., ed.), Blackwell, Oxford.

Hayami, Y., and Ruttan, V.W. (1971) *Agricultural Development: An International Perspective,* Johns Hopkins University Press, Baltimore.

Jinenez-Flores, R., and Richardson, T. (1987) in *Food Biotechnology*, vol. 1 (King, R.D., and Cheetham, P.S.J., eds.), Elsevier, London.

Klausner, A. (1985) *Bio/Technology* 3(1), 27–32.

Kloppenburg, J., Jr. (1988) *First the Seed: The Political Economy of Plant Biotechnology,* Cambridge University Press, New York.

Miller, A. [Sparse Grey Hackle, pseud.] (1971) *Fishless Days, Angling Nights,* Crown Publishers, New York.

Okun, A. (1974) *Equality and Efficiency: The Big Trade-Off,* Brookings, Washington, D.C.

Peterson, C. (1987) "Oklahoma's New Prairie." *The Washington Post,* 25 December 1987, A1, A14.
Plutarch (1988) *Life of Antony,* Cambridge University Press, Cambridge.
Pond, W.G. (1983) *Scientific American* 248(5), 102.
Popper, F.J., and Popper, D.E. (1989) "Saving the Plains: The Bison Gambit." *The Washington Post,* 6 August 1989. B3.
Powledge, T.M. (1984) *Bio/Technology* 2, 763–772.
Robertson, F.D. (1988) *Fisheries* 13(3), 22–23.
Sagoff, M. (1988a) *Tennessee Law Review* 56, 77–229.
Sagoff, M. (1988b) *Agriculture and Human Values* 5(3), 26–35.
Salquist, R. (1987) *Bio/Technology Roundtable* 1987, 130.
Stewart, G.G. (1983) in *BIOTECH 83, Conference Proceedings,* Online Publications Ltd., Northwood, UK.
Stoddart, T.T. (n.d., circa 1870) (1942) quoted in *Notable Angling Literature* (Robb, J., ed., n.d. circa 1942), p. 48, Jenkins, London.
Thoreau, H.D. (1975) in *The Portable Thoreau* (Bode, C., ed.), p. 458, Viking, New York.
Tiedje, J.M., Colwell, R.K., Grossman, Y.L., et al. (1989) *Ecology* 70, 301.
Turner, F. (1988) *Harper's Magazine* 276, 49–55.
U.S. Congress, Office of Technology Assessment (1986) Technology, Public Policy, and the Changing Structure of American Agriculture, OTA-F-285, March 1986, Washington, D.C.
van den Doel, K., and Junne, G. (1986) *Trends in Biotechnology* 4, 88–90.
Walton, I. (1975) *The Compleat Angler,* Weathervane Books, New York.
Webber, H. (1984) in *Biotechnology and the Marine Sciences* (Colwell, R., Sinskey, A., and Pariser, E., eds.), Wiley, New York.
Wheat, D. (1986) in *Biotechnology in Food Processing* (Harlander, S.K., and Labuza, T.P., eds.), pp. 279–284, Noyes Publication, Park Ridge, NJ.
Williams, T. (1987) *Audubon Magazine,* September, 75–77.
Yanchinsky, S. (1985) *Setting Genes to Work: The Industrial Era of Biotechnology,* Penguin, Hammondsworth, UK.
Zaitlin, M., Day, P., and Hollaender, A. (1985) *Biotechnology in Plant Science: Relevance to Agriculture in the Eighties*, Academic Press, Orlando, FL.

INDEX